21世纪高等职业教育计算机系列规划教材

Android程序设计实用教程

向守超　姚骏屏　主　编
朱　雷　邓书基　副主编

電子工業出版社
Publishing House of Electronics Industry
北京・BEIJING

内 容 简 介

本书面向所有对 Android SDK 在 Android 移动手机平台上创建应用程序感兴趣的读者。不管是有丰富 Java 开发经验的程序员，还是只有 Java 基础的初学者，此书都将是十分有价值的学习资料。

全书共有 10 章，分别介绍了 Android 开发环境配置、Android 应用程序、Android 常用基本控件、Android 常用高级控件、Android 游戏应用程序开发、Android 消息与广播、Service 后台服务、Android 数据存储与访问、Android 位置服务与地图应用、综合案例设计与开发等内容。

本书立足实用，实例丰富，既可作为高职高专相关专业课程的教材和教学参考书，也可供从事移动编程开发的用户学习和参考。

图书在版编目（CIP）数据

Android 程序设计实用教程 / 向守超，姚骏屏主编. —北京：电子工业出版社，2012.11
21 世纪高等职业教育计算机系列规划教材
ISBN 978-7-121-18819-0

Ⅰ. ①A… Ⅱ. ①向…②姚… Ⅲ. ①移动终端—应用程序—程序设计—高等职业教育—教材 Ⅳ. ①TN929.53

中国版本图书馆 CIP 数据核字（2012）第 254513 号

策划编辑：徐建军（xujj@phei.com.cn）
责任编辑：徐建军　　　　特约编辑：俞凌娣　赵海红
印　　刷：北京天宇星印刷厂
装　　订：三河市皇庄路通装订厂
出版发行：电子工业出版社
　　　　　北京市海淀区万寿路 173 信箱　邮编 100036
开　　本：787×1 092　1/16　印张：18.25　字数：467.2 千字
印　　次：2012 年 11 月第 1 次印刷
印　　数：3 000 册　定价：35.00 元

前言

Android 是基于 Linux 内核的软件平台和操作系统，是 Google 在 2007 年 11 月 5 日公布的手机系统平台，早期由 Google 开发，后由开放手机联盟（Open Handset Alliance）开发。它采用了软件堆层（software stack，又名以软件叠层）的架构，主要分为三部分。底层以 Linux 内核工作为基础，由 C 语言开发，只提供基本功能；中间层包括函数库 Library 和虚拟机 Virtual Machine，由 C++开发。最上层是各种应用软件，包括通话程序、短信程序等，应用软件则由各公司自行开发，以 Java 作为编写程序的主要部分。不存在任何以往阻碍移动产业创新的专有权障碍，号称是首个为移动终端打造的真正开放和完整的移动软件。Google 通过与软、硬件开发商、设备制造商、电信运营商等其他有关各方结成深层次的合作伙伴关系，希望借助建立标准化、开放式的移动电话软件平台，在移动产业内形成一个开放式的生态系统。

随着 Android 平台的发展，引发了 Android 人才荒。但符合条件的 Android 工程师屈指可数，企业招聘难度可想而知。我们相信，在未来几年内，Android 开发工程师将成为 3G 行业炙手可热的岗位之一。3G 人才全球紧缺，实用人才培养已迫在眉睫！在国内三大运营商如火如荼的 3G 营销战持续升温，再加上 3G 的推出对整个行业的巨大推动作用，无疑将引爆 3G 手机开发工程师这个黄金职位。所以我们说程序员必学 Google Android 的理由可以总结为：更快的薪酬提升通道、更好的热门就业岗位、更多的行业人才需求、最热门的新技术行业。

本书包括 10 章内容，分别介绍了以下内容：

第 1 章 Android 开发环境配置，对 Android 的起源、优点和系统架构进行了介绍；详细介绍了 Android 开发环境的配置，Android 应用程序的创建以及应用程序的结构分析。

第 2 章 Android 应用程序，对 Android 应用程序的四大基本组件进行了详细介绍，详细讲述了 Android 应用程序从活动状态、暂停状态、停止状态和非活动状态整个生命周期过程。详细介绍了 LogCat 工具在调试应用程序过程中，对程序错误的定位和分析。

第 3 章 Android 常用基本控件，对进行用户界面开发常用的 Android 布局管理器、基本控件、菜单、对话框、事件和动画播放技术进行了详细介绍。Android 中的布局包括线性布局、表格布局、相对布局、帧布局和绝对布局，基本控件主要包括文本框、按钮、单选按钮、复选按钮、状态开关按钮、日期时间控件和图片控件的使用。菜单包括选项菜单、子菜单和上下文菜单。Android 平台下的对话框主要包括普通对话框、选项对话框、单选多选对话框、日期和时间对话框以及进度对话框。

第 4 章 Android 常用高级控件，对自动完成文本框、滚动视图、列表视图、滑块与进度条、画廊与消息提示、下拉列表与选项卡等高级控件进行了详细介绍，并辅以详细案例。

第 5 章 Android 游戏应用程序开发，Android 平台下的应用开发，一般来说主要分为商业应用和游戏应用两种。本章重点介绍自定义 View 和 SurfaceView 类在游戏开发中的应用，游戏开发中的碰撞与检测技术，最后通过扫雷游戏的开发详细介绍了游戏开发的步骤。

第 6 章 Android 消息与广播，Intent 是轻量级的进程间通信机制，用于跨进程的组件通信和发送系统级的广播。本章让读者基本了解 Android 系统的组件通信原理，掌握利用组件通信启动其他组件的方法，以及利用组件通信信息和发送广播消息的方法。

第 7 章 Service 后台服务，Service 是 Android 系统的后台服务组件，适用于开发无界面、长时间运行的应用功能。本章让读者了解后台服务的基本原理，掌握进程内服务与跨进程服务的使用方法，有助于深入了解 Android 系统的进程间通信机制。

第 8 章 数据存储与访问，Android 平台提供了多种数据存储方法，包括易于使用的 SharedPreferences，经典的文件存储和轻量级的 SQLite 数据库。通过本章的学习，读者可以了解 Android 平台各种组件数据存储方法的特点和使用方法，掌握跨进度的数据共享方法。

第 9 章 Android 位置服务与地图应用，位置服务和地图应用是发展最为迅速，有着大量潜在需求的领域，通过本章的学习可以让读者简单地了解位置服务和地图应用的概念、方法和技巧。读者可以使用 Google 提供的地图服务，构建提供位置服务的应用程序。

第 10 章 综合案例设计与开发，本章将以“手机相册服务软件”作为示例，综合运用前面章节所学到的知识和技巧，从需求分析、界面设计、模块设计和程序设计等几个方面，详细介绍 Android 应用程序的设计思路与开发方法。本章提供的“手机相册服务软件”是 2012 年“全国软件杯”软件设计大赛 Android 开发项目的本地相册内容，是一个比较综合的案例。

本书由重庆正大职业技术学院的向守超、姚骏屏担任主编，由辽宁信息职业技术学院的朱雷和无锡工艺职业技术学院的邓书基担任副主编。本书在编写过程中得到了各方面的大力支持，在此一并表示感谢。

为了方便教师教学，本书配有电子教学课件及相关资源，请有此需要的教师登录华信教育资源网（www.hxedu.com.cn）免费注册后进行下载，如有问题可在网站留言板留言或与电子工业出版社联系（E-mail:hxedu@phei.com.cn）。

由于编者水平有限和时间仓促，书中难免存在疏漏和不足。希望同行专家和读者能给予批评和指正。

编　者

目 录

第 1 章

Android 开发环境配置

Android 是一个优秀的开源手机平台，我们需要对 Android 平台的起源、发展、特征和体系结构有个初步地了解。Android 开发环境的安装与配置是开发 Android 应用程序的第一步，也是深入 Android 平台的一个非常好的入口。通过本章的学习，我们可以完全了解 Android，掌握安装、配置 Android 开发环境的步骤和注意事项，熟悉 Android SDK 和 ADT 插件的用途，了解其在应用程序开发过程中可能会使用到的各种工具。

1.1 Android 简介

Android 一词的本义指“机器人”，同时也是 Google 于 2007 年 11 月 5 日宣布的基于 Linux 平台的开源手机操作系统的名称，该平台由操作系统、中间件、用户界面和应用软件组成，号称是首个为移动终端打造的真正开放和完整的移动开发软件。

1.1.1 Android 起源

Android 本是一家公司的名称，这家公司的创始人名叫 Andy Rubin。Andy Rubin 创立了两个手机操作系统公司：Danger 和 Android。Danger 以 5 亿美元卖给微软，成为了今天的 Kin，Android 以 4 千万美元卖给 Google。Android 作为 Google 企业战略的重要组成部分，将进一步推进“随时随地为每个人提供信息”这一企业目标的实现。全球为数众多的移动电话用户正在使用各种基于 Android 的电话。谷歌的目标是让移动通信不依赖于设备甚至平台。出于这个目的，Android 将被补充，而不会替代谷歌长期以来奉行的移动发展战略：通过与全球各地的手机制造商和移动运营商结成合作伙伴，开发既有用又有吸引力的移动服务，并推广这些产品。

Android 手机就是使用 Android 操作系统或 OMS 操作系统的手机，2008 年 9 月 22 日，美国运营商 T-Mobile 在纽约正式发布第一款 Android 手机——T-Mobile G1，如图 1-1 所示。该款手机为台湾宏达电代工制造，是世界上第一部使用 Android 操作系统的手机，支持 WCDMA/HSPA 网络，理论下载速率 7.2Mbps，并支持 Wi-Fi。2009 年 9 月初，摩托罗拉坐镇主场在旧金山举办的 Giga OM 2009 大会上携手 T-Mobile 正式发布了旗下首款搭载 Android 操作系统的智能手机——MOTO CLIQ（见图 1-2），使其在沉寂许久后的首次爆发吸引了全球无数用户的目光。如果说 T-Mobile G1 的出世开辟了 Android 领域先河的话，那么摩托罗拉 CLIQ 的发布则更多地被视为昔日手机霸主的强势回归！2009 年 10 月 28 日摩托罗拉和网络运营商 Verizon 共同宣布了首款采用 Android 2.0 的手机 Droid。2010 年 1 月索尼爱立信首款 Android 机型 X10 上市。2010 年 1 月 7 日，Google 在其美国总部正式向外界发布了旗下首款合作品牌手机 Nexus One（HTC G5），并同时开始对外发售。2010 年 7 月 9 日，美国 NDP 集团调查显示，Android 系统已占据了美国移动系统市场 28%的份额。我们相信未来将有越来越多的 Android 手机握在消费者的手中。

图 1-1 第一款 Google 手机 T-Mobile G1

图 1-2 MOTO CLIQ

1.1.2 Android 的优点

目前市场上的手机操作系统除了 Android，还有 Symbian、iPhone 等，与这些手机系统相比，Android 具有如下的优点：

第一，真正开放性。Android 是一个真正意义上的开放性移动开发平台，它同时包含底层操作系统以及上层的用户界面和应用程序——移动电话工作所需的全部软件，而且不存在任何以往阻碍移动产业创新的专有权障碍。Google 与 OHA 合作开发 Android，目的就是通过与运营商、设备厂商、开发商等结成深层次的合作伙伴关系，来建立标准化、开放式的移动电话软件平台，在移动产业内形成一个开放式的生态系统，这样应用程序之间的通用性和互联性将在最大程度上得到保持。另一方面，Android 平台的开放性还体现在不同的厂商可以根据自己的需求对平台进行定制和扩展，以及使用这个平台无须任何授权许可费用等。显著的开放性可以使其拥有更多的开发者，随着用户和应用的日益丰富，一个崭新的平台也将很快走向成熟。开放性对于 Android 的发展而言，有利于积累人气，这里的人气包括消费者和厂商，而对于消费者来讲，最大的收益正是丰富的软件资源。开放的平台也会带来更大竞争，如此一来，消费者将可以用更低的价位购得心仪的手机。

第二，应用程序相互平等。所有的 Android 应用程序之间是完全平等的，所有的应用程序都运行在一个核心引擎上面，这个核心引擎就是一个虚拟机，它提供了一系列用于应用程序和硬件资源间通信的 API。抛开这个核心引擎，Android 所有其他的东西，包括系统的核心应用和第三方应用都是完全平等的。

第三，应用程序之间沟通无界限。在 Android 平台下开发应用程序，能方便实现应用程序之间的数据共享，只需要经过简单的声明或操作，应用程序即可访问或调用其他应用程序的功能，或者将自己的部分数据和功能提供给其他应用程序使用。

第四，快速方便的应用程序开发。Android 平台为开发人员提供了大量的实用库和工具，开发人员可以快速创建自己的应用程序。如今叱咤互联网的 Google 已经走过 10 年，从搜索巨人到全面的互联网渗透，Google 服务如地图、邮件、搜索等已经成为连接用户和互联网的重要纽带，而 Android 平台手机将无缝结合这些优秀的 Google 服务。

1.1.3 Android 的系统架构

Android 是基于 Linux 内核的软件平台和操作系统，采用了软件堆层（Software Stack，又名软件叠层）的架构，主要分为四部分，如图 1-3 所示。第一层以 Linux 内核工作为基础，由 C 语言开发，只提供由操作系统内核管理的底层基本功能；第二层为中间件层，包括函数库 Library 和虚拟机 Virtual Machine，由 C++开发；第三层为应用程序框架层，提供了 Android 平台基本的管理功能和组件重用机制；第四层为应用程序层，提供了一系列核心应用程序，包括通话程序、短信程序等，应用软件则由各公司自行开发，以 Java 作为编写程序的一部分。

Linux Kernel Android 基于 Linux 2.6 提供核心系统服务，例如：安全、内存管理、进程管理、网络堆栈、驱动模型。Linux Kernel 也作为硬件和软件之间的抽象层，它隐藏具体硬件细节而为上层提供统一的服务。

Android Runtime Android 包含一个核心库的集合，提供大部分在 Java 编程语言核心类库中可用的功能。每一个 Android 应用程序都是 Dalvik 虚拟机中的实例，运行在它们自己的进程中。Dalvik 虚拟机设计成一个设备可以高效地运行多个虚拟机。Dalvik 虚拟机可执行的文件格式是.dex，dex 格式是专为 Dalvik 设计的一种压缩格式，适合内存和处理器速度有限的系统。大多数虚拟机包括 JVM 都是基于栈的，而 Dalvik 虚拟机则是基于寄存器的。两种架构各有优劣，一般而言，基于栈的机器需要更多指令，而基于寄存器的机器指令更大。dx 是一套工具，可以将 Java .class 转换成 .dex 格式的工具。一个 dex 文件通常会有多个.class。由于 dex 有时必须进行最佳化，会使文件大小增加 1～4 倍，以 ODEX 结尾。 Dalvik 虚拟机依赖于 Linux 内核提供基本功能，如线程和底层内存管理。

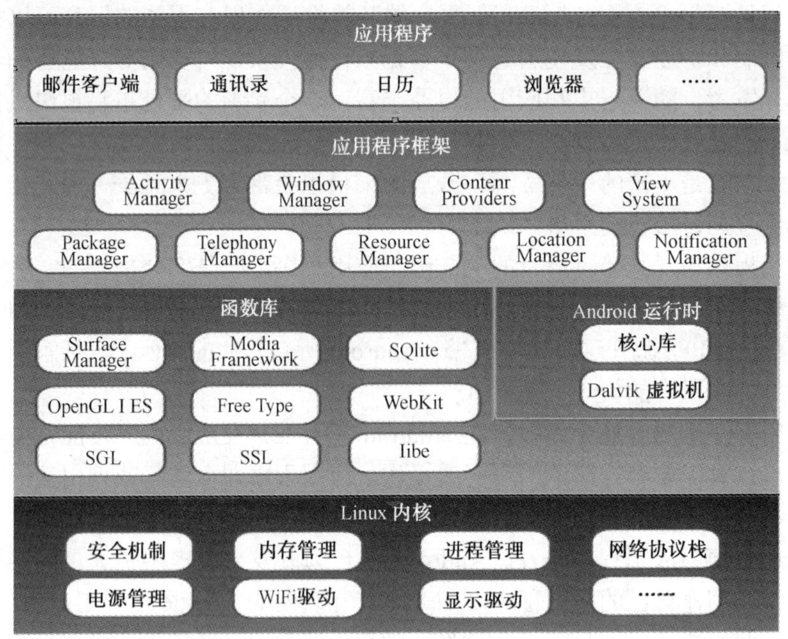

图 1-3　Android 体系结构图

Libraries Android 包含一个 C/C++库的集合，供 Android 系统的各个组件使用。这些功能通过 Android 的应用程序框架（Application Framework）暴露给开发者。下面列出一些核心库：

系统 C 库——标准 C 系统库（libc）的 BSD 衍生，调整为基于嵌入式 Linux 设备。

媒体库——基于 PacketVideo 的 OpenCore。这些库支持播放和录制许多流行的音频和视频格式，以及静态图像文件，包括 MPEG4、H.264、MP3、AAC、AMR、JPG、PNG。

界面管理——管理访问显示子系统和无缝组合多个应用程序的二维和三维图形层。

LibWebCore——新式的 Web 浏览器引擎，驱动 Android 浏览器和内嵌的 Web 视图。

SGL——基本的 2D 图形引擎。

3D 库——基于 OpenGL ES 1.0 APIs 的实现。库使用硬件 3D 加速或包含高度优化的 3D 软件光栅。

FreeType ——位图和矢量字体渲染。

SQLite ——所有应用程序都可以使用的强大而轻量级的关系数据库引擎。

Application Framework 通过提供开放的开发平台，Android 使开发者能够编制极其丰富和新颖的应用程序。开发者可以自由地利用设备硬件优势、访问位置信息、运行后台服务、设置闹钟、向状态栏添加通知等。开发者可以完全使用核心应用程序所使用的框架 APIs。应用程序的体系结构旨在简化组件的重用，任何应用程序都能发布它的功能且任何其他应用程序都可以使用这些功能（需要服从框架执行的安全限制）。这一机制允许用户替换组件。所有的应用程序其实是一组服务和系统，包括：

视图（View）——丰富的、可扩展的视图集合，可用于构建一个应用程序。包括列表、网格、文本框、按钮，甚至是内嵌的网页浏览器。

内容提供者（Content Providers）——使应用程序能访问其他应用程序（如通讯录）的数据，或共享自己的数据。

资源管理器（Resource Manager）——提供访问非代码资源，如本地化字符串、图形和布局文件。

通知管理器（Notification Manager）——使所有的应用程序能够在状态栏显示自定义警告。

活动管理器（Activity Manager）——管理应用程序生命周期，提供通用的导航回退功能。

Applications Android 装配一个核心应用程序集合，包括电子邮件客户端、SMS 程序、日历、地图、浏览器、联系人和其他设置。所有应用程序都是用 Java 编程语言写的。

在本节中，我们简单介绍了 Android 的起源、Android 的优点和 Android 的体系架构。从技术层面来讲，Android 最震撼人心之处在于其开放性和服务免费。Android 是对第三方软件完全开放的平台，开发者在为其开发程序时拥有更大的自由度，突破了 iPhone 等只能添加为数不多的固定软件的束缚。而且与 Windows Mobile、Symbian 等厂商不同，Android 操作系统可以免费使用。

1.2 Android 开发环境配置

Android 开发环境的安装和配置是开发 Android 应用程序的第一步，也是深入 Android 平台的一个非常好的机会。Eclipse 是开发 Android 应用程序的首选集成开发环境，因此我们这本书的案例都是在 Eclipse 工具中编写和调试的。由于各位读者都是有相当的 Java 基础，熟悉 Java 编程语言也对 Eclipse 集成开发环境有相当的了解，我们这里对于 Android 开发环境的安装和配置从以下几个方面简单加以介绍，相信我们的读者凭自己学习的能力，应该能融会贯通。

步骤一：安装 JDK 和 Eclipse 工具。这一步对于我们熟悉 Java 开发的读者来说，不是难事，只是告诉大家不要忘记了 JDK 环境变量的配置。这里就不赘述了。

步骤二：下载 Android SDK 工具包。Android SDK 是 Android 软件开发工具包（Android Software Development Kit）的简写。是 Google 公司为了提高 Android 应用程序开发效率、减少开发周期而提供的辅助开发工具、开发文档和程序范例。

Android SDK 可以从 Google 的中文 Android 开发网站上下载，网站的地址是 http://developer.android.com/sdk/index.html。打开 Android 开发网页，会看到如表 1-1 所示的三种版本 Android SDK 压缩文件。在开发网站，不仅能够下载最新的 Android SDK，还有许多有价值的学习内容，如开发文档、常见问题解答等。

表 1-1　Android 软件开发包 SDK 表

作业平台	Android-SDK	MD5 检查码
Windows	android-sdk_r17-windows.zip	3af1baeb39707e54df068e939aea5a79
	installer_r17-windows.exe (Recommended)	5afaf6511ebaa52bd6d1dba4afc61e41
Mac OS X (intel)	android-sdk_r17-macosx.zip	52639aae036b7c2e47cf291696b23236
Linux (i386)	android-sdk_r17-linux.tgz	14e99dfa8eb1a8fadd2f3557322245c4

如果你已经使用 Android SDK 开发环境，希望更新到新的版本或增加其他开发包，比如添加 USB 驱动程序，只要从 Android SDK 和 AVD 开发管理环境就可以取得最新的组件。不需要再下载新版的 Android SDK 开发包。Android SDK 开发包的安装这里就不赘述了，有一定 Java 基础的读者都会通过网络等资源进行学习和使用。

步骤三：ADT 插件的安装。ADT 插件是 Eclipse 集成开发环境的定制插件，为开发 Android 应用程序提供了一个强大的、完整的开发环境，可以快速建立 Android 工程，用户界面和基于 Android API 的组件，还可以使用 Android SDK 提供的工具进行程序调试，对 apk 文件进行签名等。安装 ADT 插件有两种方法，一种是手动下载 ADT 插件的压缩包，然后在 Eclipse 中进行安装；第二种是在 Eclipse 中输入插件的下载地址，由 Eclipse 自动完成下载和安装工作。第二种方法比较简单方便，但出错的概率较第一种大，我们这里用第二种方法。

启动 Eclipse，选择 Help→Install New Software，打开 Eclipse 的插件安装界面，如 Eclipse 插件安装界面如图 1-4 所示，单击 Add 按钮，进入 Add Site 界面，如图 1-5 所示，在 Add Site 界面的 Name 文本框中输入插件名称如 android，在 Location 文本框中输入 ADT 插件的下载网络路径 https://dl-ssl.google.com/android/eclipse/。

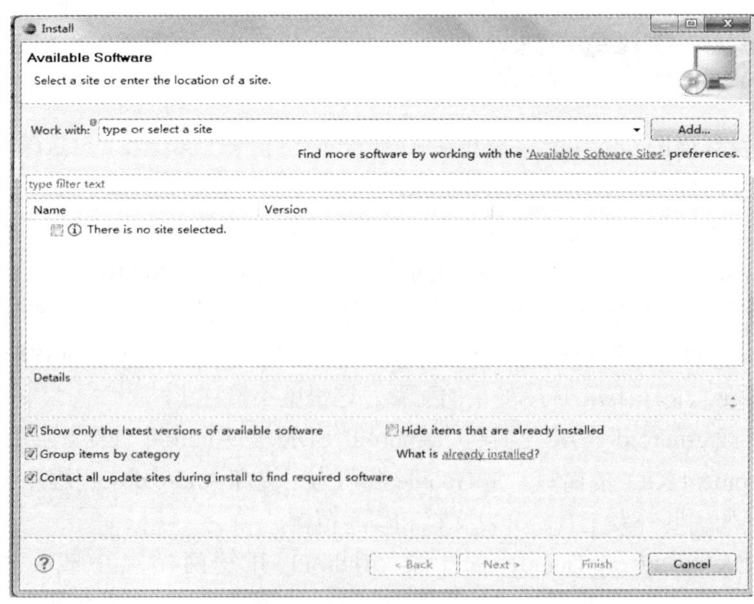

图 1-4　Eclipse 插件安装界面图

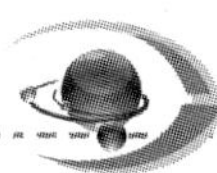

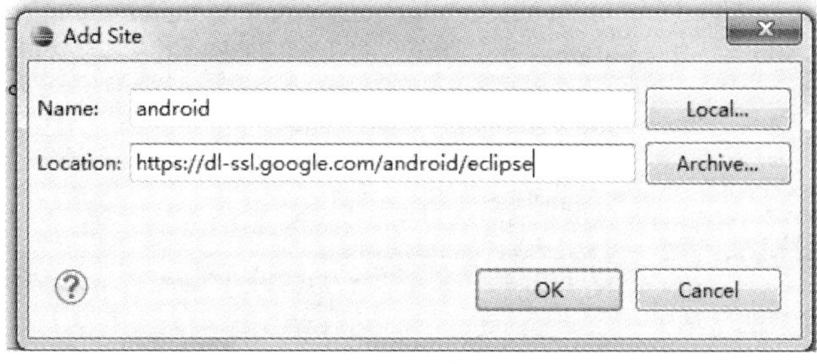

图 1-5　Add Site 界面图

正确填写 ADT 插件压缩包的下载路径后，在 Eclipse 的插件安装界面上会出现 ADT 插件的安装选项图，如图 1-6 所示，选中 Android DDMS 复选框和 Android Development Tools 复选框，然后单击 Next 按钮进入 ADT 插件许可界面，如图 1-7 所示。

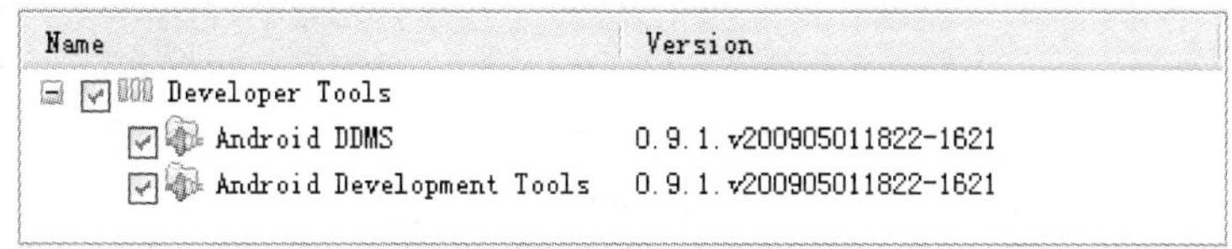

图 1-6　ADT 插件的安装选项图

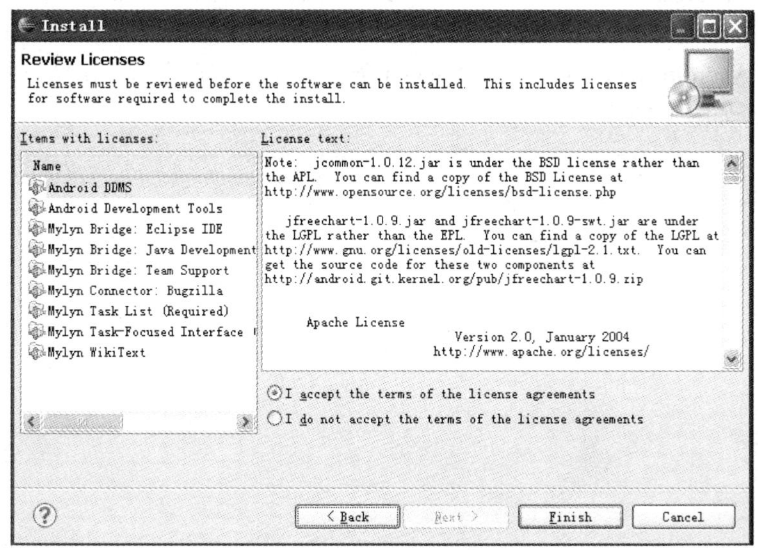

图 1-7　ADT 插件许可界面

在 ADT 插件许可界面中，选择 I accept the terms of the license agreements 单选项即可，待安装结束，重新启动 Eclipse，使 ADT 插件生效。

步骤四：配置 Android 开发环境。在 ADT 插件安装之后，开始设置 Android SDK 的保存路径。首先选择 Windows→Preferences 命令，打开 Android 配置界面，如图 1-8 所示，单击 Browse 按钮，在 SDK Location 文本框中输入 Android SDK 的保存路径，最后单击 Apply 按钮使配置生效。

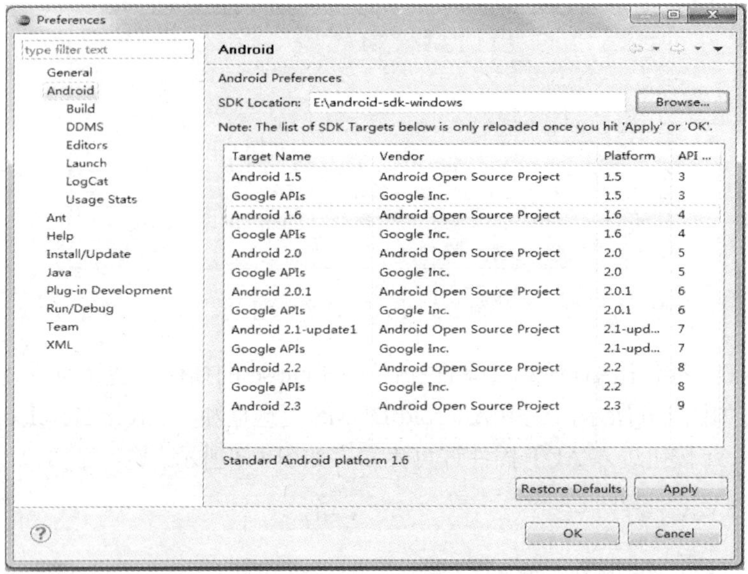

图 1-8　Android 配置界面图

步骤五：虚拟设备 AVD 的创建。使用 Android SDK 开发的 Android 应用程序需要进行测试，Android 为开发人员提供了可以在电脑上直接测试应用程序的虚拟设备 AVD（Android Virtual Device），或称做模拟器。AVD 的创建，首先启动 Eclipse，选择 Windows→Android SDK and AVD Manager，进入 Android SDK and AVD Manager 界面如图 1-9 所示，单击 New 按钮，弹出创建 AVD 的对话框，如图 1-10 所示。

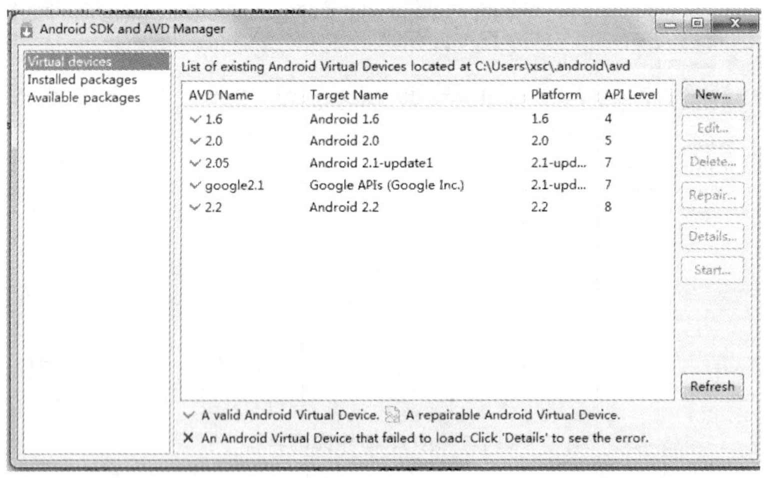

图 1-9　Android SDK and AVD Manager 界面图

在对话框中设置所要创建的 AVD 名称、API 版本、SD 卡大小以及 AVD 显示皮肤，单击 Create AVD 按钮，就完成了一个 AVD 的创建，依此类推，可以创建多个不同 API 版本的模拟器。创建成功 AVD 以后，可以启动模拟器，调试我们自己开发的 Android 应用程序了。

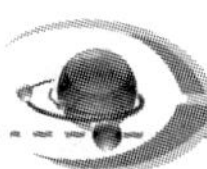

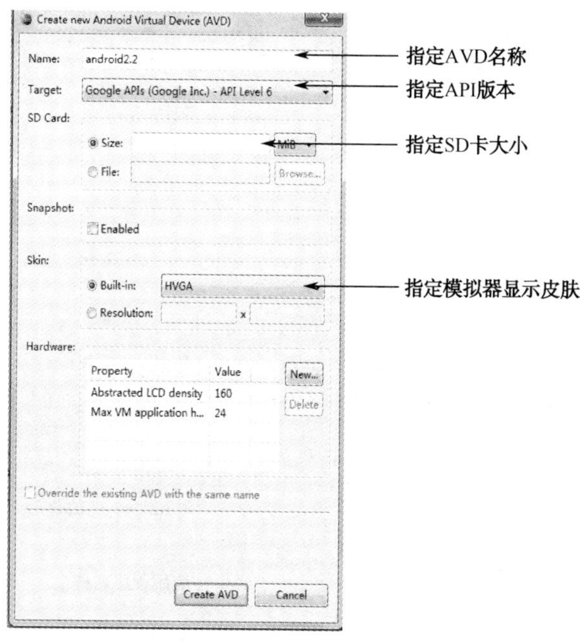

图 1-10 创建 AVD 对话框

1.3 第一个 Android 应用程序

在前面的章节中，我们已经完成了 Android 开发环境的初步搭建以及虚拟设备 AVD 的创建，在本节中我们将开发第一个 Android 应用程序——HelloAndroid。并对 Android 应用程序结构进行详细了解。Android 应用程序开发步骤如下。

步骤一：启动 Eclipse，创建 HelloAndroid 项目。打开 Android 工程向导：File→New→Project... | Android→Android Project 或 File →New→Other ... | Android→Android Project 进入 Android 工程向导对话框，如图 1-11 所示。在对话框的 Project name 文本框中填入项目名称 HelloAndroid，依次在项目界面中填入必要的信息，单击 Finish 按钮，则完成了我们的第一个项目创建。

注意：工程名称必须唯一，不能与已有的工程重名，应用程序名称，即 Android 程序在手机中显示的名称，显示在手机的顶部；包名称是包的命名空间，需遵循 Java 包的命名方法，由两个或多个标志符组成，中间用点隔开，为了包名称的唯一性，可以采用反写电子邮件地址的方式；创建 Activity 是个可选项，如需要自动生成一个 Activity 的代码文件，则选择该项。Activity 的名称与应用程序的名称不同，但为了简洁，可以让它们相同，表示这个 Activity 是 Android 程序运行时首先显示给用户的界面。应用程序版本号是可选项，可以填所选择 API 版本的版本号。

步骤二：调试项目。在 HelloAndroid 项目上，单击鼠标右键，则出现运行项目菜单选项，选择 Run As→Android Application，如图 1-12 所示。系统将自动启动虚拟设备，并将应用程序在虚拟设备中运行。观察虚拟设备屏幕，将显示我们开发的第一个 Android 应用程序项目界面，如图 1-13 所示。注意，第一次启动模拟器所用时间较长，一般需要 3～5 分钟。

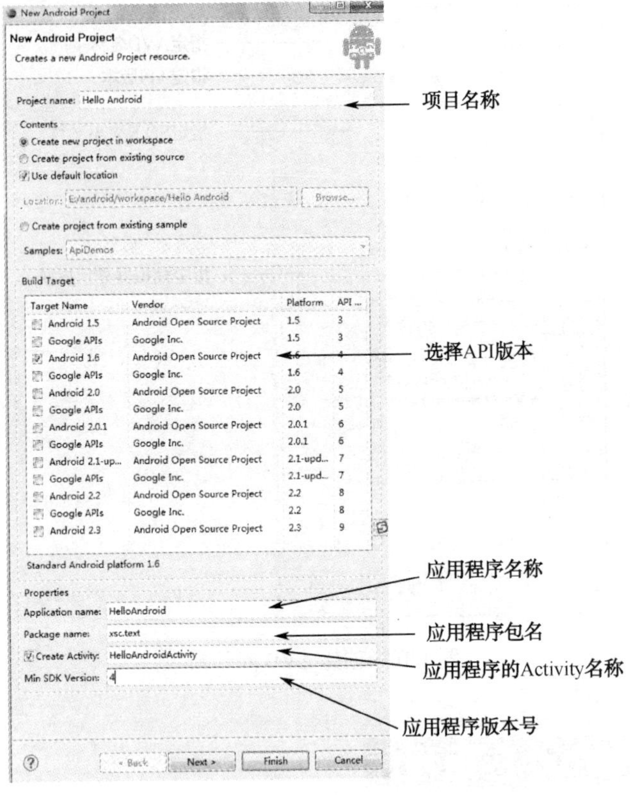

图 1-11　Android 工程向导对话框图

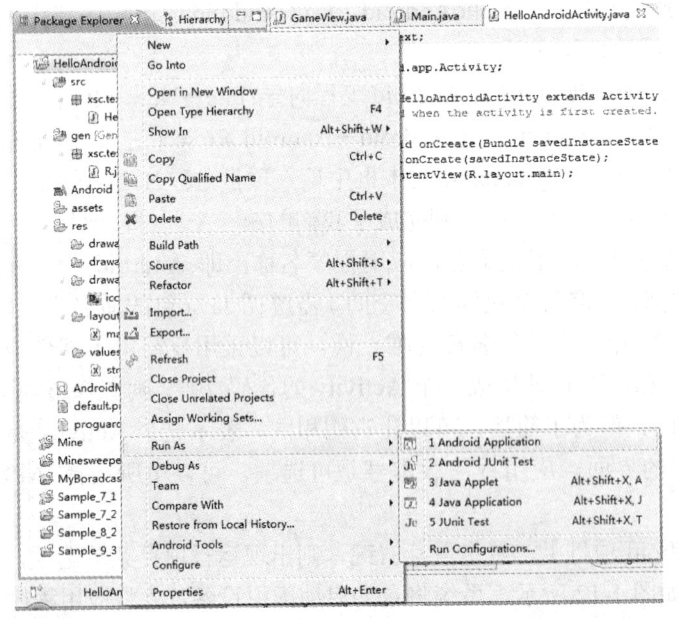

图 1-12　运行项目菜单选项图

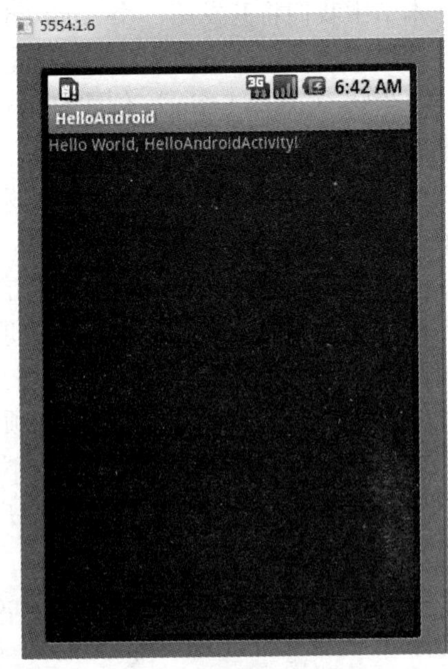

图 1-13　HelloAndroid 运行示意图

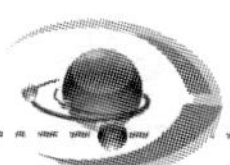

前面我们已经能够创建自己的 Android 应用程序，现在我们需要对应用程序的目录结构、资源管理以及程序权限进行更深入的理解。我们先来看我们前面建立的 HelloAndroid 项目的目录结构图，如图 1-14 所示。

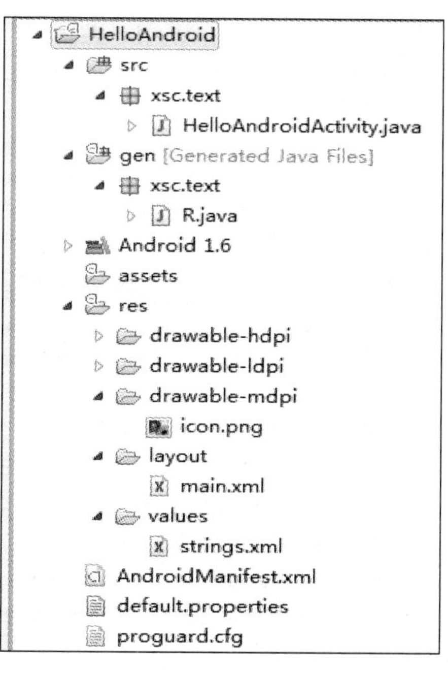

图 1-14　HelloAndroid 项目的目录结构图

src 目录中存放的是该项目的源文件，所有允许用户修改的 Java 文件和用户自己添加的 Java 文件，都保存在这个目录中。

gen 目录下的文件是 ADT 自动生成的，并不需要人为地去修改，实际上该目录下只定义了一个 R.java 文件，该文件相当于项目的字典，项目中所涉及到的用户界面、字符串、图片、声音等资源都会在该类中创建其唯一的 ID 编号，这些编号为整型，以十六进制自动生成。当项目中使用这些资源时，会通过该类得到资源的引用。

Android1.6 目录中存放的是支持该项目的 JAR 包，同时还包含项目打包是需要的 META—INF 目录。

assets 目录用于存放项目相关的资源文件，例如文本文件等。此目录中的资源不能够被 R.java 文件索引，因此只能以字节流形式进行读取，一般情况下为空。

res 目录用于存放应用程序中经常使用的资源文件，包括图片、声音、布局文件及参数描述文件等，包括多个目录，其中以 drawable 开头的三个文件夹用于存储.png、.9.png、.jpg 等图片资源，layout 文件夹存放的是应用程序的布局文件，raw 用于存放应用程序所用到的声音文件，values 存放的则是所有 XML 格式的资源描述文件，例如字符串资源描述文件 strings.xml、样式的描述文件 styles.xml、颜色描述文件 colors.xml、尺寸描述文件 dimens.xml 以及数组描述文件 arrays.xml 等。

default.properties 文件为项目配置文件，不需要人为改动，系统会自动对其进行管理。文件里面记录了 Android 工程的相关设置，例如编译目标和 apk 设置等。如果需要更改其中的设置，必须通过右击工程名称，在弹出的快捷菜单中选择 Properties 选项修改。

AndroidManifest.xml 文件为应用程序的系统配置文件也叫清单文件。该文件中包含了 Android 系统运行 Android 程序前所必须掌握的重要信息，这些信息包括应用程序名称、图标、包名称、模块组成、授权和 SDK 最低版本等。而且每个 Android 程序必须在根目录下包含一个 AndroidManifest.xml 文件。下面我们看一个 AndroidManifest.xml 文件的基本格式：

```
<?xml version="1.0" encoding="utf-8"?>
<!-在根元素里面定义命名空间 -->
<manifest xmlns:android="http://schemas.android.com/apk/res/android"
     package="xsc.text" <!- 定义应用程序包名 -->
     android:versionCode="1"<!- 定义应用程序版本号 -->
     android:versionName="1.0"><!- 定义应用程序版本名称 -->
<!- 定义应用程序的图标和标签名称 -->
   <application android:icon="@drawable/icon" android:label="@string/
   app_name">
<!- 声明需要显示的Activity类名和标签名称 -->
       <activity android:name=".HelloAndroidActivity"
                android:label="@string/app_name">
          <intent-filter><!- 定义过滤器 -->
            <action android:name="android.intent.action.MAIN" />
            <category android:name="android.intent.category.LAUNCHER" />
          </intent-filter>
       </activity>
</application>
<uses-sdk android:minSdkVersion="4" /><!- 定义版本号 -->
</manifest>
```

AndroidManifest.xml 文件中还包含其他组件，例如 Service、BroadcastReceiver、ContentProvider 等，分别使用<service></service>标签、<receiver></receiver>标签、<provider></provider>标签来声明，我们在这里就不做详细介绍，在以后的学习中再分别介绍。

在 AndroidManifest.xml 文件中，还可以为应用程序指定相应的权限，例如网络权限、发送短信的权限、打电话的权限等。应用程序的权限有很多，全部都封装到 Manifest.permission 类中，读者可以自行查阅 API。权限的使用方法是将权限声明的语句添加到 AndroidManifest.xml 文件中</manifest>标签之前，例如当某个应用程序需要添加发送短信的权限时，只需将“<uses-permission android:name="android.permission.SEND_SMS"/>”添加到 AndroidManifest.xml 文件</的 manifest>标签之前就可以了。应用程序除了声明自身应该具有的权限外，还可以声明访问本应用程序的程序应该具有的权限，这样，其他应用程序需要访问该应用程序时，必须具有该应用程序所需要的权限。例如在<activity></activity>之间添加权限<uses-permission android: name="android.permission.SEND_SMS"/>，那么当其他应用程序需要访问该程序时，就必须具有 SEND_SMS 权限。在表 1-2 中列出了部分常用的权限。

表 1-2　Android 部分常用权限表

声明语句	描述语句
android.permission.SEND_SMS	发送短信的权限
android.permission.CALL_PHONE	打电话的权限
android.permission.BLUETOOTH	使用蓝牙的权限
android.permission.WRITE_OWNER_DATA	写数据的权限
android.permission.SET_TIME_ZONE	设置时区的权限
android.permission.READ_SMS	读取短信的权限
android.permission.WRITE_GSERVICES	Google 地图服务的权限
android.permission.INTERNET	使用网络的权限
android.permission.CAMERA	允许使用照相设备的权限
android.permission.FLASHLIGHT	允许访问设备上的闪光灯的权限
android.permission.WALLPAPER	允许程序设置壁纸的权限
android.permission.VIBRATE	允许应用程序控制设备振动的权限

本章我们主要介绍了 Android 概述、Android 开发环境的配置与权限、Android 项目的建立以及项目目录结构。

第 2 章

Android 应用程序

Android 应用程序由组件组成，组件是可以被调用的基本功能模块。Android 系统利用组件实现程序内部或程序间的模块调用，以解决代码复用的问题，这是 Android 系统非常重要的特性。在程序设计时，在 AndroidManifest.xml 中声明可共享的组件，声明后其他应用程序可以直接调用这些共享组件。Android 系统有 4 个重要的组件，分别是 Activity、Service、BroadcaseReceiver 和 ContentProvider。在下面的章节中我们将进行逐一介绍。

2.1 基本组件介绍

Activity 是 Android 中最常用的组件，是应用程序的表示层，Activity 一般通过 View 来实现应用程序的用户界面，相当于一个屏幕，用户与程序的交互是通过该类实现的。Android 应用程序可以包含一个或多个 Activity，一般在程序启动后会呈现一个 Activity，用于提示用户程序已经正常启动。Activity 在界面上的表现形式一般是全体窗体，也可以是非全屏悬浮窗体或对话框。

Service 一般用于没有用户界面，但需要长时间在后台运行的应用。实际上，Service 是一个具有较长的生命周期但是并没有用户界面的程序。例如在播放 MP3 音乐时，使用 Service 播放 MP3 音乐，可以在关闭播放器界面的情况下长时间播放 MP3 音乐，并通过对外公开 Service 的通信接口，控制 MP3 音乐播放的启动、暂停和停止。

Service 一般由 Activity 启动，但是并不依赖于 Activity，即当 Activity 的生命周期结束时，Service 仍然会继续运行，直到自己的生命周期结束为止。Service 的启动方式有两种。startService 方式启动是当 Activity 调用 startService 方法启动 Service 时，会依次调用 onCreate 和 onStart 方法来启动 Service，而当调用 stopService 方法结束 Service 时，又会调用 onDestroy 方法结束 Service。Service 同样可以在自身调用 stopSelf 或 stopService 方法来结束 Service。bindService 方式启动是 Activity 调用 bindService 方法启动 Service，此时会依次调用 onCreate 和 onBind 方法启动 Service。而当通过 unbindService 方法结束 Service 时，则会依次调用 onUnbind 和 onDestroy 方法。

BroadcastReceiver 为用户接收广播通知的组件，当系统或某个应用程序发送广播时，可以使用 BroadcastReceiver 组件来接收广播信息并做相应处理。在信息发送时，需要将信息封装后添加到一个 Intent 对象中，然后通过调用 Content.sendBroadcast()、sendOrderedBroadcast()或 sendStickyBroadcast()方法将 Intent 对象广播出去，然后接收者会检查注册的 IntentFilter 是否与收到的 Intent 相同，当相同时便会调用 onReceive()方法来接收信息。三个发送方法的不同之处是使用 sendBroadcast()或者 sendStickyBroadcast()方法发送广播时，所有满足条件的接收者都会随时地执行，而使用 sendOrderedBroadcast()方法发送的广播接受者会根据 IntentFilter 中设置的优先级顺序来执行。

BroadcastReceiver 的使用过程如下：

（1）将需要广播的消息封装到 Intent 中。

（2）通过三种发送方法中的一种将 Intent 广播出去。

（3）通过 IntentFilter 对象来过滤所发送的实体 Intent。

（4）实现一个重写了 onReceive 方法的 BroadcastReceiver。

需要注意的是，注册 BroadcastReceiver 对象的方式有两种，一种是在 AndroidManifest.xml 中声明，另一种是在 Java 代码中设置。在 AndroidManifest.xml 中声明时，将注册的信息包裹在<receiver></receiver>标签中，并通过<intent-filter>标签来设置过滤条件；在 Java 代码中设置时，需要先创建 IntentFilter 对象，并为 IntentFilter 对象设置 Intent 的过滤条件，并通过 Content.register Receiver 方法来注册监听，然后通过 Content.unregisterReceiver 方法来取消监听，此种注册方式的缺点是当 Content 对象被销毁时，该 BroadcastReceiver 也就随之被摧毁了。

ContentProvider 是用来实现应用程序之间数据共享的类。当需要进行数据共享时，一般利用 ContentProvider 为需要共享的数据定义一个 URI，然后其他应用程序通过 Content 获得 Content Resolver 并将数据的 URI 传入即可。Android 系统已经为一些常用的数据创建了 ContentProvider，这些 ContentProvider 都位于 android.provider 下，只要有相应的权限，自己开发的应用程序便可以轻松地访问这些数据。

对于 ContentProvider 最重要的就是数据模型（data model）和 URI，接下来分别对其进行介绍。数据模型就是 ContentProvider 为所有需要共享的数据创建一个数据表，在表中，每一行表示一条记录，而每一列代表某个数据，并且其中每一条数据记录都包含一个名为“_ID”的字段类标志每条数据。URI 就是每个 ContentProvider 都会对外提供一个公开的 URI 来标识自己的数据集，当管理多个数据集时，将会为每个数据集分配一个独立的 URI，所有的 URI 都以“content://”开头。需要注意的是，使用 ContentProvider 访问共享资源时，需要为应用程序添加适当的权限才可以。权限为“<uses-permission.android:name=”android.permission.READ_CONTACTS”/>”。

Android 系统通过组件机制，有效地降低了应用程序的耦合性，使向其他应用程序共享私有数据（ContentProvider）和调用其他应用程序的私有模块（Service）成为可能。所有 Android 组件都具有自己的生命周期，称为组件生命周期，是从组件建立到组件销毁的整个过程。在这个过程中，组件会在可见、不可见、活动、非活动等状态中不断变化。

2.2 Activity 生命周期

Activity 生命周期指 Activity 从启动到销毁的过程，在这个过程中，Activity 一般表现为 4 种状态，分别是活动状态、暂停状态、停止状态和非活动状态。活动状态是指当 Activity 在用户界面中处于最上层，用户完全看得到，能够与用户进行交互，则这时 Activity 处于活动状态；暂停状态是指当 Activity 在界面上被部分遮挡，该 Activity 不再处于用户界面的最上层，且不能够与用户进行交互，则这个 Activity 处于暂停状态；停止状态是指 Activity 在界面上完全不能被用户看到，也就是说这个 Activity 被其他 Activity 全部遮挡，则这个 Activity 处于停止状态；非活动状态是指 Activity 所处的不在以上三种状态中的另一种状态。

Activity 的 4 种状态的变换如图 2-1 所示。Activity 启动后处于活动状态，此时的 Activity 处于最上层，是与用户正在进行交互的组件，因此 Android 系统会努力保证处于活动状态 Activity 的资源需求，资源紧张时可终止其他状态的 Activity；如果用户启动了新的 Activity，部分遮挡了当前的 Activity，或新的 Activity 是半透明的，则当前的 Activity 转换为暂停状态，Android 系统仅在为处于活动状态的 Activity 释放资源时才终止处于暂停状态的 Activity；如果用户启动新的 Activity 完全遮挡了当前的 Activity，则当前的 Activity 转变为停止状态，停止状态的 Activity 将优先被终止；活动状态的 Activity 被用户关闭后，或暂停状态或停止状态的 Activity 被系统终止后，Activity 便进入了非活动状态。

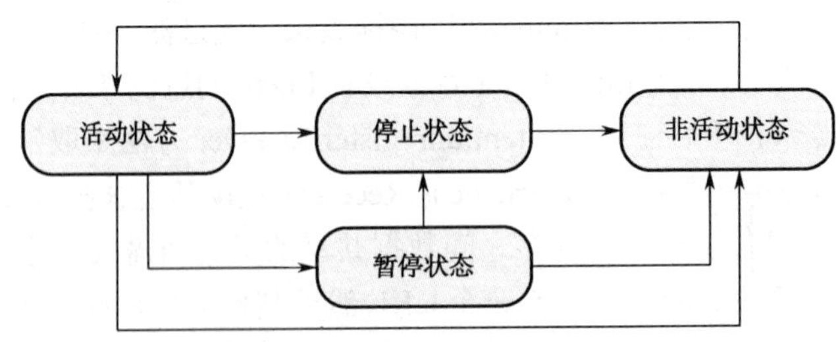

图 2-1　Activity 状态变换图

为能够更好地理解 Activity 的生命周期，还需要对 Activity 栈做一个简要的介绍。Activity 栈保存了已经启动且没有终止的所有 Activity，并遵循“后进先出”的原则。如图 2-2 所示，栈顶的 Activity 处于活动状态，除栈顶以外的其他 Activity 处于暂停状态或停止状态，而被终止的 Activity 或已经出栈的 Activity 则不在栈内。

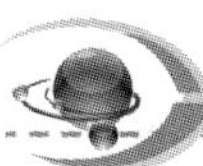

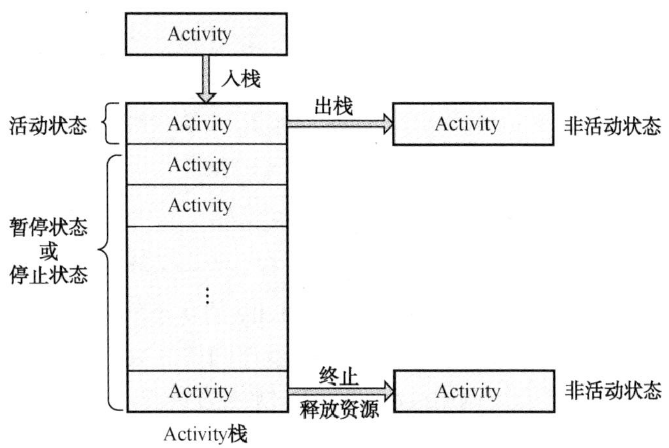

图 2-2　Activity 栈图

Activity 的状态与其在 Activity 栈的位置有着密切的关系，不仅如此，Android 系统在资源不足时，也是通过 Activity 栈来选择哪些 Activity 是可以被终止的。一般来讲，Android 系统会优先选终止处于停止状态，且位置靠近栈底的 Activity，因为这些 Activity 启动顺序最靠前，而且在界面上用户是看不到的。

随着用户在界面的操作和 Android 系统对资源的管理，Android 不断变化在 Activity 栈的位置，其状态也不断在 4 种状态中转变。为了能够让 Android 程序了解自身状态的变化，Android 系统提供了多个事件回调函数，在事件回调函数中添加相关代码，就可以在 Activity 状态变化时完成适当的工作。下面的代码给出了 Activity 的主要事件回调函数。

```
public class MyActivity extends Activity {
    protected void onCreate(Bundle savedInstanceState);
    protected void onStart();
    protected void onRestart();
    protected void onResume();
    protected void onPause();
    protected void onStop();
    protected void onDestroy(); }
```

这些事件回调函数何时被调用，具体用途是什么，以及是否可以被 Android 系统终止，可以参考表 2-1。

表 2-1　Activity 生命周期的事件回调函数表

函　　数	是否可终止	说　　明
onCreate()	否	Activity 启动后第一个被调用的函数，常用来进行 Activity 的初始化，例如创建 View、绑定数据或恢复信息等
onStart()	否	当 Activity 显示在屏幕上时，该函数被调用
onRestart()	否	当 Activity 从停止状态进入活动状态前，调用该函数
onResume()	否	当 Activity 能够与用户交互，接受用户输入时，该函数被调用。此时的 Activity 位于 Activity 栈的栈顶

续表

函　数	是否可终止	说　明
onPause()	是	当 Activity 进入暂停状态时，该函数被调用。一般用来保存持久的数据或释放占用的资源
onStop()	是	当 Activity 进入停止状态时，该函数被调用
onDestroy()	是	在 Activity 被终止前，即进入非活动状态前，该函数被调用

除了 Activity 生命周期的事件回调函数以外，还有 onRestoreInstanceState()和 onSaveInstanceState()两个函数经常会被使用，用于保存和恢复 Activity 的状态信息，例如用户在界面中选择的内容或输入的数据等。这两个函数不是生命周期的事件回调函数，不会因为 Activity 的状态变化而被调用，但在下述情况下会被调用：Android 系统因为资源紧张需要终止某个 Activity，但这个 Activity 在未来的某一时刻还会显示在屏幕上。

Activity 状态保存和恢复函数 onRestoreInstanceState()和 onSaveInstanceState ()的说明参见表 2-2。

表 2-2　Activity 状态保存/恢复的事件回调函数说明表

函　数	是否可终止	说　明
onSaveInstanceState()	否	Android 系统因资源不足终止 Activity 前调用该函数，用以保存 Activity 的状态信息，供 onRestoreInstanceState()或 onCreate()恢复之用
onRestoreInstanceState()	否	恢复 onSaveInstanceState()保存的 Activity 状态信息，在 onStart()和 onResume ()之间被调用

举个例子说明这两个函数是如何调用的。如用户启动 Activity A，然后又直接启动 Activity B，这时系统资源不足，需要终止 Activity A，则会调用 Activity A 的 onSaveInstanceState()来保存 Activity A 的状态信息；当用户关闭了 Activity B 时，Activity B 的 onSaveInstanceState()不会被调用，因为是用户主动关闭 Activity B 而不是系统终止，而且 Activity B 不会再显示在屏幕上；当 Activity A 重新显示在屏幕上后，Activity A 会调用 onRestoreInstanceState()用以恢复之前保存的 Activity A 的状态信息。

函数 onSaveInstanceState()保存的 Activity 状态信息，也可以在 onCreate()函数中进行恢复，但 onRestoreInstanceState()使用起来更加方便，也使得代码更加整洁。因为使用者可以在 onCreate()函数中完成初始化工作结束后，再在 onRestoreInstanceState()中决定是否使用默认设置或恢复保存的数据。

在 Activity 的生命周期中，并不是所有的事件回调函数都会被执行，但如果被调用则遵循一定的调用顺序。那么，这个调用顺序如图 2-3 所示。

由上图可知，Activity 的生命周期可分为全生命周期、可视生命周期和活动生命周期。每种生命周期中包含不同的事件回调函数。

全生命周期是从 Activity 建立到销毁的全部过程，始于 onCreante()，结束于 onDestroy()。一般情况下，使用者在 onCreate()中初始化 Activity 所能使用的全局资源和状态，并在 onDestroy()中释放这些资源。例如，Activity 中使用后台线程下载网络数据，则在 onCreate()中创建线程，在 onDestroy()中停止并销毁线程。在一些极端的情况下，Android 系统会不调用 onDestroy()函数，而是直接终止进程。

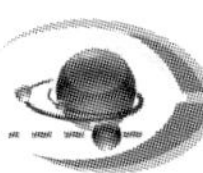

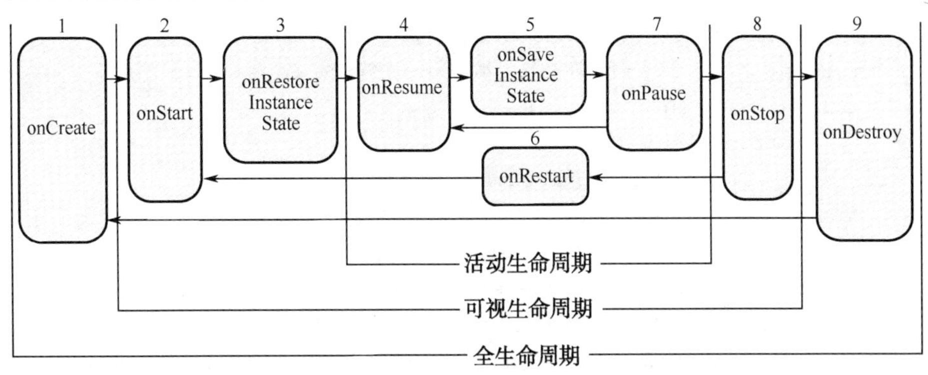

图 2-3　Activity 事件回调函数的调用顺序图

可视化生命周期是Activity在界面上从可见到不可见的过程，开始于onStart()，结束于onStop()。onStart()一般用来初始化或启动与更新界面相关的资源。onStop()一般用来暂停或停止一切与更新用户界面相关的线程、计时器和服务，因为在调用onStop()后，Activity对用户不再可见，更新用户界面也就没有任何实际意义。onRestart()函数在onStart()前被调用，用来在Activity从不可见变为可见的过程中，进行一些特定的处理过程。因为Activity不断从可见变为不可见，再从不可见变为可见，所以onStart()和onStop()会被多次调用。另外，onStart()和onStop()也经常被用来注册和注销BroadcastReceive，例如使用者可以在onStart()注册一个BroadcastReceive，用来监视某些重要的广播信息，使用这些消息更新用户界面中的相关内容，并可以在onStop()中注销BroadcastReceive。

活动生命周期是Activity在屏幕的最上层，并能够与用户交互的阶段，开始于onResume()，结束于onPause()。例如，在手机进入休眠状态时，处于活动生命周期的Activity会调用onPause()，当手机从休眠状态被唤醒后，则会调用onResume()。因为在Activity的状态变换过程中onResume()和onPause()经常被调用，因此这两个函数中应使用更为简单、高效的代码。

为了能够更好地理解Activity事件回调函数的调用顺序，下面建立一个新的Android工程来进行说明。工程名称为ActivityLifeCycle，下面的代码是ActivityLifeCycle.java文件的代码。

```
public class ActivityLifeCycle extends Activity {
    private static String TAG = "LIFTCYCLE";
   @Override  //完全生命周期开始时被调用，初始化Activity
    public void onCreate(Bundle savedInstanceState) {
        super.onCreate(savedInstanceState);
        setContentView(R.layout.main);
        Log.i(TAG, "(1) onCreate()");    }
 @Override  //可视生命周期开始时被调用，对用户界面进行必要的更改
   public void onStart() {
        super.onStart();
        Log.i(TAG, "(2) onStart()");}
  @Override //在onStart()后被调用，用于恢复onSaveInstanceState()保存的用户界面信息
   public void onRestoreInstanceState(Bundle savedInstanceState) {
        super.onRestoreInstanceState(savedInstanceState);
```

```
        Log.i(TAG, "(3) onRestoreInstanceState()");   }
    @Override //在活动生命周期开始时被调用，恢复被onPause()停止的用于界面更新的资源
      public void onResume() {
        super.onResume();
        Log.i(TAG, "(4) onResume()"); }
      @Override  // 在onResume()后被调用，保存界面信息
      public void onSaveInstanceState(Bundle savedInstanceState) {
  super.onSaveInstanceState(savedInstanceState);
        Log.i(TAG, "(5) onSaveInstanceState()");    }
    @Override //在重新进入可视生命周期前被调用，载入界面所需要的更改信息
    public void onRestart() {
       super.onRestart();
       Log.i(TAG, "(6) onRestart()");    }
    @Override  //在活动生命周期结束时被调用，用来保存持久的数据或释放占用的资源
    public void onPause() {
      super.onPause();
       Log.i(TAG, "(7) onPause()");  }
  @Override //在可视生命周期结束时被调用，一般用来保存持久的数据或释放占用的资源
  public void onStop() {
       super.onStop();
        Log.i(TAG, "(8) onStop()");   }
     @Override //在完全生命周期结束时被调用，释放资源，包括线程、数据连接等
     public void onDestroy() {
         super.onDestroy();
         Log.i(TAG, "(9) onDestroy()");    }
  }
```

上面的程序主要通过在生命周期函数中添加“日志点”的方法进行调试，程序的运行结果将会显示在LogCat中。为了显示结果易于观察和分析，在LogCat设置过滤器LifeCycleFilter，过滤方法选择by Log Tag，过滤关键字为LIFTCYCLE。

为了观察Android程序启动和关闭时调用生命周期函数的顺序，首先正常启动ActivityLifeCycle，然后按下模拟器的返回键，关闭ActivityLifeCycle。我们会看到LogCat的输出结果如图2-4所示。从图中我们可以得知，函数调用顺序为：onCreate()→onStart()→onResume()→onPause()→onStop()→onDestroy()。在Activity启动时，系统首先调用调用onCreate()函数分配资源，然后调用onStart()将Activity显示在屏幕上，之后调用onResume()获取屏幕焦点，使Activity能接受用户的输入，这时用户能够正常使用这个Android程序了。用户单击回退键或Activity调用finish()函数时，会导致Activity关闭，系统会相继调用onPause()、onStop()和onDestroy()，释放资源并销毁进程。因为Activity关闭后，除非用户重新启动应用程序，否则这个Activity不会出现在屏幕上，因此系统直接调用onDestroy()销毁了进程，且没有调用onSaveInstanceState()函数来保存Activity状态。

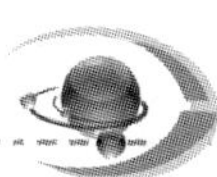

Time	pid	tag	Message
06-07 01:22:40.741 I	779	LIFECYCLE	(1) onCreate()
06-07 01:22:40.753 I	779	LIFECYCLE	(2) onStart()
06-07 01:22:40.760 I	779	LIFECYCLE	(4) onResume()
06-07 01:23:30.192 I	779	LIFECYCLE	(7) onPause()
06-07 01:23:30.473 I	779	LIFECYCLE	(8) onStop()
06-07 01:23:30.490 I	779	LIFECYCLE	(9) onDestroy()

图 2-4　全生命周期的 LogCat 输出图

在 Activity 启动后，如果启动其他的程序，原有的 Activity 会被新启动程序的 Activity 完全遮挡，因此原有 Activity 会进入停止状态；如果将新启动的程序关闭，则原有 Activity 会从停止状态恢复到活动状态。为了能够分析上述状态转换过程中的函数调用顺序，首先正常启动 ActivityLifeCycle，然后通过呼出/接听键启动内置的拨号程序，再通过返回键退出拨号程序，使 ActivityLifeCycle 重新显示在屏幕上。LogCat 的输出结果如下可视化生命周期的 LogCat 输出图如图 2-5 所示，ActivityLifeCycle 启动时调用的函数没有在 LogCat 中显示。

Time	pid	tag	Message
06-07 02:23:01.530 I	779	LIFECYCLE	(5) onSaveInstanceState()
06-07 02:23:01.530 I	779	LIFECYCLE	(7) onPause()
06-07 02:23:01.609 I	779	LIFECYCLE	(8) onStop()
06-07 02:23:04.483 I	779	LIFECYCLE	(6) onRestart()
06-07 02:23:04.499 I	779	LIFECYCLE	(2) onStart()
06-07 02:23:04.499 I	779	LIFECYCLE	(4) onResume()

图 2-5　可视化生命周期的 LogCat 输出图

从可视化生命周期的 LogCat 输出图可以得知，函数调用顺序为：onSaveInstanceState()→onPause()→onStop()→onRestart()→onStart()→onResume()。Activity 启动后，当内置的拨号程序被启动时，原有的 Activity 被完全覆盖，系统首先调用 onSaveInstanceState()函数保存 Activity 状态，因为界面上不可见的 Activity 可能被系统终止；系统然后调用 onPause()和 onStop()，停止对不可见 Activity 的更新。在用户关闭拨号程序后，系统调用 onRestart()恢复界面上需要更新的信息，然后调用 onStart()和 onResume()重新显示 Activity，并接受用户交互。系统调用了 onSave InstanceState()保存 Activity 的状态，因为进程没有被终止，所以没有调用 onRestoreInstanceState()恢复保存的 Activity 状态。

Activity 启动后，如果被其他程序部分遮挡，或手机进入休眠状态，这个 Activity 进入暂停状态。为了分析活动生命周期中的函数调用顺序，首先启动 ActivityLifeCycle，然后通过“挂断键”使模拟器进入休眠状态，然后再通过“挂断键”唤醒模拟器。此时 LogCat 的输出结果如图 2-6 所示。

Time	pid	tag	Message
06-07 04:47:13.673 I	779	LIFECYCLE	(5) onSaveInstanceState()
06-07 04:47:13.673 I	779	LIFECYCLE	(7) onPause()
06-07 04:47:15.038 I	779	LIFECYCLE	(4) onResume()

图 2-6　活动生命周期的 LogCat 输出图

从图 2-5 可以得知，函数调用顺序如下：onSaveInstanceState()→onPause()→onResume()。在 Activity 进入休眠状态前，系统首先调用 onSaveInstanceState()保存 Activity 的状态，然后调用 onPause()停止与用户交互；在系统被唤醒后，系统调用 onResume()恢复与用户的交互。

2.3 Android 程序调试

在 Android 程序开发过程中，出现错误（Bug）是不可避免的事情。一般情况下，语法错误会被集成开发环境监测到，并提示使用者错误的位置以及修改方法。但逻辑错误就不那么容易发现了，只有将程序在模拟器或硬件设备上运行才能够发现。逻辑错误的定位和分析是件困难的事情，尤其是代码量较大且结构复杂的应用程序，仅凭直觉很难快速找到并解决问题。因此，Android 系统提供了几种调试工具，用于定位、分析及修复程序中出现的错误，这些工具包括 LogCat 和 DevTools，我们这里仅对 LogCat 工具进行介绍。

LogCat 是用来获取系统日志信息的工具，并可以显示在 Eclipse 集成开发环境中。LogCat 能够捕获的信息包括 Dalvik 虚拟机产生的信息、进程信息、ActivityManager 信息、PackagerManager 信息、Homeloader 信息、WindowsManager 信息、Android 运行时信息和应用程序信息等。在 Eclipse 的默认开发模式下没有 LogCat 的显示页，用户可以使用 Window→Show View→Other 打开 Show View 的选择菜单，然后在 Android→LogCat 中选择 LogCat，如图 2-7 所示。

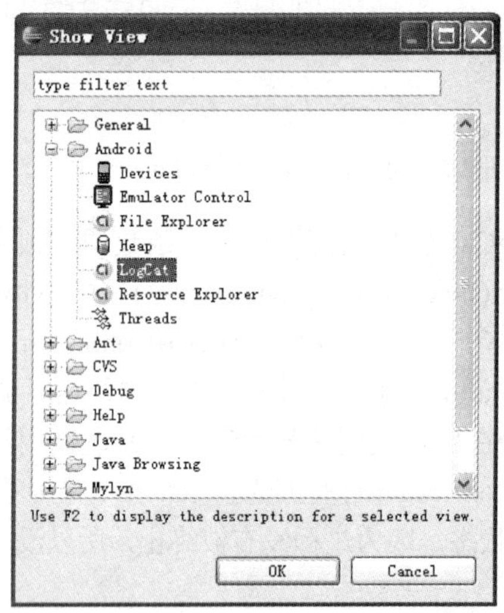

图 2-7　在 Show View 中选择 LogCat 图

这样，LogCat 便会在 Eclipse 的下方区域显示出来。如图 2-8 所示。

LogCat 的右上方的 5 个字母 V、D、I、W、E，表示 5 种不同的日志信息，分别是详细（Verbose）信息、调试（Debug）信息、通告（Info）信息、警告（Warn）信息、错误（Error）信息。不同类型日志信息的级别是不相同的，级别最高的是错误信息其次是警告信息，然后是通知信息和调试信息，最后是详细信息。在 LogCat 中，用户可以通过 5 个字母图标选择显示的信息类型，级别高于所选类型的信息也会在 LogCat 中显示，但级别低于所选类型的信息则不会被显示。

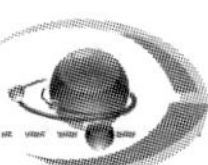

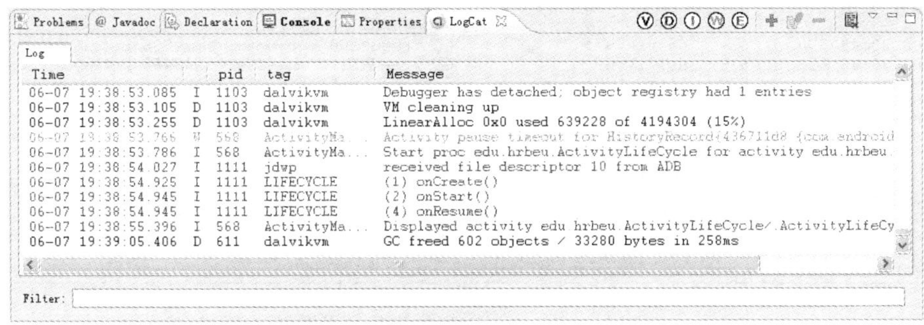

图 2-8　Eclipse 中的 LogCat 图

即使用户指定了所显日志信息的级别，仍然会产生很多日志信息，很容易让用户不知所措。LogCat 还提供了“过滤”功能，在右上角的“+”号和“-”号，分别是添加和删除过滤器。用户可以根据日志信息的标签（Tag）、产生日志的进程编号（Pid）或信息等级（Level），对显示的日志内容进行过滤。

在 Android 程序调试过程中，首先需要引入 android.util.Log 包，然后使用 Log.v()、Log.d()、Log.i()、Log.w() 和 Log.e()5 个函数在程序中设置“日志点”。每当程序运行到“日志点”时，应用程序的日志信息便被发送到 LogCat 中，使用者可以根据“日志点”信息是否与预期的内容一致，判断程序是否存在错误。之所以使用 5 个不同的函数产生日志，主要是为了区分日志信息的类型，其中，Log.v()用来记录详细信息，Log.d()用来记录调试信息，Log.i()用来记录通告信息，Log.w()用来记录警告信息，Log.e()用来记录通错误信息。

在下面的程序中，演示了 Log 类的具体使用方法。

```
package edu.hrbeu.LogCat;
import android.app.Activity;
import android.os.Bundle;
import android.util.Log;
public class LogCat extends Activity {
    final static String TAG = "LOGCAT";
    @Override
    public void onCreate(Bundle savedInstanceState) {
        super.onCreate(savedInstanceState);
        setContentView(R.layout.main);
Log.v(TAG,"Verbose");
        Log.d(TAG,"Debug");
        Log.i(TAG,"Info");
        Log.w(TAG,"Warn");
        Log.e(TAG,"Error");
}
}
```

程序运行后，LogCat 捕获得到应用程序发生的日志信息，显示结果如图 2-9 所示。在 LogCat 中显示了标签为 LOGCAT 的日志信息共 5 条，并以不同颜色加以显示。可见，LogCat 对不同类型的信息使用了不同的颜色加以区别。

Time		pid	tag	Message
06-07 00:33:43.700	D	656	dalvikvm	GC freed 2807 objects / 172736 bytes in 297ms
06-07 00:33:48.653	D	612	dalvikvm	GC freed 3111 objects / 189488 bytes in 228ms
06-07 00:34:01.919	I	570	ActivityManager	Starting activity: Intent { action=android.inte
06-07 00:34:02.000	V	714	LOGCAT	Verbose
06-07 00:34:02.000	D	714	LOGCAT	Debug
06-07 00:34:02.000	I	714	LOGCAT	Info
06-07 00:34:02.000	W	714	LOGCAT	Warn
06-07 00:34:02.000	E	714	LOGCAT	Error
06-07 00:34:02.216	W	612	IInputConnectionW...	showStatusIcon on inactive InputConnection
06-07 00:34:02.251	I	570	ActivityManager	Displayed activity edu.hrbeu.LogCat/.LogCat: 32
06-07 00:34:07.482	D	612	dalvikvm	GC freed 1842 objects / 98152 bytes in 208ms

图 2-9 LogCat 工程的运行结果图

如果能够使用 LogCat 的过滤器，则可以使显示的结果更加清晰。下面使用 LogCat 右上角的“+”号，添加一个名为 LogcatFilter 的过滤器，并设置过滤条件为“标签=LOGCAT”，具体设置方法如图 2-10 所示。

Log Filter

Filter Name: LogcatFilter

by Log Tag: LOGCAT

by pid:

by Log level: <none>

OK Cancel

图 2-10 LogCat 过滤器设置图

过滤器设置好后，LogcatFilter 过滤后的日志信息如图 2-11 所示。在这之后，无论什么类型的日志信息，属于哪一个进程，只要标签为 LOGCAT，都将显示在 LogcatFilter 区域内。

Log (3) | LogcatFilter

Time		pid	tag	Message
06-07 00:58:18.261	V	714	LOGCAT	Verbose
06-07 00:58:18.274	D	714	LOGCAT	Debug
06-07 00:58:18.279	I	714	LOGCAT	Info
06-07 00:58:18.279	W	714	LOGCAT	Warn
06-07 00:58:18.279	E	714	LOGCAT	Error

图 2-11 LogCat 过滤后的输入结果图

第 3 章

Android 常用基本控件

本章将进行用户界面开发时常用到的 Android 布局管理器和基本控件进行详细介绍。Android 中的布局包括线性布局、表格布局、相对布局、帧布局和绝对布局。Android 中的基本控件主要包括文本框、按钮、单选按钮、复选按钮、状态开发按钮、日期时间控件、图片控件等。熟练掌握了这些布局和控件的使用，对我们以后 Android 项目开发具有非常重要的意义和作用。

3.1 界面布局

在介绍 Android 的布局管理器之前，有必要让读者了解 Android 平台下的控件类。首先要了解的是 View 类，该类为所有可视化控件的基类，主要提供了控件绘制和事件处理的方法。创建用户界面所使用的控件都继承自 View，如 TextView、Button、CheckBox 等。

关于 View 及其子类的相关属性，可以通过两种方法来设置，既可以在布局 XML 文件中进行设置，也可以通过成员方法在代码中动态设置。View 类常用的属性及其对应方法如表 3-1 所示。

表 3-1　View 类常用属性及其对应方法说明表

属性名称	对应方法	描　　述
android:background	setBackgroundResource(int)	设置背景颜色
android:clickable	setClickable(boolean)	设置 View 是否响应单击事件
android:visibility	setVisibility(int)	控制 View 的可见性
android:focusable	setFocusable(boolean)	控制 View 是否可以获取焦点
android:id	setId(int)	为 View 设置标志符，可通过 findViewById 方法获取
android:longClickable	setLongClickable(boolean)	设置 View 是否响应长单击事件
android:soundEffectsEnabled	setSoundEffectsEnabled(boolean)	设置当 View 触发单击等事件时是否播放音效
android:saveEnabled	setSaveEnabled(boolean)	如果未作设置，当 View 被冻结时将不会保存其状态
android:nextFocusDown	setNextFocusDownId(int)	定义当向下检索时应该获取焦点的 View，如果该 View 不存在或不可见，则会抛出 RuntimeException 异常
android:nextFocusLeft	setNextFocusLeftId(int)	定义当向左检索时应该获取焦点的 View
android:nextFocusRight	setNextFocusRightId(int)	定义当向右检索时应该获取焦点的 View
android:nextFocusUp	setNextFocusUpId(int)	定义当向上检索时应该获取焦点的 View，如果该 View 不存在或不可见，则会抛出 RuntimeException 异常

另外一个需要了解的是 ViewGroup 类，它也是 View 类的子类，但是可以充当其他控件的容器。ViewGroup 的子控件既可以是普通的 View，也可以是 ViewGroup，实际上，这是使用了 Composite 的设计模式。Android 中的一些高级控件如 Galley、GridView 等都继承自 ViewGroup。与 Java SE 不同，Android 中并没有设计布局管理器，而是为每种不同的布局提供了一个 ViewGroup 的子类。

3.1.1　线性布局

线性布局（LinearLayout）是最简单的布局之一，它提供了控件水平或者垂直排列的模型。同时，使用此布局时可以通过设置控件的 Weight 参数控制各个控件在容器中的相对大小。LinearLayout 布局的属性既可以在布局文件（XML）中设置，也可以通过成员方法进行设置。表 3-2 给出 LinearLayout 常用的属性及这些属性的对应设置方法。

表 3-2　LinearLayout 常用属性及对应设置方法

属性名称	对应方法	描　　述
android:orientation	setOrientation(int)	设置线性布局的朝向，可取 Horizontal（水平）和 Vertical（垂直）
android:gravity	setGravity(int)	设置线性布局的内部元素的布局方式

在线性布局中可使用 gravity 属性来设置控件的对齐方式，gravity 可取的值及说明如表 3-3 所示。当需要为 gravity 设置多个值时，可以用“|”分隔符号隔开即可。

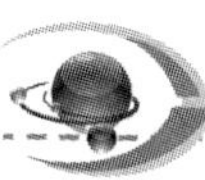

表 3-3　gravity 可取的属性及说明表

属 性 值	说　明
top	不改变控件大小，对齐到容器顶部
bottom	不改变控件大小，对齐到容器底部
left	不改变控件大小，对齐到容器左侧
right	不改变控件大小，对齐到容器右侧
center_vertical	不改变控件大小，对齐到容器纵向中央位置
center_horizontal	不改变控件大小，对齐到容器横向中央位置
center	不改变控件大小，对齐到容器中央位置
fill_vertical	若有可能，纵向拉伸以填充容器
fill_horizontal	若有可能，横向拉伸以填充容器
fill	若有可能，纵向横向同时拉伸以填充容器

接下来我们通过一个案例来学习 LinearLayout 的具体用法。该案例比较简单，其开发步骤如下。

在 Eclipse 中新建一个项目名称为 LinearExample，打开 res\layout 目录下的 main.xml 文件，将其中的代码改写为如下代码。

```
    //声明文件的处理指令名，版本号，字符集
<?xml version="1.0" encoding="utf-8"?>
<LinearLayout xmlns:android=http://schemas.android.com/apk/res/android
    //设置线性布局的朝向为竖直
    android:orientation="vertical"
    //设置布局的宽度和高度填充整个界面
    android:layout_width="fill_parent"
    android:layout_height="fill_parent"
     //设置布局的 id 编号
    android:id="@+id/lla"
    //设置线性布局的内部元素的布局方式为向上
    android:gravity="top">
     //声明一个不可编译的文本框，设置文本框的 id 编号
 <TextView  android:id="@+id/text"
    //设置文本框上显示字符内容
     android:text="用户名:"
  //设置文本框的宽度和高度为控件大小
     android:layout_width="wrap_content"
     android:layout_height="wrap_content"/>
   //声明一个可编译的文本框，设置文本框的 id 编号
 <EditText android:id="@+id/name"
     android:layout_height="wrap_content"
     android:layout_width="fill_parent" />
  //声明一个按钮
```

```
    <Button  android:text="确认"
      android:id="@+id/Button01"
      android:layout_width="wrap_content"
      android:layout_height="wrap_content"/>
</LinearLayout>
```

然后运行项目，会出现如图 3-1 所示的界面，这个界面就是一个 LinearLayout 布局的简单例子。

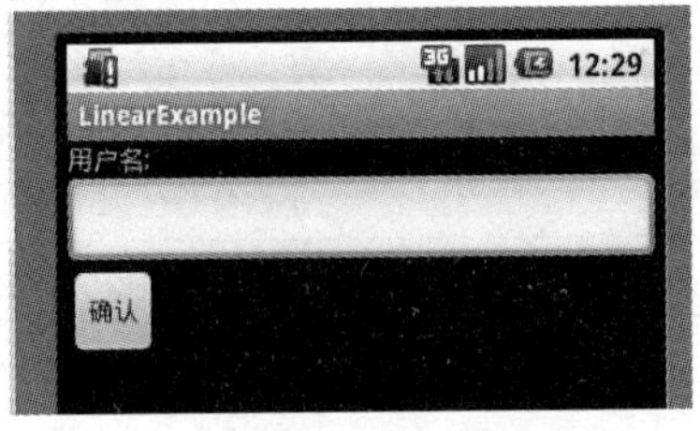

图 3-1　线性布局案例效果图

3.1.2　表格布局

本节将介绍布局管理器的表格布局，首先将对 TableLayout 类进行介绍。TableLayout 类以行和列的形式管理控件，每行为一个 TableRow 对象，也可以为一个 View 对象，当为 View 对象时，该 View 对象将跨越该行的所有列。在 TableRow 中可以添加子控件，每添加一个子控件为一列。

TableLayout 布局中并不会为每一行、每一列或每个单元格绘制边框，每一行可以有 0 或多个单元格，每个单元格为一个 View 对象。TableLayout 中可以有空的单元格，单元格也可以像 HTML 中那样跨越多个列。

在表格布局中，一个列的宽带由该列中最宽的那个单元格指定，而表格的宽带是由父容器指定的。在 TableLayout 中，可以为列设置三种属性：Shrinkable、Stretchable 和 Collapsed。如果一个列被标识为 Shrinkable，则该列的宽带可以进行收缩，以使表格能够适应其父容器的大小；如果一个列被标识为 Stretchable，则该列的宽带可以进行拉伸，以使填满表格中空闲的空间；如果一个列被标识为 Collapsed，则该列将会被隐藏；如果一个列同时具有 Shrinkable 和 Stretchable 属性，则该列的宽带将任意拉伸或收缩以适应父容器。

TableLayout 继承自 LinearLayout 类，除了继承来自父类的属性和方法，TableLayout 类中还包含表格布局所特有的属性和方法。这些属性和方法说明如表 3-4 所示。

表 3-4　TableLayout 类常用属性及其对应方法说明表

属性名称	对应方法	描　述
android:collapseColumns	setCoiumnCollapsed(int,boolean)	设置指定列号的列为 Collapsed，列号从 0 开始计算
android:shrinkColumns	setShrinkColumns(boolean)	设置指定列号的列为 Shrinkable，列号从 0 开始计算
android:stretchColumns	setStretchAllColumns(boolean)	设置指定列号的列为 Stretchable，列号从 0 开始计算

接下来我们通过一个案例来具体学习 TableLayout 布局的使用，该案例的开发步骤如下所示。

步骤一：在 Eclipse 中创建一个项目名称为 TableExample。打开项目 res\values 目录下的 string.xml 文件，修改其字符串，描述文件内容为如下所示。

```
<?xml version="1.0" encoding="utf-8"?>
<resources>
  <string name="app_name">TableExample</string>
  <string name="tv1">我自己是一行......我自己是一行</string>
  <!-- 该值用于独占一行的列，显示内容-->
  <string name="tvShort">我的内容少</string>
  <!-- 该值用于内容较少的列，显示内容-->
  <string name="tvStrech">我是被拉伸的一列</string>
  <!-- 该值用于被拉伸的列，显示内容 -->
  <string name="tvShrink">我是被收缩的一列被收缩的一列</string>      <!-- 该值
 用于被收缩的列，显示内容-->
  <string name="tvLong">我的内容比较长比较长比较长</string>
  <!-- 该值用于内容比较长的列，显示内容-->
</resources>
```

步骤二：在项目的 res\values 目录下创建一个名称为 colors.xml 文件，该文件设置颜色描述内容。编写的代码内容如下所示。

```
<?xml version="1.0" encoding="utf-8"?>
<resources>
   <!-设置颜色属性值为"red"时，显示的颜色（十六进制的 RGB 值）-->
   <color name="red">#fd8d8d</color>
   <color name="green">#9cfda3</color>
   <color name="blue">#8d9dfd</color>
   <color name="white">#FFFFFF</color>
   <color name="black">#000000</color>
   </resources>
```

步骤三：开发程序的布局文件，打开项目 res\layout 目录下的 main.xml 文件，编写其中的代码，内容如下。

```
<?xml version="1.0" encoding="utf-8"?>
<LinearLayout   android:id="@+id/LinearLayout01"
   android:layout_width="fill_parent"
   android:layout_height="fill_parent"
   xmlns:android="http://schemas.android.com/apk/res/android"
   android:orientation="vertical"
   android:gravity="top">
   <TableLayout   android:id="@+id/TableLayout01"
   android:layout_width="fill_parent"
   android:layout_height="wrap_content"
   android:background="@color/white"
xmlns:android="http://schemas.android.com/apk/res/android">
   <TextView   android:text="@string/tv1"
```

```
    android:id="@+id/TextView01"
    android:layout_width="wrap_content"
    android:layout_height="wrap_content"
    android:layout_centerInParent="true"
   <!—设置文本框的背景颜色 -->
   android:background="@color/red"
   <!—设置文本字体的颜色 -->
    android:textColor="@color/black"
    android:layout_margin="4px"/>
</TableLayout>
<TableLayout  android:id="@+id/TableLayout02"
   android:layout_width="fill_parent"
   android:layout_height="wrap_content"
   android:background="@color/white"
   android:stretchColumns="0"
   xmlns:android="http://schemas.android.com/apk/res/android">
   <!-- android:stretchColumns="0" 设置 0 号列为可伸展的列，当有多个列可伸展时用
   逗号隔开 -->
   <TableRow android:id="@+id/TableRow01"
     android:layout_width="wrap_content"
     android:layout_height="wrap_content">
        <TextView  android:text="@string/tvStrech"
         android:id="@+id/TextView02"
         android:layout_width="wrap_content"
         android:layout_height="wrap_content"
         android:layout_centerInParent="true"
         android:background="@color/green"
         android:textColor="@color/black"
         android:layout_margin="4px" />
        <TextView  android:text="@string/tvShort"
         android:id="@+id/TextView03"
         android:layout_width="wrap_content"
         android:layout_height="wrap_content"
         android:layout_centerInParent="true"
         android:background="@color/blue"
         android:textColor="@color/black"
         android:layout_margin="4px" >
        </TextView>   </TableRow>  </TableLayout>
<TableLayout android:id="@+id/TableLayout03"
   android:layout_width="fill_parent"
   android:layout_height="wrap_content"
   android:background="@color/white"
   android:collapseColumns="1"
   android:shrinkColumns="0"
```

```
xmlns:android="http://schemas.android.com/apk/res/android">
<!-- android:collapseColumns="1" 隐藏编号为 1 的列，若有多个列要隐藏，则用逗号
隔开，如 0,2 -->
<!-- android:shrinkColumns="0"
          设置 0 号列为可收缩的列，可收缩的列会纵向扩展
          若有多个列要收缩，则用逗号隔开，如 0,2 -->
<TableRow
  android:id="@+id/TableRow02"
  android:layout_width="wrap_content"
  android:layout_height="wrap_content" >
     <TextView  android:text="@string/tvShrink"
      android:id="@+id/TextView04"
      android:layout_width="wrap_content"
      android:layout_height="wrap_content"
      android:layout_centerInParent="true"
      android:background="@color/green"
      android:textColor="@color/black"
      android:layout_margin="4px"/>
     <TextView  android:text="@string/tvShort"
      android:id="@+id/TextView05"
      android:layout_width="wrap_content"
      android:layout_height="wrap_content"
      android:layout_centerInParent="true"
      android:background="@color/blue"
      android:textColor="@color/black"
      android:layout_margin="4px"/>
     <TextView   android:text="@string/tvLong"
      android:id="@+id/TextView06"
      android:layout_width="wrap_content"
      android:layout_height="wrap_content"
      android:layout_centerInParent="true"
      android:background="@color/red"
      android:textColor="@color/black"
      android:layout_margin="4px" />
    </TableRow>       </TableLayout>  </LinearLayout>
```

步骤四：运行项目，可得到如图 3-2 所示的效果。

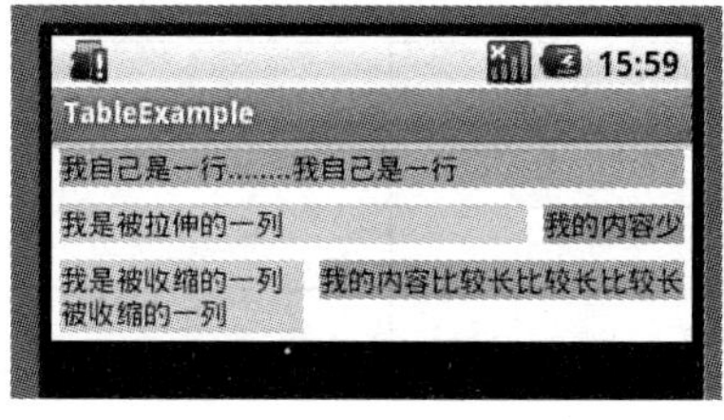

图 3-2　表格布局案例效果图

3.1.3 相对布局

在相对布局（RelativeLayout）中，子控件的位置是相对兄弟控件或父容器来决定的。出于对性能的考虑，在设计相对布局时要按照控件之间的依赖关系排列，如 View A 的位置相对于 View B 来决定，则需要保证在布局文件中 View B 在 View A 的前面。在进行相对布局时用到的属性很多，读者可参照如表 3-5～表 3-7 所示进行学习。

表 3-5 相对布局中属性值只取 true 或 false 的属性表

属性名称	属性说明
android:layout_centerHorizontal	当前控件位于父控件的横向中间位置
android:layout_centerVertical	当前控件位于父控件的纵向中间位置
android:layout_centerInParent	当前控件位于父控件的中间位置
android:layout_alignParentBottm	当前控件底部与父控件底部对齐
android:layout_alignParentLeft	当前控件左侧与父控件左侧对齐
android:layout_alignParentRight	当前控件右侧与父控件右侧对齐
android:layout_alignParentTop	当前控件顶部与父控件顶部对齐
android:layout_alignWithParentIfMissing	参照控件不存在或不可见时参照父控件

表 3-6 相对布局中属性值为其他控件 id 的属性及说明表

属性名称	属性说明
android:layout_toRightOf	使当前控件位于给出 id 控件的右侧
android:layout_toLeftOf	使当前控件位于给出 id 控件的左侧
android:layout_above	使当前控件位于给出 id 控件的上方
android:layout_below	使当前控件位于给出 id 控件的下方
android:layout_alignTop	使当前控件的上边界与给出 id 控件的上边界对齐
android:layout_alignBottom	使当前控件的下边界与给出 id 控件的下边界对齐
android:layout_alignLeft	使当前控件的左边界与给出 id 控件的左边界对齐
android:layout_alignRight	使当前控件的右边界与给出 id 控件的右边界对齐

表 3-7 相对布局中取值为像素的属性及说明表

属性名称	属性说明
android:layout_marginLeft	当前控件左侧留白
android:layout_marginRight	当前控件右侧留白
android:layout_marginTop	当前控件顶部留白
android:layout_marginBottom	当前控件底部留白

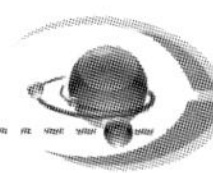

需要注意的是，在进行相对布局时要避免出现循环依赖，例如设置相对布局在父容器中的排列方式为 WRAP_CONTENT，就不能再将相对布局的子控件设置为 ALIGN_RARENT_BOTTOM。因为这样会造成子控件和父控件相互依赖和参照的错误。

接下来我们通过一个相对布局的案例来具体学习 RelativeLayout 类的使用。该案例开发比较简单，具体步骤为：在 Eclipse 中创建一个项目名称为 RelativeExample。打开项目 res\layout 目录下的 main.xml 文件，修改其布局，文件内容如下所示。

```
<?xml version="1.0" encoding="utf-8"?>
<RelativeLayout
    android:id="@+id/RelativeLayout01"
    android:layout_width="fill_parent"
    android:layout_height="fill_parent"
    xmlns:android="http://schemas.android.com/apk/res/android">
<!-- 声明一个相对布局 -->
    <ImageView android:id="@+id/ImageView01"
        android:background="@drawable/center"
        android:layout_width="wrap_content"
        android:layout_height="wrap_content"
        android:layout_centerInParent="true"/>
<!-- 声明一个 ImageView 控件 -->
    <ImageView android:id="@+id/ImageView02"
        android:background="@drawable/down"
        android:layout_width="wrap_content"
        android:layout_height="wrap_content"
        android:layout_toRightOf="@id/ImageView01"
        android:layout_alignTop="@id/ImageView01" />
    <ImageView android:id="@+id/ImageView03"
        android:background="@drawable/up"
        android:layout_width="wrap_content"
        android:layout_height="wrap_content"
        android:layout_above="@id/ImageView01"
        android:layout_alignLeft="@id/ImageView01"/>
</RelativeLayout>
```

运行项目，可以看到案例的效果如图 3-3 所示。

图 3-3　相对布局案例效果图

3.1.4 帧布局

FrameLayout 帧布局在屏幕上开辟出了一块区域，在这块区域中可以添加多个子控件，但是所有的子控件都被对齐到屏幕的左上角。帧布局的大小由子控件尺寸最大的子控件来决定。在 FrameLayout 中，子控件是通过栈来绘制的，所以后添加的子控件会被绘制在上层。如果子控件一样大，同一时刻只能看到最上面的子控件。

FrameLayout 继承自 ViewGroup，除了继承自父类的属性和方法，FrameLayout 类中包含了自己特有的属性和方法，见表 3-8。

表 3-8 FrameLayout 属性及对应方法表

属性名称	对应方法	描　述
android:foreground	setForeground(Drawable)	设置绘制在所有子控件之上的内容
android:foregroundGravity	setForegroundGravity(Drawable)	设置绘制在所有子控件之上的内容的 gravity 属性

接下来我们通过一个相对布局的案例来具体学习 FrameLayout 类的使用。该案例开发比较简单，具体步骤为：在 Eclipse 中创建一个项目名称为 FrameExample。打开项目 res\layout 目录下的 main.xml 文件，修改其布局，文件内容如下所示。

```
<?xml version="1.0" encoding="utf-8"?>
<FrameLayout  android:id="@+id/FrameLayout01"
     android:layout_width="fill_parent"
     android:layout_height="fill_parent"
     xmlns:android="http://schemas.android.com/apk/res/android">
     <TextView  android:text="大的"
        android:id="@+id/TextView01"
        android:layout_width="wrap_content"
        android:layout_height="wrap_content"
        android:textSize="120px" />
     <TextView  android:text="中的"
        android:id="@+id/TextView02"
        android:layout_width="wrap_content"
        android:layout_height="wrap_content"
        android:textSize="60px"/>
     <TextView  android:text="小的"
        android:id="@+id/TextView03"
        android:layout_width="wrap_content"
        android:layout_height="wrap_content"
        android:textSize="20px" />
</FrameLayout>
```

运行项目，可以看到案例的效果如图 3-4 所示。

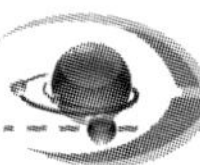

图 3-4　帧布局案例效果图

3.1.5　绝对布局

所谓绝对布局，是指屏幕中所有控件的摆放都由开发人员通过设置控件的坐标来指定，控件容器不再负责管理其子控件的位置。由于子控件的位置和布局都通过坐标来指定，因此AbsoluteLayout 类中并没有开发特有的属性和方法。

我们通过一个绝对布局的案例来具体学习 AbsoluteLayout 类的使用。该案例开发比较简单，具体步骤为：在 Eclipse 中创建一个项目名称为 AbsoluteExample。打开项目 res\layout 目录下的 main.xml 文件，修改其布局，文件内容如下所示。

```
<?xml version="1.0" encoding="utf-8"?>
<AbsoluteLayout   android:id="@+id/AbsoluteLayout01"
   android:layout_width="fill_parent"
   android:layout_height="fill_parent"
   xmlns:android="http://schemas.android.com/apk/res/android">
   <TextView   android:layout_x="20dip"
       android:layout_y="20dip"
       android:layout_height="wrap_content"
       android:layout_width="wrap_content"
       android:id="@+id/TextView01"
       android:text="用户名"/>
   <TextView   android:layout_x="20dip"
       android:layout_y="80dip"
       android:layout_height="wrap_content"
       android:layout_width="wrap_content"
       android:id="@+id/TextView02"
       android:text="密码"/>
   <EditText   android:layout_x="80dip"
      android:layout_y="20dip"
      android:layout_height="wrap_content"
      android:layout_width="180dip"
      android:id="@+id/EditText01"/>
   <EditText   android:layout_x="80dip"
      android:layout_y="80dip"
```

```
        android:layout_height="wrap_content"
        android:layout_width="180dip"
        android:id="@+id/EditText02"
        android:password="true"/>
    <!-- android:password 设置是否为密码框 -->
      <Button   android:layout_x="155dip"
         android:layout_y="140dip"
         android:layout_height="wrap_content"
         android:id="@+id/Button01"
         android:layout_width="wrap_content"
         android:text="确定"  />
      <Button  android:layout_x="210dip"
         android:layout_y="140dip"
         android:layout_height="wrap_content"
         android:id="@+id/Button02"
         android:layout_width="wrap_content"
         android:text="取消"  />
      <ScrollView    android:layout_x="10dip"
        android:layout_y="200dip"
        android:layout_height="150dip"
        android:layout_width="250dip"
        android:id="@+id/ScrollView01">
         <EditText   android:layout_width="fill_parent"
          android:layout_height="wrap_content"
         android:id="@+id/EditText03"
         android:singleLine="false"
         android:gravity="top"  />
       </ScrollView>
    </AbsoluteLayout>
```

运行项目，可以看到案例的效果如图3-5所示。

图3-5　绝对布局案例效果图

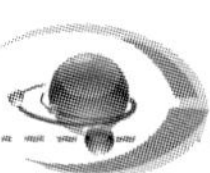

3.2 界面控件

3.2.1 文本控件

文本控件主要包括 TextView 控件和 EditView 控件。其中 TextView 控件继承自 View 类，其主要功能是向用户显示文本内容，同时可选择性地让用户编辑文本。从功能上来说，一个 TextView 就是一个完整的文本编辑器，只不过其本身被设置为不允许编辑，其子类 EditView 被设置为允许用户对内容进行编辑。TextView 控件中包含很多可以在 XML 文件中设置的属性，这些属性同样可以在 Java 代码中动态声明。表 3-9 为 TextView 常用属性及对应方法说明。

表 3-9 TextView 常用属性及对应方法说明

属性名称	对应方法	说　明
android:autoLink	setAutoLinkMask(int)	设置是否将指定格式的文本转换为可单击的超链接显示。传入的参数值可取 ALL、EMAIL_ADDRESSES、MAP_ADDRESSES、PHONE_NUMBERS 和 WEB_URLS
android:gravity	setGravity(int)	定义 TextView 在 X 轴和 Y 轴方向上显示的方式
android:height	setHeight(int)	定义 TextView 的准确高度，以像素为单位
android:minHeight	setMinHeight(int)	定义 TextView 的最小高度，以像素为单位
android:maxHeight	setMaxHeight(int)	定义 TextView 的最大高度，以像素为单位
android:width	setWidth(int)	定义 TextView 的准确宽度，以像素为单位
android:minWidth	setMinWidth(int)	定义 TextView 的最小宽度，以像素为单位
android:maxWIndth	setMaxWidth(int)	定义 TextView 的最大宽度，以像素为单位
android:hint	setHint(int)	当 TextView 中显示的内容为空时，显示该文本
android:text	setText(CharSequence)	为 TextView 设置显示的文本内容
android:textColor	setTextColor (ColorStateList)	设置 TextView 的文本颜色
android:textSize	setTextSize (float)	设置 TextView 的文本大小
android:typeface	setTypeface(Typeface)	设置 TextView 的文本字体
android:ellipsize	setEllipsize(TextUtils.TruncateAt)	如果设置了该属性，当 TextView 中要显示的内容超过了 TextView 的长度时，会对内容进行省略。可取的值有 start、middle、end 和 marquee

EditText 类继承自 TextView。EditText 和 TextView 最大的不同就是用户可以对 EditText 控件进行编辑，同时还可以为 EditText 控件设置监听器，用来检测用户的输入是否合法等。表 3-10 为 EditText 常用属性及对应方法说明。

表 3-10　EditText 常用属性及对应方法说明

属性名称	对应方法	说　明
android:cursorVisible	setCursorVisible(boolean)	设置光标是否可见，默认可见
android:lines	setLines(int)	通过设置固定的行数来决定 EditText 的高度
android:maxLines	setMaxLines(int)	设置最大的行数
android:minLines	setMinLines(int)	设置最小的行数
android:password	setTransformationMethod(TransformationMethod)	设置文本框中的内容是否显示为密码
android:phoneNumber	setKeyListener(KeyListener)	设置文本框中的内容只能是电话号码
android:scorllHorizontally	setHorizontallyScorlling(boolean)	设置文本框是否可以进行水平滚动
android:selectAllOnFocus	setSelectAllOnFocus(boolean)	如果文本内容可选中，当文本框获得焦点时自动选中全部文本内容
android:shadowColor	setShadowLayer(float,float,float,int)	为文本框设置指定颜色的阴影
android:shadowDx	setShadowLayer(float,float,float,int)	为文本框设置阴影的水平偏移，为浮点数
android:shadowDy	setShadowLayer(float,float,float,int)	为文本框设置阴影的垂直偏移，为浮点数
android:shadowRadius	setShadowLayer(float,float,float,int)	为文本框设置阴影的半径，为浮点数
android:singleLine	setTransformationMethod(TransformationMethod)	设置文本框的单行模式
android:maxLength	setFilters(InputFilter)	设置最大显示长度

3.2.2　按钮控件

Android 中的按钮主要包括 Button 控件、ImageButton 控件、ToggleButton 控件、RadioButton 控件和 CheckBox 控件。接下来我们将对这些按钮控件进行详细介绍。

Button 控件继承自 TextView 类，用户可以对 Button 控件执行按下或单击等操作。Button 控件的用法很简单，主要是为 Button 控件设置 View.OnClickListener 监听器并在监听器的实现代码中开发按钮按下事件的处理代码。

ImageButton 控件继承自 ImageView 类，ImageButton 控件与 Button 控件的主要区别是 ImageButton 中没有 text 属性，即按钮中将显示图片而不是文本。ImageButton 控件中设置按钮显示的图片可以通过 android:src 属性实现，也可以通过 setImageResource(int)方法来实现。

默认情况下，ImageButton 控件与 Button 控件一样具有背景色，当按钮处于不同的状态时，背景色也会随之改变。如果当 ImageButton 控件所显示的图片不能完全覆盖掉背景色时，效果非常差，所以一般使用 ImageButton 控件要将背景色设置为其他图片或直接设置为透明。

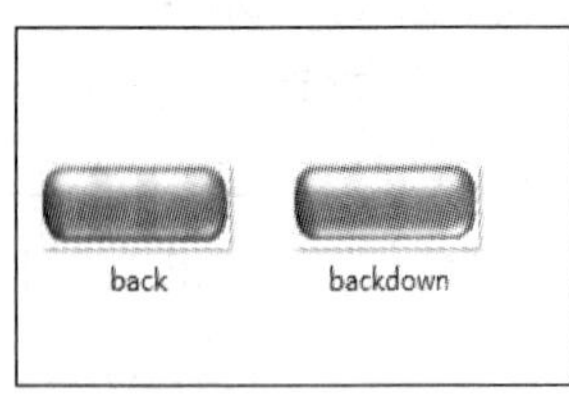

图 3-6　背景图片

接下来我们通过一个个人信息登记表来对前面所述的文本控件、Button 控件和 ImageButton 控件的具体应用。该案例的开发步骤如下。

步骤一：建立项目，准备资源。打开 Eclipse 工具，建立一个名为 Example_EditText 的项目，准备如图 3-6 所示图片，存放在项目的 res\drawable-mdpi 目录下。

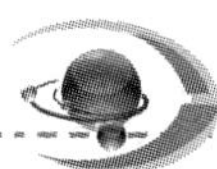

步骤二：修改字符串资源描述文件。打开项目 res\values 目录下的 string.xml 文件，在<resources>和</resources>标记之间加入如下的代码：

```
<string name="app_name">个人信息登记表一</string>
<string name="tvEmail">邮箱地址\n(如: xsc78@163.com)</string>
<string name="etEmail">请输入电子邮件地址</string>
<string name="tvPhone">电话号码\n(如:  1234567890)</string>
<string name="etPhone">请输入电话号码</string>
<string name="etInfo">此处显示登记信息</string>
```

步骤三：修改颜色资源描述文件。打开项目 res/values 目录下的 colors.xml 文件，此文件在项目建立时，并不自动生成，需要我们读者自己去建立一个名为 colors.xml 的文件。在<resources>和</resources>标记之间加入如下的代码：

```
<color name="shadow">#fd8d8d</color><!-- 声明名为 shadow 的资源 -->
<color name="back">#000000</color><!-- 声明名为 back 的颜色资源 -->
```

步骤四：修改布局文件。打开项目 res/layout 目录下的 main.xml 文件，界面布局效果如图 3-7 所示。

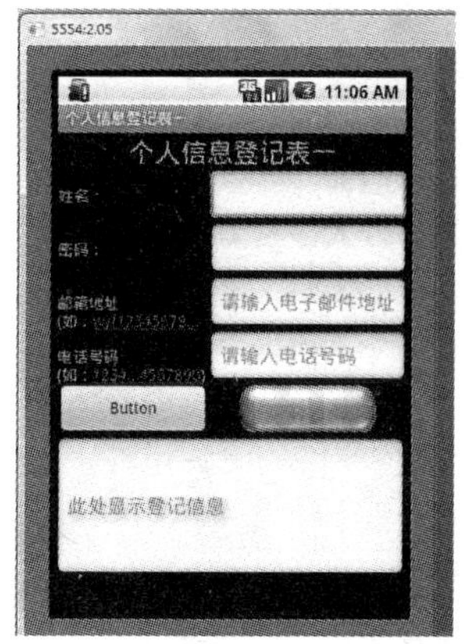

图 3-7　个人信息登记表效果图

修改代码如下：

```
<?xml version="1.0" encoding="utf-8"?>
 <TableLayout xmlns:android="http://schemas.android.com/apk/res/android"
  android:layout_width="fill_parent" android:layout_height="fill_parent"
  android:shrinkColumns="0,2">
 <TextView android:id="@+id/textinfo"
  android:layout_width="wrap_content"
  android:layout_height="wrap_content" android:text="个人信息登记表一"
  android:textSize="24px" android:textColor="@color/shadow"
  android:gravity="center_horizontal" />
 <TableRow android:layout_width="fill_parent"
   android:layout_height="wrap_content"><!-- 声明一个 TableRow 控件 -->
  <TextView android:id="@+id/tvName"
    android:layout_width="wrap_content"
```

```
    android:layout_height="wrap_content" android:text="姓名: "
    android:ellipsize="end"
     android:autoLink="email" />      <!-- 声明一个 TextView 控件 -->
   <EditText android:id="@+id/etName"
    android:layout_width="wrap_content"
    android:layout_height="wrap_content" />
  </TableRow>        <!-- 声明一个 TableLayout -->
  <TableRow android:layout_width="fill_parent"
   android:layout_height="wrap_content">
  <TextView android:id="@+id/tvPassword"
    android:layout_width="wrap_content"
     android:layout_height="wrap_content" android:text="密码: " />
   <EditText android:id="@+id/etPassword"
    android:layout_width="wrap_content"
    android:layout_height="wrap_content" android:password="true" />
  </TableRow>   <!-- 声明一个 TableLayout -->
  <TableRow android:layout_width="fill_parent"
   android:layout_height="wrap_content"><!-- 声明一个 TableRow 控件 -->
  <TextView android:id="@+id/tvEmail"
   android:layout_width="wrap_content"
   android:layout_height="wrap_content"
   android:text="@string/tvEmail"
   android:ellipsize="end" android:autoLink="email" />      <EditText
   android:id="@+id/etEmail" android:hint="@string/etEmail"
   android:layout_width="wrap_content"
   android:layout_height="wrap_content"
   android:selectAllOnFocus="true" />   <!-- 声明一个 EditText 控件 -->
  </TableRow>
  <TableRow android:layout_width="fill_parent"
    android:layout_height="wrap_content">
      <!-- 声明一个 TableRow -->
   <TextView android:id="@+id/tvPhone"
    android:layout_width="wrap_content"
    android:layout_height="wrap_content"
    android:text="@string/tvPhone"
    android:ellipsize="middle" android:autoLink="phone" />
   <EditText android:id="@+id/etPhone"
     android:hint="@string/etPhone"
     android:layout_width="wrap_content"
     android:layout_height="wrap_content"
     android:selectAllOnFocus="true" android:maxWidth="160px"
     android:phoneNumber="true" android:singleLine="true" />
      </TableRow>
   <TableRow android:layout_width="fill_parent"
```

```
    android:layout_height="wrap_content">
    <Button android:id="@+id/btn" android:layout_width="wrap_content"
android:layout_height="wrap_content" android:text="Button" />
<!-- 声明一个 Button 控件 -->
     <ImageButton android:id="@+id/imagebtn"
        android:layout_width="wrap_content"
android:layout_height="wrap_content"
             android:src="@drawable/back"
android:background="@color/back" />
    <!-- 声明一个 ImageButton 控件 -->
    </TableRow>
    <EditText android:id="@+id/etInfo"
android:layout_width="wrap_content"
  android:layout_height="wrap_content" android:editable="false"
android:hint="@string/etInfo" android:cursorVisible="false"
  android:lines="5" android:shadowColor="@color/shadow"
  android:shadowDx="2.5" android:shadowDy="2.5"
  android:shadowRadius="5.0" /><!-- 声明一个 EditText 控件 -->
</TableLayout>
```

步骤五：开发主逻辑界面文件。打开项目中的 Example_EditText.java 文件，编写代码如下：

```
public class Example_EditText extends Activity implements OnClickListener{
    //声明文本框、按钮等相应的对象
    EditText etName=null;
    EditText etPassword=null;
    EditText etInfo=null;
    EditText etEmail=null;
    EditText etPhone=null;
    Button button;
    ImageButton imageButton;
   @Override
   public void onCreate(Bundle savedInstanceState) {   //重写onCreate方法
       super.onCreate(savedInstanceState);
       setContentView(R.layout.main);   //设置当前屏幕
      //对所声明的文本框和按钮依次实例化
       etName=(EditText)findViewById(R.id.etName);
       etPassword=(EditText)findViewById(R.id.etPassword);
       etEmail = (EditText)findViewById(R.id.etEmail);
       etInfo = (EditText)findViewById(R.id.etInfo);
       etPhone=(EditText)findViewById(R.id.etPhone);
       button=(Button)findViewById(R.id.btn);
       imageButton=(ImageButton)findViewById(R.id.imagebtn);
     //为 EditText 控件设置 OnKeyListner 监听器
       etEmail.setOnKeyListener(myOnKeyListener);
     //为两个按钮控件绑定监听器
```

```
        button.setOnClickListener(this);
        imageButton.setOnClickListener(this); }
      //自定义的 OnKeyListner 对象
    private OnKeyListener myOnKeyListener = new OnKeyListener(){
        @Override
    //重写 onKey 方法
    public boolean onKey(View v, int keyCode, KeyEvent event){
    //设置 EditText 控件的显示内容
   etInfo.setText("您输入的邮箱地址为："+etEmail.getText());
            return false;}};
   @Override
   public void onClick(View v) {
        // TODO Auto-generated method stub
        if(v==button){
            etInfo.setText("");
            etInfo.append("你的姓名为："+etName.getText()+"\n");
            etInfo.append("你的密码为："+etPassword.getText()+"\n");
etInfo.append("你的电话号码为："+etPhone.getText()+"\n");
            imageButton.setImageResource(R.drawable.back);}
        if(v==imageButton){etInfo.setText("");
        imageButton.setImageResource(R.drawable.backdown);}}}
```

步骤六：调试案例，测试运行结果。

当用户在邮箱地址文本框中输完信息以后，光标焦点发生改变，则将邮箱地址信息自动显示在下方不可编辑的文本框中，如图 3-8 所示；当用户点击 Button 控件，则将姓名、密码和电话号码文本框中的信息显示在下方不可编辑的文本框中，如图 3-9 所示；当用户点击 ImageButton 控件，在 ImageButton 控件图片发生改变，并清空下方不可编辑的文本框中的内容，如图 3-10 所示。

图 3-8 显示邮箱

图 3-9 显示姓名、密码、电话号码

图 3-10 清空内容

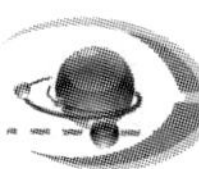

CheckBox 控件和 RadioButton 控件继承自 CompoundButton 类，存放在 android.widget 包中。CheckBox 控件和 RadioButton 控件的状态只能是选中和未选中两种，不同的是 RadioButton 控件是单选按钮，需要编制到一个 RadioGroup 中，同一时刻一个 RadioGroup 中只能有一个按钮处于选中状态。表 3-11 所示为 CheckBox 控件和 RadioButton 控件常用方法。

表 3-11 CheckBox 控件和 RadioButton 控件常用方法及说明

属性名称	说　明
isChecked()	判断是否被选中，如果被选中返回 true，否则返回 false
performClick()	调用 OnClickListener 监听器，即模拟一次单击
setChecked(boolean checked)	通过传入的参数设置控件状态
toggle()	置反控件当前的状态
setOnCheckedChangeListener(CompoundButton.OnCheckedChangeListener listener)	为控件设置 OnCheckedChangeListener 监听器

接下来我们通过个人信息登记表二来具体学习 CheckBox 控件和 RadioButton 控件的应用。项目的开发步骤如下。

步骤一：创建项目。打开 Eclipse 工具，建立一个名为 Example_infomation 的项目。

步骤二：修改字符串资源描述文件。打开项目 res\values 目录下的 string.xml 文件，在<resources>和</resources>标记之间加入如下的代码：

```
<?xml version="1.0" encoding="utf-8"?>
<resources>
  <string name="hello">Hello World, Example_infomation!</string>
  <string name="app_name">个人信息登记表二</string>
  <string name="sex">性别：</string>
  <string name="like">爱好：</string>
  <string name="man">男</string>
  <string name="woman">女</string>
  <string name="basketball">篮球</string>
  <string name="football">足球</string>
  <string name="volleyball">排球</string>
</resources>
```

步骤三：修改布局文件。打开项目 res\layout 目录下的 main.xml 文件，修改代码如下：

```
<?xml version="1.0" encoding="utf-8"?>
<LinearLayout xmlns:android="http://schemas.android.com/apk/res/android"
    android:orientation="vertical" android:layout_width="fill_parent"
    android:layout_height="fill_parent">
    <LinearLayout
    xmlns:android="http://schemas.android.com/apk/res/android"
    android:orientation="horizontal"
    android:layout_width="fill_parent"
    android:layout_height="wrap_content">
    <TextView android:layout_width="wrap_content"
        android:layout_height="wrap_content"
```

```
android:text="@string/sex"  android:textSize="20px" />
    <RadioGroup android:id="@+id/sexmenu"
 android:layout_width="fill_parent"
        android:layout_height="wrap_content"
android:orientation="horizontal" android:checkedButton="@+id/man">
<RadioButton android:id="@+id/man"android:text="@string/man" />
    <RadioButton android:id="@+id/woman" android:text="@string/woman" />
    </RadioGroup> </LinearLayout>
    <LinearLayout
    xmlns:android="http://schemas.android.com/apk/res/android"
        android:orientation="horizontal"
        android:layout_width="fill_parent"
        android:layout_height="wrap_content">
    <TextView android:layout_width="wrap_content"
        android:layout_height="wrap_content"
android:text="@string/like" android:textSize="20px" />
    <CheckBox android:id="@+id/basketball"
 android:text="@string/basketball"
        android:layout_width="wrap_content"
android:layout_height="wrap_content" />
    <CheckBox android:id="@+id/football"
android:text="@string/football"
        android:layout_width="wrap_content"
android:layout_height="wrap_content" />
    <CheckBox android:id="@+id/volleyball"
android:text="@string/volleyball"
        android:layout_width="wrap_content"
 android:layout_height="wrap_content" /></LinearLayout>
    <Button android:id="@+id/editButton" android:text="点击提交"
        android:layout_width="fill_parent"
android:layout_height="wrap_content"
        android:gravity="center_horizontal" />
    <TextView android:id="@+id/edittext"
android:layout_width="fill_parent"
        android:layout_height="wrap_content" android:textSize="20px" />
</LinearLayout>
```

获得如图 3-11 所示界面布局图。

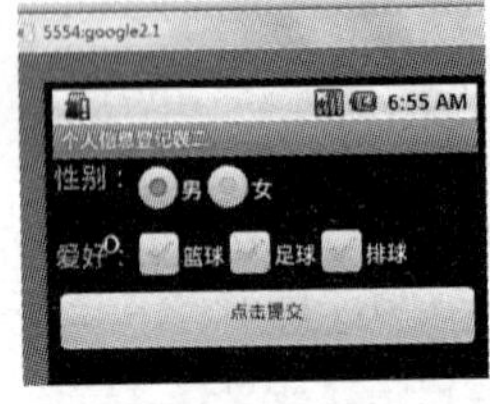

图 3-11　个人信息登记表二界面布局图

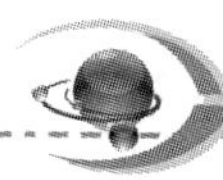

步骤四：开发界面文件。打开项目中的 Example_infomation.java 文件，编写代码如下：

```
public class Example_infomation extends Activity
    implements OnClickListener{
// RadioGroup rGroup=null;
   RadioButton rbman=null;
  RadioButton rbwoman=null;  CheckBox cbbasket=null;
  CheckBox cbfoot=null;      CheckBox cbvolley=null;
  Button btn=null;           TextView tView=null;
  @Override
  public void onCreate(Bundle savedInstanceState) {
      super.onCreate(savedInstanceState);
      setContentView(R.layout.main);
     // rGroup=(RadioGroup)findViewById(R.id.sexmenu);
      rbman=(RadioButton)findViewById(R.id.man);
      rbwoman=(RadioButton)findViewById(R.id.woman);
      cbbasket=(CheckBox)findViewById(R.id.basketball);
      cbfoot=(CheckBox)findViewById(R.id.football);
      cbvolley=(CheckBox)findViewById(R.id.volleyball);
      btn=(Button)findViewById(R.id.editButton);
      tView=(TextView)findViewById(R.id.edittext);
      btn.setOnClickListener(this);}
  @Override
  public void onClick(View v) {
       // TODO Auto-generated method stub
       int num=0;
       if(v==btn){
         tView. setText ("你的性别为：");
       if(rbman.isChecked()){tView.append ("男\n");}
       if(rbwoman.isChecked()){tView.append
           (rbwoman.getText().toString()+"\n");}
       tView.append("你的爱好为：");
       if(cbbasket.isChecked()){num++;tView.append("  篮球");}
       if(cbfoot.isChecked()){num++;tView.append("   足球");}
       if(cbvolley.isChecked()){num++;tView.append("   排球");}
       if(num>0){tView.append("\n 你的爱好有"+num+"种，谢谢参与");} }}}}
```

步骤五：运行程序，查看效果，如图 3-12 所示。

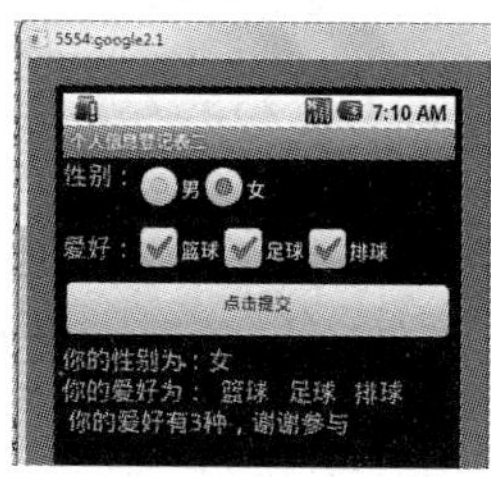

图 3-12 运行效果图

ToggleButton 控件继承也自 CompoundButton 类，存放在 android.widget 包中。ToggleButton 控件的状态只能是选中和未选中两种，并且需要为不同的状态设置不同的显示文本。ToggleButton 控件除了继承自父类的一些属性和方法之外，也具有一些自己的属性。表 3-12 所示为 ToggleButton 控件常用属性说明。

表 3-12　ToggleButton 控件常用属性说明

属性名称	说　明
android:textOff	按钮未被选中时显示的文本内容
android:textOn	按钮被选中时显示的文本内容

接下来我们再用一个电灯开关的案例（见图 3-13）来具体学习 ToggleButton 控件、CheckBox 控件和 RadioButton 控件的使用方法。该案例的开发步骤如下。

图 3-13

步骤一：建立项目，准备资源。打开 Eclipse 工具，建立一个名为 Example_ButtonText 的项目，准备相应的图片存放到项目的 res/drawable-mdpi 目录下。

步骤二：修改字符串资源描述文件。打开项目 res\values 目录下的 string.xml 文件，在<resources>和</resources>标记之间加入如下的代码：

```
<string name="app_name">电灯开关</string>
<string name="on">开灯</string> <!-- 声明名为 on 的字符串资源 -->
<string name="off">关灯</string>     <!-- 声明名为 off 的字符串资源 -->
```

步骤三：修改布局文件。打开项目 res\layout 目录下的 main.xml 文件，修改代码如下：

```
<?xml version="1.0" encoding="utf-8"?>
<LinearLayout xmlns:android="http://schemas.android.com/apk/res/android"
 android:orientation="vertical" android:layout_width="fill_parent"
 android:layout_height="fill_parent"><!-- 声明一个线性布局 -->
 <ImageView android:id="@+id/ImageView01" android:src="@drawable/bulb_off"
 android:layout_width="wrap_content" android:layout_height="wrap_content"
 android:layout_gravity="center_horizontal">
 </ImageView><!-- 声明一个 ImageView 控件 -->
 <RadioGroup android:id="@+id/RadioGroup01"
 android:orientation="horizontal" android:layout_width="wrap_content"
 android:layout_height="wrap_content"
 android:layout_gravity="center_horizontal">    <!-- 声明一个 RadioGroup 控
 件 -->
 <RadioButton android:text="@string/off" android:id="@+id/off"
 android:checked="true" android:layout_width="wrap_content"
 android:layout_height="wrap_content">
 </RadioButton>      <!-- 声明一个 RadioButton 控件 -->
 <RadioButton android:text="@string/on" android:id="@+id/on"
 android:layout_width="wrap_content"
 android:layout_height="wrap_content">
```

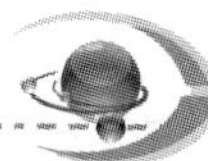

```
 </RadioButton></RadioGroup>  <!-- 声明一个 RadioButton 控件 -->
<CheckBox android:text="@string/on" android:id="@+id/CheckBox01"
android:layout_width="wrap_content"
 android:layout_height="wrap_content"
android:layout_gravity="center_horizontal">
</CheckBox>
<ToggleButton android:textOn="@string/off"
android:textOff="@string/on" android:id="@+id/ToggleButton01"
android:layout_width="140dip" android:layout_height="wrap_content"
android:layout_gravity="center_horizontal" />
<!--声明一个 ToggleButton 控件--> <!-- 声明一个 CheckBox 控件 -->
</LinearLayout>
```

步骤四：开发界面文件。打开项目中的 Example_ButtonText.java 文件，编写代码如下：

```
public class Example_ButtonText extends Activity {
  @Override
  public void onCreate(Bundle savedInstanceState){// 重写 onCreate 方法
      super.onCreate(savedInstanceState);
      setContentView(R.layout.main);
      CheckBox cb = (CheckBox) this.findViewById(R.id.CheckBox01);

      cb.setOnCheckedChangeListener(new OnCheckedChangeListener() {// 为
      CheckBox 添加监听器及开关灯业务代码
  @Override
  public void onCheckedChanged(CompoundButton buttonView,boolean isChecked) {
      setBulbState(isChecked);}});


      RadioButton rb = (RadioButton) findViewById(R.id.off);
      rb.setOnCheckedChangeListener(new OnCheckedChangeListener() { // 为
      RadioButton 添加监听器及开关灯业务代码
              @Override
 public void onCheckedChanged(CompoundButton buttonView, boolean isChecked) {
 setBulbState(!isChecked);}});
 ToggleButton tb = (ToggleButton) this.findViewById(R.id.ToggleButton01);
 tb.setOnCheckedChangeListener(new OnCheckedChangeListener() {// 为
 ToggleButton 添加监听器
    @Override
 public void onCheckedChanged(CompoundButton buttonView,
 boolean isChecked) { // 重写 onCheckedChanged 方法
 setBulbState(isChecked); }});}// 设置控件状态
 // 设置程序状态的方法
 public void setBulbState(boolean state) {
      // 设置图片状态
      ImageView iv = (ImageView) findViewById(R.id.ImageView01);
```

```
        iv.setImageResource((state) ? R.drawable.bulb_on :
        R.drawable.bulb_off);
        CheckBox cb = (CheckBox) this.findViewById(R.id.CheckBox01);
    cb.setText((state) ? R.string.off : R.string.on);// 设置复选框文字状态
        cb.setChecked(state); // 设置复选框状态
        RadioButton rb = (RadioButton) findViewById(R.id.off);
        rb.setChecked(!state);
        rb = (RadioButton) findViewById(R.id.on);
        rb.setChecked(state); // 设置单选按钮状态
        ToggleButton tb = (ToggleButton) this.findViewById(R.id.ToggleButton01);
        tb.setChecked(state);// 设置 ToggleButton 状态
    }
}
```

步骤五：运行项目，查看效果，如图 3-14 所示。启动模拟器，调试该项目。在项目运行中，当我们选中关灯单选按钮或者关灯的开关按钮或者不选中复选框，则出现关灯状态，否则出现开灯状态。

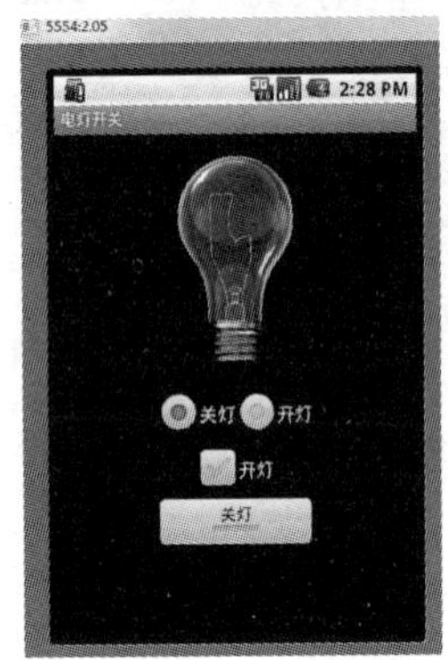

图 3-14　电灯开关案例开关状态图

3.2.3　图片控件

图片控件（ImageView）负责显示图片，其图片的来源既可以是资源文件的 id，也可以是 Drawable 对象或 Bitmap 对象，还可以是 ContentProvider 的 Uri。ImageView 控件中常用的属性和方法如表 3-13 所示。

表 3-13　ImageView 控件中常用的属性和方法表

属性名称	对应方法	说　明
android:adjustViewBounds	set AdjustViewBounds(boolean)	设置是否需要 ImageView 调整自己的边界来保证所显示的图片的长宽比例
android:maxHeight	set MaxHeight(int)	ImageView 的最大高度，可选
android:maxWidth	set MaxWidth(int)	ImageView 的最大宽度，可选

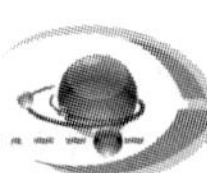

续表

属性名称	对应方法	说　　明
android:scaleType	setScaleType(ImageView.ScaleType)	控制图片应如何调整或移动来适合 ImageView 的尺寸
android:src	setImageResource(int)	设置 ImageVIew 要显示的图片

同时，ImageView 类中还有一些成员方法比较常用，其方法说明见表 3-14。

表 3-14　ImageView 类常用成员方法表

方法名称	说　　明
setAlpha(int alpha)	设置 ImageView 的透明度
setImageBitmap(Bitmap bm)	设置 ImageView 所显示的内容为指定 Bitmap 对象
setImageDrawable(Drawable drawable)	设置 ImageView 所显示的内容为指定 drawable
setImageResource(int resId)	设置 ImageView 所显示的内容为指定 id 的资源
setImageURI(Uri uri)	设置 ImageView 所显示的内容为指定 Uri
setSelected(boolean selected)	设置 ImageView 的选中状态

下面我们通过一个简单的图片查看器来介绍 ImageView 控件的具体用法，案例开发步骤如下。

步骤一：建立项目，准备资源。打开 Eclipse 工具，建立一个名为 Example_ImageView 的项目，准备如图 3-15 所示图片，并将其存放在项目的 res\drawable-mdpi 目录下。

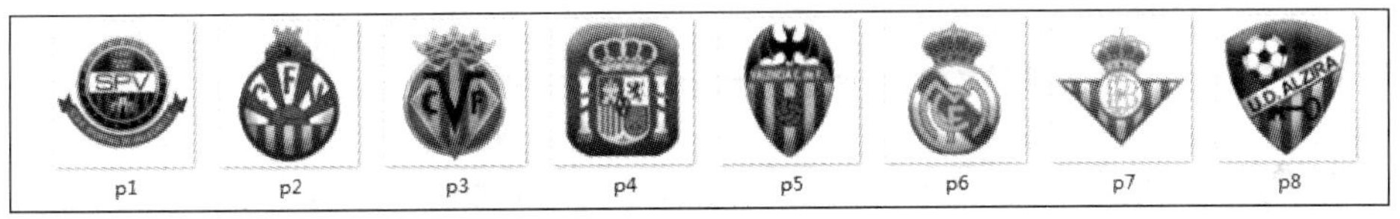

图 3-15

步骤二：修改字符串资源描述文件。打开项目 res\values 目录下的 string.xml 文件，在<resources>和</resources>标记之间加入如下代码：

```
<string name="next">下一张</string>
<string name="previous">上一张</string>
<string name="alpha_plus">透明度增加</string>
<string name="alpha_minus">透明度减少</string>
```

步骤三：修改布局文件。打开项目 res\layout 目录下的 main.xml 文件，修改代码如下：

```
<?xml version="1.0" encoding="utf-8"?>
<LinearLayout xmlns:android="http://schemas.android.com/apk/res/android"
    android:orientation="vertical"
    android:layout_width="fill_parent"
    android:layout_height="fill_parent"
    >    <!-- 声明了一个垂直分布的线性布局 -->
    <ImageView  android:id="@+id/iv"
        android:layout_width="wrap_content"
```

```
            android:layout_height="wrap_content"
            android:layout_gravity="center_horizontal"
            android:src="@drawable/p1"
            />  <!-- 声明了 ImageView 控件  -->
        <LinearLayout
        xmlns:android="http://schemas.android.com/apk/res/android"
        android:orientation="horizontal"
        android:layout_width="fill_parent"
        android:layout_height="wrap_content"
        android:layout_gravity="center_horizontal"
        >              <!-- 声明了一个水平分布的线性布局 -->
        <Button  android:id="@+id/previous"
            android:text="@string/previous"
            android:layout_width="wrap_content"
            android:layout_height="wrap_content"
            android:layout_gravity="center_horizontal"
            />        <!-- 声明了一个上一张 Button 控件按钮 -->
        <Button  android:id="@+id/alpha_plus"
            android:text="@string/alpha_plus"
            android:layout_width="wrap_content"
            android:layout_height="wrap_content"
            android:layout_gravity="center_horizontal"
            />        <!-- 声明了一个增加透明度 Button 控件按钮 -->
        <Button  android:id="@+id/alpha_minus"
            android:text="@string/alpha_minus"
            android:layout_width="wrap_content"
            android:layout_height="wrap_content"
            android:layout_gravity="center_horizontal"
            />  <!-- 声明了一个减少透明度 Button 控件按钮 -->
        <Button  android:id="@+id/next"
            android:text="@string/next"
            android:layout_width="wrap_content"
            android:layout_height="wrap_content"
            android:layout_gravity="center_horizontal"
            />        <!-- 声明了一个下一张 Button 控件按钮 -->
    </LinearLayout></LinearLayout>
```

步骤四：开发界面文件。打开项目中的 Example_ImageView.java 文件，编写代码如下：

```
public class Example_ImageView extends Activity {
    ImageView iv;   //ImageView 对象引用
    Button btnNext, btnPrevious, btnAlphaPlus, btnAlphaMinus;
    int currImgId = 0;              //记录当前 ImageView 显示的图片 id
    int alpha=255;                  //记录 ImageView 的透明度
    int [] imgId = {                //ImageView 显示的图片数组
        R.drawable.p1,R.drawable.p2,R.drawable.p3,R.drawable.p4,
```

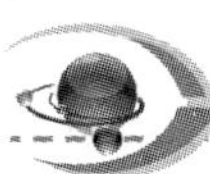

```
        R.drawable.p5,R.drawable.p6,R.drawable.p7,R.drawable.p8,};
    private View.OnClickListener myListener = new View.OnClickListener(){//
    自定义的 OnClickListener 监听器
        @Override
      public void onClick(View v) {   //判断按下的是哪个 Button
      if(v == btnNext){            //下一张图片按钮被按下
      currImgId = (currImgId+1)%imgId.length;
      iv.setImageResource(imgId[currImgId]); }//设置 ImageView 的显示图片
       else if(v == btnPrevious){ //上一张图片按钮被按下
       currImgId = (currImgId-1+imgId.length)%imgId.length;
       iv.setImageResource(imgId[currImgId]);}//设置 ImageView 的显示图片
        else if(v == btnAlphaPlus){ //增加透明度按钮被按下
            alpha += 25;
        if(alpha >255){ alpha =0;}
            iv.setAlpha(alpha); }   //设置 ImageView 的透明度
        else if(v == btnAlphaMinus){    //减少透明度按钮被按下
            alpha -= 25;
        if(alpha <0){alpha = 255;}
            iv.setAlpha(alpha); }}};    //设置 ImageView 的透明度
      @Override
      public void onCreate(Bundle savedInstanceState) {   //重写onCreate方法
        super.onCreate(savedInstanceState);
        setContentView(R.layout.main);
      iv = (ImageView)findViewById(R.id.iv);  //获得 ImageView 对象引用
      btnNext = (Button)findViewById(R.id.next);//获得 ImageView 对象引用
    btnPrevious = (Button)findViewById(R.id.previous);
    btnAlphaPlus = (Button)findViewById(R.id.alpha_plus);       btnAlphaMinus
= (Button)findViewById(R.id.alpha_minus); btnNext.setOnClickListener(myListener);
//为 Button 对象设 OnClickListener 监听器
     btnPrevious.setOnClickListener(myListener);
    btnAlphaPlus.setOnClickListener(myListener);
btnAlphaMinus.setOnClickListener(myListener); }}
```

步骤五：运行项目，查看效果。启动模拟器，调试该项目。在项目运行中，当我们单击“上一张”或者“下一张”按钮时，看到模拟器界面所显示的图片不断更换，循环显示。当我们单击“透明度增加”或者“透明度减少”按钮，发现界面图片的明暗程度会发生相应的改变，如图 3-16 所示。

图 3-16　图片查看器效果图

3.2.4 时钟控件

时钟控件是Android用户界面中比较简单的控件，时钟控件包括AnalogClock和DigitalClock控件。这两种控件都负责显示时钟，所不同的是 AnalogClock 控件显示模拟时钟，它只能显示时针和分针，而 DigitalClock 显示数字时钟，可精确到秒。

下面通过一个案例来介绍时钟控件的用法，时钟控件的用法比较简单，只需要在布局文件中声明控件即可。该案例的开发步骤如下。

在 Eclipse 中创建一个项目名称为 ClockExample。打开项目 res\layout 目录下的 main.xml 文件，修改其布局，文件内容如下所示。

```
<?xml version="1.0" encoding="utf-8"?>
<LinearLayout xmlns:android="http://schemas.android.com/apk/res/android"
    android:orientation="vertical"
    android:layout_width="fill_parent"
    android:layout_height="fill_parent"
    ><!-- 声明一个垂直分布 LinearLayout 布局 -->
    <AnalogClock
        android:id="@+id/analog"
        android:layout_width="wrap_content"
        android:layout_height="wrap_content"
        android:layout_gravity="center_horizontal"
        />  <!-- 声明一个 AnalogClock 控件 -->
    <DigitalClock android:id="@+id/digital"
        android:layout_width="wrap_content"
        android:layout_height="wrap_content"
        android:layout_gravity="center_horizontal"
        />  <!-- 声明一个 DigitalClock 控件 -->
</LinearLayout>
```

运行文件，其效果如图 3-17 所示。

图 3-17　时钟控件案例效果图

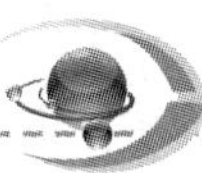

3.2.5　日期与时间选择控件

日期选择控件（DatePicker）继承自 FrameLayout 类，日期选择控件的主要功能是向用户提供包含年、月、日的日期数据并允许用户对其进行选择。如果要捕获用户修改日期选择控件中数据的事件，需要为 DatePicker 添加 onDateChangedListener 监听器。DatePicker 类主要的成员方法如表 3-15 所示。

表 3-15　DatePicker 类主要的成员方法及说明

方法名称	方法说明
getDayOfMonth()	获取日期天数
getMonth()	获取日期月份
getYear()	获取日期年份
init(int year, int monthOfYear, int dayOfMonth, DatePicker.onDateChangedListener onDateChangedListener)	初始化 DatePicker 控件的属性，参数 onDateChangedListener 为监听器对象，负责监听日期数据的变化
setEnabled(boolean enabled)	根据传入的参数设置日期选择控件是否可用
updateDate(int year, int monthOfYear, int dayOfMonth)	根据传入的参数更新日期选择控件的各个属性值

TimePicker 同样继承自 FrameLayout 类。时间选择控件向用户显示一天中的时间（可以为 24 小时制，也可以为 AM/PM），并允许用户进行选择。如果要捕获用户修改时间数据的事件，便需要为 TimePicker 添加 OnTimeChangedListener 监听器。TimePicker 类的主要成员方法如表 3-16 所示。

表 3-16　TimePicker 类的主要成员方法及说明

方法名称	方法说明
getCurrentHour()	获取时间选择控件的当前小时，返回 Integer 对象
getCurrentMinute()	获取时间选择控件的当前分钟，返回 Integer 对象
is24HourView()	判断时间选择控件是否为 24 小时制
setCurrentHour(Integer　currentHour)	设置时间选择控件的当前小时，传入 Integer 对象
setCurrentMinute(Integer　currentMinute)	设置时间选择控件的当前分钟，传入 Integer 对象
setEnabled(boolean　enabled)	根据传入的参数设置时间选择控件是否可用
setIs24HourView(boolean　is24HourView)	根据传入的参数设置时间选择控件是否为 24 小时制
setOnTimeChangedListener(TimePicker.OnTimeChangedListener onTimeChangedListener)	为时间选择控件添加 OnTimeChangedListener 监听器

接下来我们通过一个案例来学习日期和时间选择控件的具体使用方法，该案例的开发步骤如下。

在 Eclipse 中创建一个项目名称为 DateTimeExample。打开项目 res\layout 目录下的 main.xml 文件，修改其布局，文件内容如下所示。

```
<?xml version="1.0" encoding="utf-8"?>
<LinearLayout xmlns:android="http://schemas.android.com/apk/res/android"
   android:orientation="vertical"
   android:layout_width="fill_parent"
   android:layout_height="fill_parent"
   >  <!-- 声明了一个垂直分布的线性布局 -->
   <DatePicker android:id="@+id/datepicker"
      android:layout_width="wrap_content"
      android:layout_height="wrap_content"
      android:layout_gravity="center_horizontal"
      />  <!-- 声明了一个 DataPicker 控件 -->
   <EditText android:id="@+id/etDate"
      android:layout_width="fill_parent"
      android:layout_height="wrap_content"
      android:cursorVisible="false"
      android:editable="false"
      /><!-- 声明了一个用于显示日期的 EditText 控件 -->
   <TimePicker    android:id="@+id/timepicker"
      android:layout_width="wrap_content"
      android:layout_height="wrap_content"
      android:layout_gravity="center_horizontal"
      /><!-- 声明了一个 TimePicker 控件 -->
   <EditText android:id="@+id/etTime"
      android:layout_width="fill_parent"
      android:layout_height="wrap_content"
      android:cursorVisible="false"
      android:editable="false"
      /><!-- 声明了一个用于显示时间的 EditText 控件 -->
</LinearLayout>
```

开发 Activity 界面的代码。打开项目 src 目录下的 DateTimeExample.java 文件，编写代码如下：

```
public class DateTimeExample extends Activity {
 @Override
 public void onCreate(Bundle savedInstanceState) {   //重写onCreate方法
    super.onCreate(savedInstanceState);
    setContentView(R.layout.main);        //设置当前屏幕
    DatePicker dp = (DatePicker)findViewById(R.id.datepicker);
    TimePicker tp = (TimePicker)findViewById(R.id.timepicker);
    Calendar c = Calendar.getInstance(); //获得 Calendar 对象
    int year = c.get(Calendar.YEAR);
    int monthOfYear = c.get(Calendar.MONTH);
    int dayOfMonth = c.get(Calendar.DAY_OF_MONTH);
dp.init(year, monthOfYear, dayOfMonth, new
OnDateChangedListener(){    //初始化 DatePicker
       @Override
```

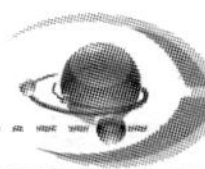

```
        public void onDateChanged(DatePicker view, int year,
      int monthOfYear, int dayOfMonth) {
        flushDate(year,monthOfYear,dayOfMonth);//更新 EditText 所显示内容
                } });
    //为 TimePicker 添加监听器
       tp.setOnTimeChangedListener(new OnTimeChangedListener(){
                @Override
    public void onTimeChanged(TimePicker view,int hourOfDay,int minute)
    {flushTime(hourOfDay,minute);//更新 EditText 所显示内容
                }   }); }
        //方法：刷新 EditText 所显示的内容
        public void flushDate(int year, int monthOfYear,int dayOfMonth){
        EditText et = (EditText)findViewById(R.id.etDate);
        et.setText("您选择的日期是："+year+"年"+(monthOfYear+1)+"月
    "+dayOfMonth+"日。"); }
        //方法：刷新时间 EditText 所显示的内容
        public void flushTime(int hourOfDay,int minute){
        EditText et = (EditText)findViewById(R.id.etTime);
        et.setText("您选择的时间是："+hourOfDay+"时"+minute+"分。");}
    }
```

最后运行案例，效果如图 3-18 所示。

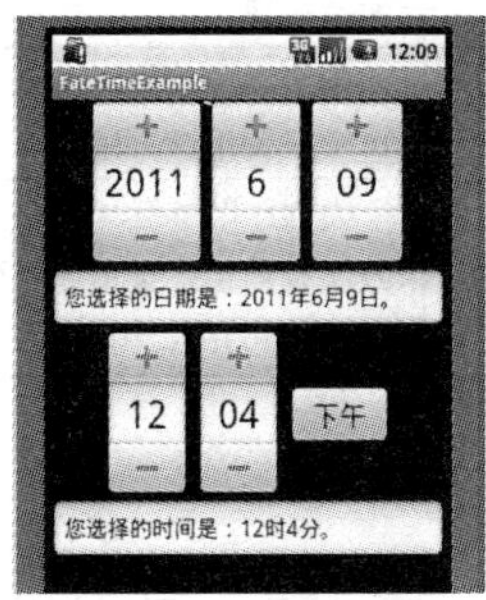

图 3-18　日期时间选择控件案例效果图

3.3 菜单

要想让 Android 应用程序有更完善的用户体验，除了设计人性化的用户界面以外，添加一些菜单可以让应用程序在功能上更加完备。Android 在平台下所提供的菜单大体上可以分为三类：选项菜单（Options Menu）、上下文菜单（Context Menu）和子菜单（Submenu）。在用户界面中，除了经常用到菜单之外，对话框也是程序与用户进行交互的主要途径之一，Android 平台下的对话框主要包括普通对话框、选项对话框、单选多选对话框、日期与时间对话框、进度对话框等。

3.3.1　选项菜单和子菜单

本节将介绍选项菜单及子菜单，当 Activity 在前台运行时，如果用户按下手机上的 Menu 键，此时就会在屏幕底部弹出相应的选项菜单。但这个功能是需要开发人员编程来实现的，如果在开发应用程序时没有实现该功能，那么程序运用时按下手机上的 Menu 键是不会起作用的。

对于携带图标的选项菜单，每次最多只能显示 6 个，当菜单选项多于 6 个时，将只显示前 5 个和一个扩展菜单选项，单击扩展菜单选项将会弹出其余的菜单项。扩展菜单项中将不会显示图标，但是可以显示单选按钮及复选框。

在 Android 中通过回调方法来创建菜单并处理菜单按下的事件，除了开发回调方法 onOptionsItemSelected 来处理用户选中事件，还可以为每个菜单项 MenuItem 对象添加 onOptionsItemClickListener 监听器来处理菜单选中事件。开发选项菜单主要用到 Menu、MenuItem 及 SubMenu。表 3-17 所示为选项菜单相关的回调方法。

表 3-17　选项菜单相关的回调方法及说明

方　法　名	描　　述
onCreateOptionsMenu(Menu menu)	初始化选项菜单，该方法只在第一次显示菜单时调用，如果每次显示菜单时更新菜单项，则需要重写 OnPrepareOptionsMenu(Menu)
public boolean onOptionsItemSelected(MenuItem item)	当选项菜单中某个选项被选中时调用该方法，默认的是一个返回 false 的空实现
public void onOptionsMenuClosed(Menu menu)	当选项菜单关闭时（或者由于用户按下了返回键或者是选择了某个菜单选项）调用该方法
public boolean OnPrepareOptionsMenu(Menu menu)	为程序准备选项菜单，每次选项菜单显示前会调用该方法。可以通过该方法设置某些菜单项可用或不可用或者修改菜单项的内容。重写该方法时需要返回 true，否则选项菜单将不会显示

1. Menu 类。

一个 Menu 对象代表一个菜单，Menu 对象中可以添加菜单项 MenuItem，也可以添加子菜单。SubMenu. Menu 中常用的方法如表 3-18 所示。

表 3-18　Menu 的常用方法及说明

方法名称	参数说明	方法说明
（1）MenuItem add(int groupId,int itemId,int order, Charsequence title); （2）MenuItem add(int groupId,int itemId,int order,int titleRes); （3）MenuItem add(Charsequence title); （4）MenuItem add(int titleRes)	groupId：菜单项所在的组 id，通过分组可以对菜单项进行通过分组可以对菜单项进行批量操作，如果菜单项不需要属于任何组，传入 NONE； itemId：唯一标识菜单项的 id，可传入 NONE； order：菜单项的顺序，可传入 NONE； title：菜单项显示的文本内容； titleRes：String 对象的资源标识符	向 Menu 添加一个菜单项，返回 MenuItem 对象

续表

方法名称	参数说明	方法说明
（1）SubMenu addSubMenu(int titleRes); （2）SubMenu addSubMenu(int groupId,int itemId,int order,int titleRes); （3）SubMenu addSubMenu(Charsequence title); （4）SubMenu addSubMenu(int groupId,int itemId,int order,Charsequence title)	groupId：子菜单项所在的组 id，通过分组可以对菜单项进行通过分组可以对菜单项进行批量操作，如果菜单项不需要属于任何组，传入 NONE； itemId：唯一标识子菜单项的 id，可传入 NONE； order：子菜单项的顺序，可传入 NONE； title：子菜单项显示的文本内容； titleRes：String 对象的资源标识符	向 Menu 添加一个子菜单项，返回 SubMenu 对象
void clear()	——	移除菜单中所有的子项
void close()	——	如果菜单正显示，关闭菜单
MenuItem findItem(int id)	id：MenuItem 的标识符	返回指定 id 的 MenuItem 对象
void removeGroup(int gronpld)	groupId：组 id	如果指定 id 的组不为空，从菜单中移除改组
void remove Item(int id)	id：MenuItem 的 id	移除指定 id 的 MenuItem
int size()	——	返回 Menu 中菜单项的个数

2. Menuitem。

MenuItem 对象代表一个菜单项，通常 Menuitem 实例通过 Menu 的 add 方法获得，Menuitem 中常用的成员方法及说明如表 3-19 所示

表 3-19 选项菜单相关的回调方法及说明

方法名称	参数说明	方法说明
setAlphabeticShortcut(char alphaChar)	alphaChar：字母快捷键	设置 MenuItem 的字母快捷键
MenuItem setNumericShortcut(char numericChar)	numericChar：数字快捷键	设置 MenuItem 的数字快捷键
MenuItem setIcon(Drawable icon)	icon：图标 Drawable 对象	设置 MenuItem 的图标
MenuItem setIntent(Intent intent)	intent：与 MenuItem 绑定的 Intent 对象	为 MenuItem 绑定 intent 对象，当被选中时将会调用 startActivity 方法处理动作相应的 Intent
setOnMenuItemClickListener (MenuItem. OnMenuItemClickListener menuItemClickListener)	menuItemClickListener：监听器	为 MenuItem 设置自定义的监听器，一般情况下，使用回调方法 onOptionsItemSelected 会更有效率

续表

方法名称	参数说明	方法说明
setShortcut(char numericChar ,char alphaChar)	numericChar：数字快捷键；alphaChar：字母快捷键	为 MenuItem 设置数字快捷键和字母快捷键，当按下快捷键或在按住 Alt 键的同时按下快捷键时将会触发 MenuItem 的选中事件
setTitle(int title)	title：标题的资源 id	为 MenuItem 设置标题
setTitle(charSequence title)	title：标题的名称	
setTitleCondensed(charSequence title)	title：MenuItem 的缩略标题	设置 MenuItem 的缩略标题，当 MenuItem 不能显示全部的标题时，显示缩略标题

3. SubMenu。

SubMenu 继承自 Menu，每个 SubMenu 实例代表一个子菜单，SubMenu 中常用的方法及说明如表 3-20 所示。

表 3-20　SubMenu 中常用方法及说明

方法名称	参数说明	方法说明
setHeaderIcon(Drawable icon)	icon：标题图标 Drawable 对象	设置子菜单的标题图标
setHeaderIcon(int iconRes)	iconRes：标题图标的资源 id	
setHeaderTitle(int titleRes)	titleRes：标题文本的资源 id	设置子菜单的标题
setHeaderTitle(charSequence title)	title：标题文本对象	
setIcon(Drawable icon)	icon：图标 Drawable 对象	设置子菜单在父菜单中显示的标题
setIcon(int iconRes)	iconRes：图标资源 id	
SetHeaderView(View view)	view：用于子菜单标题的 View 对象	设置指定 View 对象作为子菜单图标

接下来我们通过一个选项菜单和子菜单使用案例，来具体学习这两种菜单的使用方法。本案例的主要功能是接收用户在菜单中的选项并输出到文本框控件中显示出来。该案例的开发步骤如下。

步骤一：创建项目，准备资源。打开 Eclipse 开发工具，创建一个新的 Android 项目，名为 Example_Menu。复制本程序将会用到的图片到 res\drawable-mdpi 目录下，如图 3-19 所示。

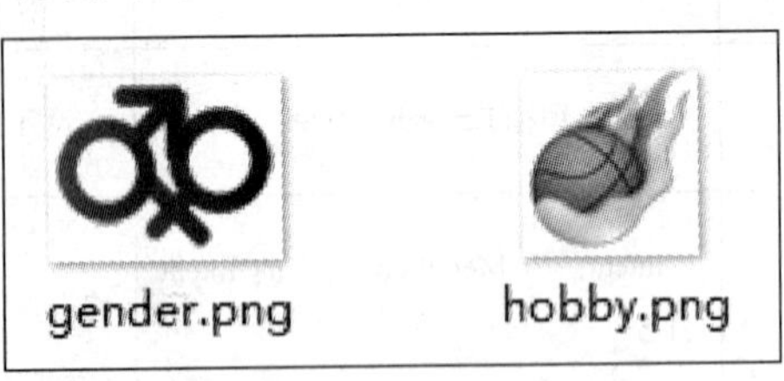

图 3-19　图片

步骤二：准备字符串资源文件。打开项目 res\values 目录下的 strings.xml 文件，在其中的<resources></resources>标记之间添加如下代码：

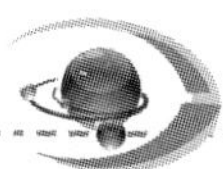

```
<string name="app_name">选项菜单与子菜单的应用</string>
<string name="label">您的选择为\n</string>
<string name="gender">性别</string>
<string name="male">男</string>
<string name="female">女</string>
<string name="hobby">爱好</string>
<string name="hobby1">游泳</string>
<string name="hobby2">唱歌</string>
<string name="hobby3">写 Java 程序</string>
<string name="hobby4">Android 开发</string>
<string name="ok">确定</string>
```

步骤三：布局文件的编写。打开项目 res\layout 目录下的 main.xml 文件，添加如下代码：

```
<?xml version="1.0" encoding="utf-8"?>
<LinearLayout
    android:id="@+id/LinearLayout01"
    android:layout_width="fill_parent"
    android:layout_height="fill_parent"
    android:orientation="vertical"
    xmlns:android="http://schemas.android.com/apk/res/android">
<EditText android:id="@+id/EditText01"
    android:layout_width="fill_parent"
    android:layout_height="wrap_content"
    android:editable="false"
    android:cursorVisible="false" ></EditText>
</LinearLayout>
```

步骤四：编写主逻辑文件。打开 Example_Menu.java 文件，编写代码如下：

```
public class Example_Menu extends Activity {
    //声明程序中菜单选项及子菜单的编号，编号必须是唯一的
    final int MENU_GENDER_MALE=0;     //性别为男选项编号
    final int MENU_GENDER_FEMALE=1; //性别为女选项编号
    final int MENU_HOBBY1=2;          //爱好 1 选项编号
    final int MENU_HOBBY2=3;          //爱好 2 选项编号
    final int MENU_HOBBY3=4;          //爱好 3 选项编号
    final int MENU_HOBBY4=5;          //爱好 4 选项编号
    final int MENU_OK=5;          //确定菜单选项编号
    final int MENU_GENDER=6;          //性别子菜单编号
    final int MENU_HOBBY=7;          //爱好子菜单编号每个菜单项目的编号
    //声明程序中菜单选项组的编号，编号必须是唯一的
    final int GENDER_GROUP=0;        //性别子菜单项组的编号
    final int HOBBY_GROUP=1;         //爱好子菜单项组的编号
    final int MAIN_GROUP=2;        //外层总菜单项组的编号
    MenuItem[] miaHobby=new MenuItem[4];//创建爱好菜单项组
    MenuItem male=null;//男性性别菜单项
```

```
MenuItem female=null;//女性性别菜单项
@Override
public void onCreate(Bundle savedInstanceState) {   //重写onCreate方法
    super.onCreate(savedInstanceState);
    setContentView(R.layout.main); }  //设置当前屏幕
@Override
public boolean onCreateOptionsMenu(Menu menu){
//性别单选菜单项组   对其进行编组成为单选菜单项组
SubMenu  subMenuGender  =  menu.addSubMenu(MAIN_GROUP,MENU_GENDER,0,
R.string.gender);
subMenuGender.setIcon(R.drawable.gender);//设置子菜单图标
subMenuGender.setHeaderIcon(R.drawable.gender);//设置子菜单标题图标
//向子菜单中添加MenuItem，即男、女性别选项
male=subMenuGender.add(GENDER_GROUP, MENU_GENDER_MALE, 0, R.string.
male);
female=subMenuGender.add(GENDER_GROUP, MENU_GENDER_FEMALE, 0, R.string.
female);
//subMenuGender.add(GENDER_GROUP, MENU_GENDER_FEMALE, 0, R.string.
female);
    male.setChecked(true);//设置默认选项
    //设置GENDER_GROUP组是可选择的，互斥的，否则出现多选现象
    subMenuGender.setGroupCheckable(GENDER_GROUP, true,true);
    //爱好复选菜单项组
SubMenu subMenuHobby = menu.addSubMenu(MAIN_GROUP,MENU_HOBBY,0,R.
string.hobby);
subMenuHobby.setIcon(R.drawable.hobby); //设置子菜单图标
miaHobby[0]=subMenuHobby.add(HOBBY_GROUP, MENU_HOBBY1, 0, R.string.
hobby1);
miaHobby[1]=subMenuHobby.add(HOBBY_GROUP, MENU_HOBBY2, 0, R.string.
hobby2);
miaHobby[2]=subMenuHobby.add(HOBBY_GROUP, MENU_HOBBY3, 0, R.string.
hobby3);
miaHobby[3]=subMenuHobby.add(HOBBY_GROUP, MENU_HOBBY4, 0, R.string.
hobby4);
miaHobby[0].setCheckable(true);//设置菜单项为复选菜单项
miaHobby[1].setCheckable(true);
miaHobby[2].setCheckable(true);
miaHobby[3].setCheckable(true);
//确定菜单项
MenuItem ok=menu.add(GENDER_GROUP+3,MENU_OK,0,R.string.ok);
OnMenuItemClickListener lsn=new OnMenuItemClickListener(){//实现菜单项
单击事件监听接口
        @Override
        public boolean onMenuItemClick(MenuItem item) {
```

```
            appendStateStr();
            return true;}    };
    ok.setOnMenuItemClickListener(lsn);//给确定菜单项添加监听器
    //给确定菜单项添加快捷键
    ok.setAlphabeticShortcut('o');//为确定菜单设置字符快捷键
    ok.setNumericShortcut('1');//为确定菜单设置数字快捷键
    //ok.setShortcut('2', 'a');//为确定菜单同时设置两种快捷键
    //要注意，同时设置多次时只有最后一个设置起作用
        return true;}
    @Override  //单选或复选菜单项选中状态变化后的回调方法
    public boolean onOptionsItemSelected(MenuItem mi){
    switch(mi.getItemId()){
      case MENU_GENDER_MALE://单选菜单项状态的切换要自行写代码完成
      case MENU_GENDER_FEMALE:
          mi.setChecked(true);
          appendStateStr();break;  //当有效项目变化时记录在文本区中
      case MENU_HOBBY1://复选菜单项状态的切换要自行写代码完成
      case MENU_HOBBY2:  case MENU_HOBBY3: case MENU_HOBBY4:
         mi.setChecked(!mi.isChecked());
         appendStateStr();break; } //当有效项目变化时记录在文本区中
        return true; }
    //获取当前选择状态的方法
    public void appendStateStr(){
        String result="您选择的性别为：";
        if(male.isChecked()){ result=result+"男"; }
        if(female.isChecked()){ result=result+female.getTitle();}
        String hobbyStr="";
        for(MenuItem mi:miaHobby){
           if(mi.isChecked()){
              hobbyStr=hobbyStr+mi.getTitle()+"、";}}
        if(hobbyStr.length()>0){
           result=result+"\n您的爱好为："+hobbyStr.substring(0, hobbyStr.
           length()-1)+"。\n";}
        else{ result=result+"。\n"; }
        EditText et=(EditText) Example_Menu.this.findViewById(R.id.
        EditText01);
        et.setText(""); et.append(result); }
}
```

步骤五：运行程序，测试效果。该案例的运行效果图如下，当案例启动时，运行效果如3-20所示；当用户单击Menu菜单时，出现如图3-21所示效果；当用户单击图3-21的性别菜单，则出现如图3-22所示效果；当用户单击图3-22的爱好菜单，则出现如图3-23所示效果。当用户选择完以后，或单击图3-21的“确定”菜单，则出现如图3-24所示效果。

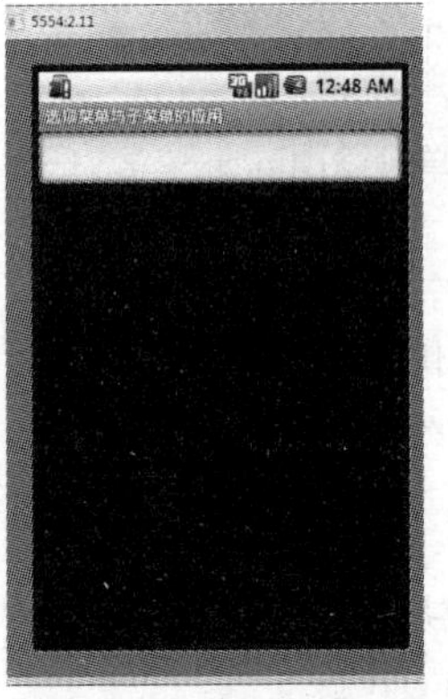

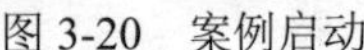

图 3-20　案例启动

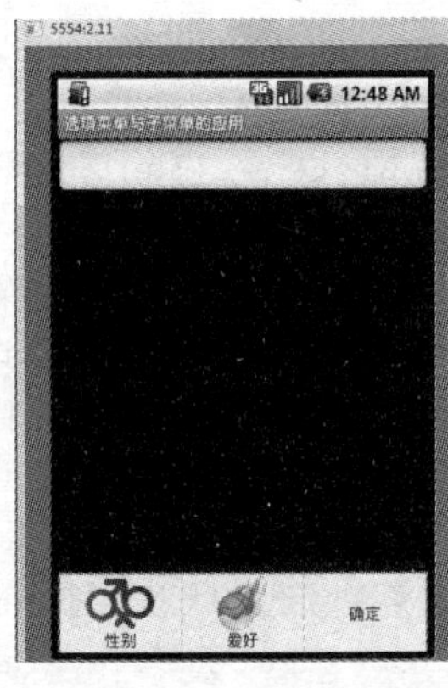

图 3-21　Menu 菜单

图 3-22　性别菜单

图 3-23　爱好菜单

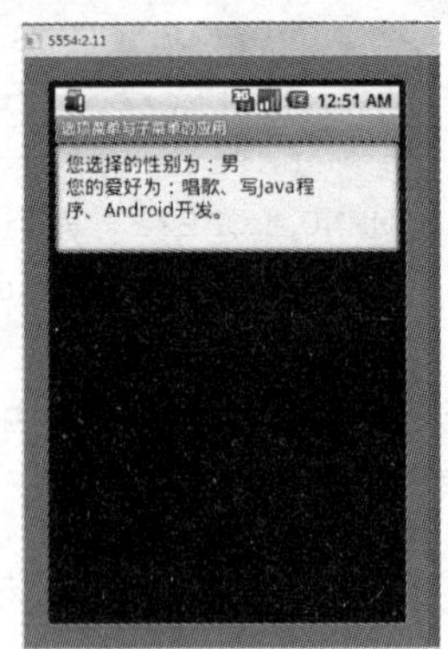

图 3-24　确定菜单

3.3.2　上下文菜单

本节将介绍上下文菜单（ContextMenu）的使用，ContextMenu 继承自 Menu。上下文菜单不同与选项菜单，选项菜单服务于 Activity，而上下文菜单是注册到某个 View 对象上的。如果一个 View 对象注册了上下文菜单，用户可以通过长按（约 2 秒）该 View 对象以呼出上下文菜单。

上下文菜单不支持快捷键（shortcut），其菜单选项也不能附带图标，但是可以为上下文菜单的标题指定图标。使用上下文菜单时常用到 Activity 类的成员方法，如表 3-21 所示。

表 3-21　Activity 类中与 ContextMenu 相关的方法及说明

方法名称	参数说明	方法说明
onCreateContextMenu(ContextMenu menu, View v, ContextMenu. ContextMenuInfo menuInfo)	menu：创建的上下文菜单 v：上下文菜单依附的 View 对象； menuInfo：上下文菜单需要额外显示的信息	每次为 View 对象呼出上下文菜单时都将调用该方法

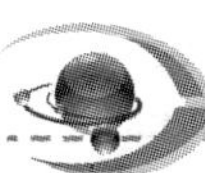

续表

方法名称	参数说明	方法说明
onContextItemSelected(MenuItem item)	Item：被选中的上下文菜单选项	当用户选择了上下文菜单选项后调用该方法进行处理
onContextMenuClosed(Menu menu)	menu：被关闭的上下文菜单	当上下文菜单被关闭时调用该方法
registerForContextMenu(View view)	View：要显示上下文菜单的 View 对象	为指定的 View 对象注册一个上下文菜单

接下来我们还是通过一个个人爱好和性别选择的案例来具体学习上下文菜单的使用方法，该案例的开发步骤如下。

步骤一：创建项目，准备资源。打开 Eclipse 开发工具，创建一个新的 Android 项目，名为 Example_ContextMenu。将如图 3-25 所示本程序会用到的图片资源复制到 res/drawable-mdpi 目录下。

图 3-25 图片资源

步骤二：准备字符串资源文件。打开项目 res\values 目录下的 strings.xml 文件，在其中的<resources></resources>标记之间添加如下代码：

```
<string name="mi1">篮球</string>
<string name="mi2">排球</string>
<string name="mi3">足球</string>
<string name="mi4">男</string>
<string name="mi5">女</string>
<string name="et1">长按进行爱好选择</string>
<string name="et2">长按进行性别选择</string>
```

步骤三：布局文件的编写。打开项目 res\layout 目录下的 main.xml 文件，添加如下代码：

```
<?xml version="1.0" encoding="utf-8"?>
<LinearLayout
    android:id="@+id/LinearLayout01"
    android:layout_width="fill_parent"
    android:layout_height="fill_parent"
    android:orientation="vertical"
    xmlns:android="http://schemas.android.com/apk/res/android">
<EditText android:text="@string/et1"
    android:id="@+id/EditText01"
 android:layout_width="fill_parent"
android:layout_height="wrap_content" />
<EditText android:text="@string/et2"
android:id="@+id/EditText02"
android:layout_width="fill_parent"
android:layout_height="wrap_content" />
</LinearLayout>
```

步骤四：编写主逻辑文件。打开 Example_Menu.java 文件，编写代码如下：

```
public class Example_ContextMenu extends Activity {
final int MENU1=1;//每个菜单项目的编号，编号必须唯一
final int MENU2=2;    final int MENU3=3;
final int MENU4=4;    final int MENU5=5;
@Override
public void onCreate(Bundle savedInstanceState) {        //重写onCreate方法
    super.onCreate(savedInstanceState);
    setContentView(R.layout.main);
    //为两个文本框注册上下文菜单
this.registerForContextMenu(findViewById(R.id.EditText01));
this.registerForContextMenu(findViewById(R.id.EditText02));}
    @Override
    public void onCreateContextMenu (ContextMenu menu, View v,
            ContextMenu.ContextMenuInfo menuInfo){//此方法在每次调出上下文菜单
            时都会被调用一次
     menu.setHeaderIcon(R.drawable.header);
     if(v==findViewById(R.id.EditText01)){//若是第一个文本框
         menu.add(0, MENU1, 0, R.string.mi1);
         menu.add(0, MENU2, 0, R.string.mi2);
         menu.add(0, MENU3, 0, R.string.mi3);    }
     else if(v==findViewById(R.id.EditText02)){//若是第二个文本框
         menu.add(0, MENU4, 0, R.string.mi4);
         menu.add(0, MENU5, 0, R.string.mi5);    }}
    @Override  //菜单项选中状态变化后的回调方法
    public boolean onContextItemSelected(MenuItem mi){
         switch(mi.getItemId()){
           case MENU1: case MENU2:  case MENU3:
          EditText et1=(EditText)this.findViewById(R.id.EditText01);
           et1.setText("你的爱好：");
           et1.append("\n"+mi.getTitle());   break;
       case MENU4:  case MENU5:
       EditText et2=(EditText)this.findViewById(R.id.EditText02);
         et2.setText("你的性别：");
         et2.append("\n"+mi.getTitle());  break;}
    return true; }
}
```

步骤五：运行程序，测试效果。程序启动运行，显示如图 3-26 所示效果，当用户长按爱好选择文本框时，弹出如图 3-27 所示爱好选择上下文菜单，当用户长按性别选择文本框时，弹出如图 3-28 所示性别选择上下文菜单。当用户在图 3-27 和图 3-28 进行选择后，出现如图 3-29 所示选择效果。

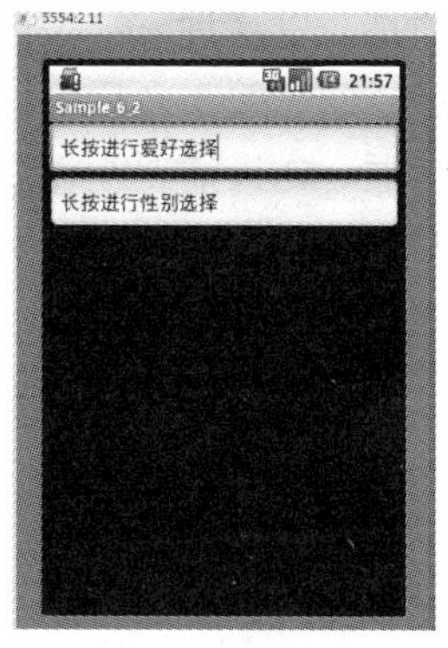

图 3-26　程序启运

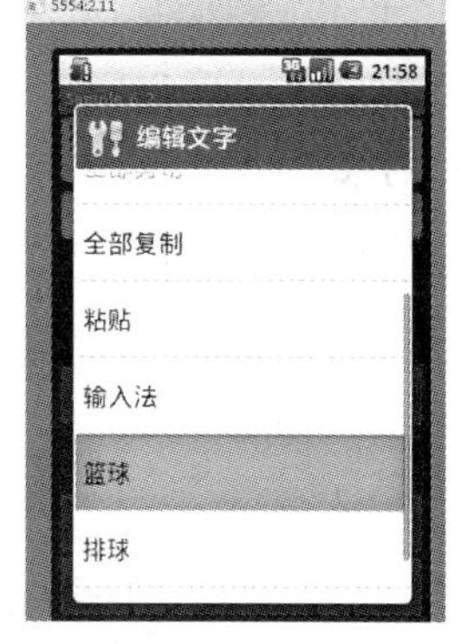

图 3-27　爱好选择菜单

图 3-28　性别选择菜单

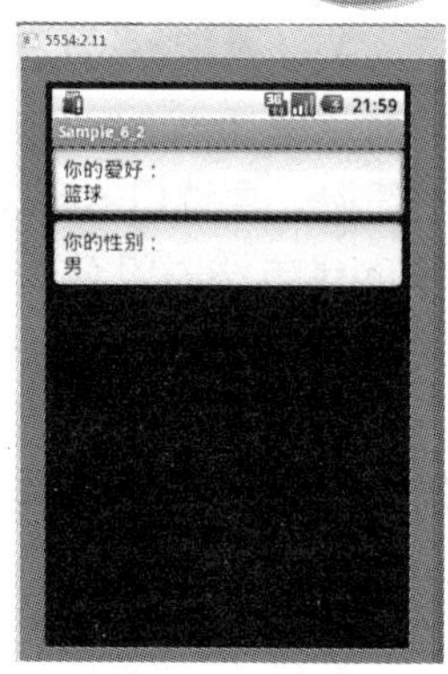

图 3-29　选择结果

3.4　对话框

在用户界面中，除了经常用到菜单之外，对话框也是程序与用户进行交互的主要途径之一。Android 平台下的对话框主要包括普通对话框、选项对话框、单选多选对话框、日期与时间对话框等。本节对 Android 平台下对话框功能的开发进行简单的介绍。

3.4.1　对话框简介

对话框是 Activity 运行时显示的小窗口，当显示对话框时，当前 Activity 失去焦点而由对话框负责所有的人机交互。一般来说，对话框用于提示消息或弹出一个与程序主进程直接相关的小程序。在 Android 平台下主要支持以下几种对话框。

（1）提示对话框 AlertDialog。AlertDialog 对话框可以包含若干按钮（0～4 个不等）和一些可选的单选按钮和复选框。一般来说，AlertDialog 能够满足常见的对话框用户界面的需求。

（2）进度对话框 ProgressDialog。ProgressDialog 可以显示进度轮（wheel）和进度条（bar），由于 ProgressDialog 继承自 AlertDialog，所以在进度对话框中也可以添加按钮。

（3）日期选择对话框 DatePickerDialog。DatePickerDialog 对话框可以显示并允许用户选择日期。

（4）时间选择对话框 TimePickerDialog。TimePickerDialog 对话框可以显示并允许用户选择时间。

对话框是作为 Activity 的一部分被创建和显示的，在程序中通过开发回调方法 onCreateDialog 来完成对话框的创建，该方法需要传入代表对话框 id 参数。如果需要显示对话框，则调用 showDialog 方法传入对话框的 id 来显示指定的对话框。

当对话框第一次被显示时，Android 会调用 onCreateDialog 方法来创建对话框实例，之后将不再重复创建该实例，这一点和选项菜单比较类似。同时，每次对话框在被显示之前都会调用 onPrepareDialog，如果不重写该方法，那么每次显示的对话框将会是最初创建的那个。

关闭对话框可以调用 Dialog 类的 dismiss 方法来实现。但是要注意的是，以这种方式关闭的对话框并不会彻底消失，Android 会在后台保留其状态。如果需要让对话框在关闭之后彻底被清除，要调用 removeDialog 方法并传入 Dialog 的 id 值来彻底释放对话框。如果需要在调用 dismiss 方法关闭对话框时执行一些特定的工作，则可以为对话框设置 OnDismissListener 并重写其中的 onDismiss 方法来开发特定的功能。

3.4.2 普通对话框

普通对话框中只显示提示信息和一个“确定”按钮，通过 AlertDialog 来实现。下面介绍一个普通对话框的案例来具体学习普通对话框的使用。该案例的具体开发步骤如下。

在 Eclipse 中创建一个项目名称为 AlertDialogExample。打开项目 src 目录下的 AlertDialogExample.java 文件，修改文件，内容如下所示。

```
public class AlertDialogExample extends Activity {
    final int COMMON_DIALOG = 1;                      //普通对话框 id
    @Override
    public void onCreate(Bundle savedInstanceState) {
      super.onCreate(savedInstanceState);
      setContentView(R.layout.main);      //设置当前屏幕
     Button btn = (Button)findViewById(R.id.Button01);//获得 Button 对象
     btn.setOnClickListener(new View.OnClickListener(){//为 Button 设置
     OnClickListener 监听器
            @Override
public void onClick(View v) {   //重写 onClick 方法
        showDialog(COMMON_DIALOG); //显示普通对话框
        }}); }
    @Override
protected Dialog onCreateDialog(int id) {//重写 onCreateDialog 方法
        Dialog dialog = null;        //声明一个 Dialog 对象用于返回
        switch(id){      //对 id 进行判断
        case COMMON_DIALOG:
            Builder b = new AlertDialog.Builder(this);
            b.setIcon(R.drawable.icon);//设置对话框的图标
            b.setTitle("显示普通对话框");//设置对话框的标题
            b.setMessage("感谢你学习 Android! ");//设置对话框的显示内容
            b.setPositiveButton("确定", //添加按钮
                new OnClickListener() { //为按钮添加监听器
                    @Override
        public void onClick(DialogInterface dialog, int which) {
            EditText et = (EditText)findViewById(R.id.EditText01);
            et.setText("游戏学院欢迎你！Android 欢迎你！");//设置 EditText 内容
                    }});
            b.setNegativeButton("取消", //设置“取消”按钮
```

```
                new DialogInterface.OnClickListener() {
            public void onClick(DialogInterface dialog,int whichButton){
                //单击"取消"按钮之后退出程序
                AlertDialogExample.this.finish(); } });
            dialog = b.create();                //生成 Dialog 对象
            break;
        default: break; }
        return dialog; }//返回生成 Dialog 的对象
    }
```

最后，运行项目。项目启动运行效果如图 3-30 所示，用户单击“显示普通对话框”按钮，则跳转到效果图 3-31，当用户单击图 3-31 的“确定”按钮，则会跳转到效果图 3-32，如果用户单击“取消”按钮，则程序关闭。

图 3-30 项目启动

图 3-31 普通对话框效果

图 3-32 确定显示普通对话框

3.4.3 列表对话框

列表对话框也属于 AlertDialog。我们通过一个爱好选择的案例来说明列表对话框的具体用法。该案例的开发步骤如下。

在 Eclipse 中创建一个项目名称为 ListDialogExample。打开项目 src 目录下的 ListDialogExample.java 文件，修改文件，内容如下所示。

```
public class ListDialogExample extends Activity {
    final int LIST_DIALOG = 2;  //声明列表对话框的 id
    public void onCreate(Bundle savedInstanceState){     //重写onCreate方法
        super.onCreate(savedInstanceState);
        setContentView(R.layout.main);        //设置当前屏幕
        Button btn = (Button)findViewById(R.id.Button01);
        //为按钮添加 OnClickListener 监听器
btn.setOnClickListener(new View.OnClickListener() {
        public void onClick(View v) {
            showDialog(LIST_DIALOG);     //显示列表对话框
            }}
);}
     protected Dialog onCreateDialog(int id){//重写的 onCreateDialog 方法
```

```
        Dialog dialog = null;
        switch(id){      //对id进行判断
        case LIST_DIALOG:
    Builder b = new AlertDialog.Builder(this);//创建Builder对象
            b.setIcon(R.drawable.icon);                          //设置图标
            b.setTitle("列表对话框");                            //设置标题
            b.setItems( //设置列表中的各个属性
            R.array.msa,         //字符串数组
          //为列表设置OnClickListener监听器
            new DialogInterface.OnClickListener() {
                    @Override
        public void onClick(DialogInterface dialog, int which) {
            EditText et = (EditText)findViewById(R.id.EditText01);
            et.setText("您选择了: "
+getResources().getStringArray(R.array.msa)[which]);
            }});
            dialog=b.create();//生成Dialog对象
            break;
        default: break;     }
        return dialog;  }//返回Dialog对象
}
```

在res/values目录下新建一个数组的XML文件名称为array.xml，并在其中输入以下代码。该文件定义的字符串可以通过R.array.msa获取到，作为列表对话框里面的值。

```
<?xml version="1.0" encoding="utf-8"?>
<resources>
   <string-array name="msa">     <!-- 声明一个字符串数组 -->
     <item>游泳</item>   <!-- 向数组中加入元素 -->
     <item>打篮球</item> <!-- 向数组中加入元素 -->
     <item>写Java程序</item><!-- 向数组中加入元素 -->
   </string-array>
</resources>
```

最后，运行项目。当项目启动运行效果如图3-33所示，用户单击“显示列表对话框”按钮，则跳转到效果图3-34，当用户单击图3-34的某一个列表信息，则会获取该选择项信息并跳转到效果图3-35。

图3-33　项目启动

图3-34　显示列表对话框

图3-35　任意列表信息

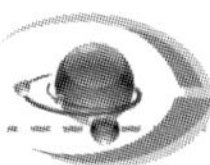

3.4.4 单选按钮和复选框对话框

单选按钮对话框和复选框同样是通过 AlertDialog 来实现的。本节我们仍然通过一个案例来学习单选按钮对话框的使用。该案例开发步骤与列表对话框案例有很多雷同的地方，这里仅进行简单说明。

```
public class RadioDialog extends Activity {
    final int LIST_DIALOG_SINGLE = 3;        //记录单选列表对话框的id
    @Override
    public void onCreate(Bundle savedInstanceState){    //重写onCreate方法
        super.onCreate(savedInstanceState);
        setContentView(R.layout.main);         //设置当前屏幕
        Button btn = (Button)findViewById(R.id.Button01);
         //为Button设置OnClickListener监听器
        btn.setOnClickListener(new View.OnClickListener() {
            @Override
            public void onClick(View v) {
        showDialog(LIST_DIALOG_SINGLE);        //显示单选按钮对话框
            }}); }
    @Override
    protected Dialog onCreateDialog(int id){   //重写onCreateDialog方法
        Dialog dialog = null;//声明一个Dialog对象用于返回
        switch(id){ //对id进行判断
        case LIST_DIALOG_SINGLE:
      Builder b = new AlertDialog.Builder(this);//创建Builder对象
            b.setIcon(R.drawable.icon);        //设置图标
            b.setTitle("单选列表对话框");       //设置标题
            b.setSingleChoiceItems( //设置单选列表选项
                    R.array.msa,
                    0,
            new DialogInterface.OnClickListener() {
                        @Override
        public void onClick(DialogInterface dialog, int which) {
        EditText et = (EditText)findViewById(R.id.EditText01);
        et.setText("您选择了："
        + getResources().getStringArray(R.array.msa)[which]);
                }});
            b.setPositiveButton(          //添加一个按钮
            "确定",                //按钮显示的文本
            new DialogInterface.OnClickListener() {
                @Override
        public void onClick(DialogInterface dialog, int which){}
```

```
            });
        dialog = b.create();          //生成 Dialog 对象
            break;
        default: break; }
        return dialog; } //返回生成的 Dialog 对象
}
```

复选框对话框案例与前面单选按钮对话框案例相似，这里仅把案例的 Activity 代码编写出来。代码如下：

```
public class CheckBoxDialog extends Activity {
    final int LIST_DIALOG_MULTIPLE = 4;//记录多选按钮对话框的 id
boolean[] mulFlags=new boolean[]{true,false,true};//初始复选情况
    String[] items=null;//选项数组
    @Override
    public void onCreate(Bundle savedInstanceState) {   //重写 onCreate 方法
        super.onCreate(savedInstanceState);
        setContentView(R.layout.main);          //设置当前屏幕
//获得 XML 文件中的字符串数组
        items=getResources().getStringArray(R.array.msa);
        Button btn = (Button)findViewById(R.id.Button01);
        btn.setOnClickListener(new View.OnClickListener() {      @Override
         public void onClick(View v) {
         showDialog(LIST_DIALOG_MULTIPLE);   //显示多选按钮对话框
            }}); }
    @Override
 protected Dialog onCreateDialog(int id){//重写 onCreateDialog 方法
         Dialog dialog = null;
         switch(id){      //对 id 进行判断
         case LIST_DIALOG_MULTIPLE:
       Builder b = new AlertDialog.Builder(this);//创建 Builder 对象
            b.setIcon(R.drawable.icon); //设置图标
            b.setTitle(R.string.title);     //设置标题
            b.setMultiChoiceItems( //设置多选选项
            R.array.msa,
            mulFlags,    //传入初始的选中状态
            new DialogInterface.OnMultiChoiceClickListener() {
                @Override
          public void onClick(DialogInterface dialog, int which,
boolean isChecked) {
            mulFlags[which] = isChecked;//设置选中标志位
            String result = "您选择了：";
            for(int i=0;i<mulFlags.length;i++){
                    if(mulFlags[i]){          //如果该选项被选中
```

```
                result = result+items[i]+"、"; } }
        EditText et = (EditText)findViewById(R.id.EditText01);
                        //设置 EditText 显示的内容
      et.setText(result.substring(0,result.length()-1));
                        }  });
            b.setPositiveButton(    //添加按钮
                    "确定",
                    new DialogInterface.OnClickListener(){
                        @Override
        public void onClick(DialogInterface dialog, int which) {}
            });
            dialog = b.create();    //生成 Dialog 方法
            break;
        default: break; }
        return dialog;  }//返回 Dialog 方法
}
```

3.4.5　日期及时间选择对话框

实现日期及时间选择对话框的开发分别需要使用 DatePickerDialog 类和 TimePickerDialog 类。下面我们通过一个案例来说明如何基于这两个类开发日期和时间选择对话框。该案例的开发步骤如下。

```
public class DateTimeDialog extends Activity {
final int DATE_DIALOG=0;      //日期选择对话框 id
final int TIME_DIALOG=1;      //时间选择对话框 id
Calendar c=null;      //声明一个日历对象
@Override
public void onCreate(Bundle savedInstanceState) {
   super.onCreate(savedInstanceState);
   setContentView(R.layout.main);
   //打开日期对话框的按钮
   Button bDate =(Button) this.findViewById(R.id.Button01);
   bDate.setOnClickListener(new OnClickListener(){
            @Override
        public void onClick(View v){//重写 onClick 方法
        showDialog(DATE_DIALOG);//打开单选列表对话框
            }  });
   //打开时间对话框的按钮
   Button bTime =(Button) this.findViewById(R.id.Button02);
   bTime.setOnClickListener(new OnClickListener(){
            @Override
        public void onClick(View v){        //重写 onClick 方法
```

```
                showDialog(TIME_DIALOG);           //打开单选列表对话框
                        } }); }
        @Override
        public Dialog onCreateDialog(int id){//重写 onCreateDialog 方法
         Dialog dialog=null;
         switch(id){      //对 id 进行判断
           case DATE_DIALOG://生成日期对话框的代码
                c=Calendar.getInstance();//获取日期对象
                dialog=new DatePickerDialog( //创建 DatePickerDialog 对象
                  this,
                //创建 OnDateSetListener 监听器
                new DatePickerDialog.OnDateSetListener(){
                        @Override
            public void onDateSet(DatePicker dp, int year, int month,
    int dayOfMonth) {
            EditText et = (EditText)findViewById(R.id.et);
        et.setText("您选择了："+year+"年"+(month+1)+"月"+dayOfMonth+"日");
                        }   },
                  c.get(Calendar.YEAR),                          //传入年份
                  c.get(Calendar.MONTH),                         //传入月份
                  c.get(Calendar.DAY_OF_MONTH)    );            //传入天数
        break;
          case TIME_DIALOG://生成时间对话框的代码
              c=Calendar.getInstance();//获取日期对象
              dialog=new TimePickerDialog(  //创建 TimePickerDialog 对象
                 this,
               //创建 OnTimeSetListener 监听器
                 new TimePickerDialog.OnTimeSetListener(){
                        @Override
        public void onTimeSet(TimePicker tp,int hourOfDay,int minute){
            EditText et = (EditText)findViewById(R.id.et);
    //设置 EditText 控件的属性
        et.setText("您选择了："+hourOfDay+"时"+minute+"分");
                        }   },
             c.get(Calendar.HOUR_OF_DAY),//传入当前小时数
             c.get(Calendar.MINUTE),     //传入当前分钟数
                  false );  break;    }
        return dialog;    }
    }
```

最后，运行项目，效果如图 3-36 所示。

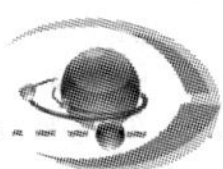

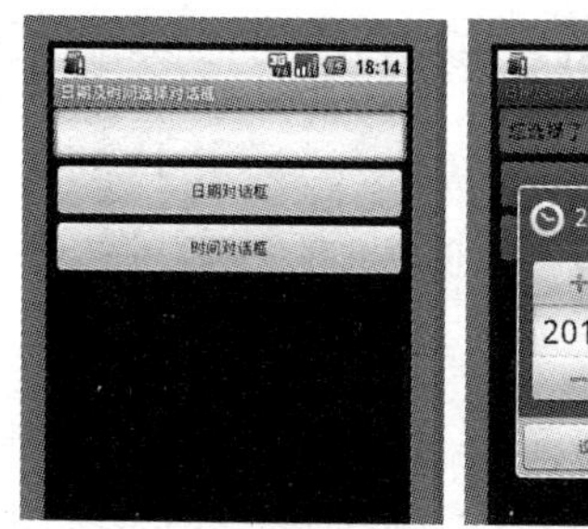
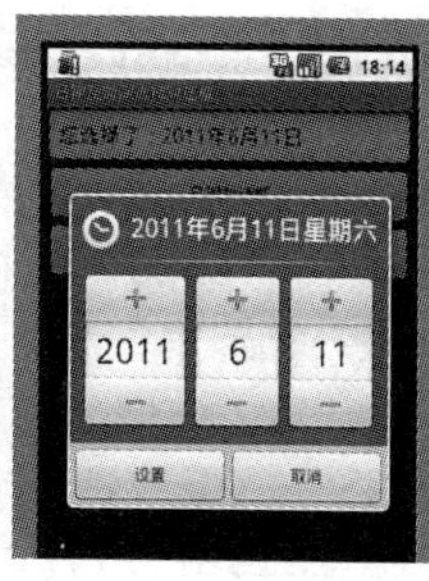
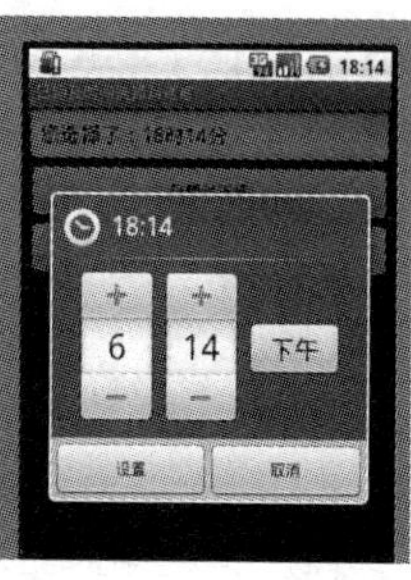

图 3-36　日期时间选择对话框案例效果图

3.4.6　进度对话框

本节我们通过一个案例来介绍进度对话框的应用。该案例的布局文件中声明了一个垂直分布的线性布局，该布局中包含一个按钮。在程序运行时按下该按钮会弹出进度对话框。该案例的主逻辑文件代码如下。

```
    public class ProgressDialogExample extends Activity {
        final int PROGRESS_DIALOG=0;//声明进度对话框 id
        final int INCREASE=0;    //Handler 消息类型
        ProgressDialog pd;
        Handler myHandler;        //Handler 对象引用
        @Override
        public void onCreate(Bundle savedInstanceState) {//重写 onCreate 方法
            super.onCreate(savedInstanceState);
            setContentView(R.layout.main);        //设置屏幕
           Button bok=(Button)this.findViewById(R.id.Button01); bok.
setOnClickListener( //设置 OnClickListener 监听器
                new OnClickListener(){
                    @Override
            public void onClick(View v) {        //重写 onClick 方法
           showDialog(PROGRESS_DIALOG); //显示进度对话框
                    }   }  );
           myHandler=new Handler(){//创建 Handler 对象
            @Override
            public void handleMessage(Message msg){
            switch(msg.what){
                case INCREASE:
               pd.incrementProgressBy(1);//进度每次加 1
            if(pd.getProgress()>=100){  //判断是否结束进度
               pd.dismiss();    }  //如果进度条走完则关闭窗口
               break;}
            super.handleMessage(msg);}  };  }
       @Override
```

```
    public Dialog onCreateDialog(int id){//重写onCreateDialog方法
     switch(id){            //对id进行判断
       case  PROGRESS_DIALOG:      //创建进度对话框
          pd=new ProgressDialog(this);//创建进度对话框
          pd.setMax(100);//设置最大值
            pd.setProgressStyle(ProgressDialog.STYLE_HORIZONTAL);
           pd.setTitle("安装进度");//设置标题
           pd.setCancelable(false);//设置进度对话框不能用回退按钮关闭
          break;}
     return pd; }
 //每次弹出对话框时被回调以动态更新对话框内容的方法
    public void onPrepareDialog(int id, Dialog dialog){
     super.onPrepareDialog(id, dialog);
     switch(id){
         case  PROGRESS_DIALOG:
         pd.incrementProgressBy(-pd.getProgress());//对话框进度清零
         new Thread(){               //创建一个线程
     public void run(){
         while(true){
         myHandler.sendEmptyMessage(INCREASE);   //发送Handler消息
         if(pd.getProgress()>=100){break;}
         try{Thread.sleep(40);   }     //线程休眠
         catch(Exception e){e.printStackTrace();//捕获并打印异常
            } }   }}.start();    //启动线程
            break;}   }
 }
```

运行该案例，运行效果如图 3-37 所示。

图 3-37　进度对话框案例效果图

3.5　界面事件

本节将对 Android 平台用户界面的常用事件响应进行详细介绍，以加深读者对 Android 平台的事件处理模型的理解，熟练掌握控件的各种事件处理方法。Android 平台的事件处理机制有两种，一种是基于回调机制的事件处理，一种是基于监听接口的事件处理。在 Android 平台中，每

个 View 都有自己的处理事件的回调方法，开发人员可以通过重写 View 中的这些回调方法来实现需要的响应事件。当某个事件被任何一个 View 处理时，便会调用 Activity 中相应的回调方法。基于回调机制的事件主要有 onKeyDown 方法、onKeyUp 方法、onTouchEvent 方法、onTrackBallEvent 方法、onFocusChanged 方法。基于监听接口的事件主要有 OnClickListener 接口、OnLongClickListener 接口、OnFocusChangeListener 接口、OnKeyListener 接口、OnTouchListener 接口、OncreateContextMenuListener 接口。

3.5.1 onKeyDown 方法简介

onKeyDown 方法是接口 KeyEvent.Callback 中的抽象方法，所有的 View 全部实现了该接口并重写了该方法，该方法用来捕捉手机键盘被按下的事件。方法的声明格式如下所示：

```
public boolean onKeyDown (int keyCode, KeyEvent event)
```

参数 keyCode，未被按下的键值即键盘码，手机键盘中每个按钮都会有其单独的键盘码，在应用程序中，都是通过键盘码才知道用户按下的是哪个键。参数 event，为按键事件的对象，其中包含了触发事件的详细信息，例如事件的状态、事件的类型、事件发生的时间等。当用户按下按键时，系统会自动将事件封装成 KeyEvent 对象供应用程序使用。返回值，该方法的返回值为一个 boolean 类型的变量，当返回 true 时，表示已经完整地处理了这个事件，并不希望其他的回调方法再次进行处理，而当返回 false 时，表示并没有完全处理完该事件，更希望其他回调方法继续对其进行处理，例如 Activity 中的回调方法。

接下来通过一个简单的例子来介绍该方法的使用方法及原理。该例子中自定义一个 Button 并显示到窗口中，然后对键盘进行监听，根据不同情况打印相关信息。该案例的开发步骤如下：

步骤一：创建一个新的 Android 项目，名为 Example_onKeyDown。

步骤二：编写 Example_onKeyDown.java 文件，用下列代码替换原有代码。

```
public class Example_onKeyDown extends Activity {
  MyButton myButton;//自定义的 Button
    public void onCreate(Bundle savedInstanceState) {
        super.onCreate(savedInstanceState);
        myButton = new MyButton(this);
        myButton.setText("全屏按钮");//设置按钮上的文字
        myButton.setTextSize(30);//设置文字的大小
        setContentView(myButton); }//将按钮显示在屏幕上
  public boolean onKeyDown(int keyCode, KeyEvent event) {//重写键盘按下监听
      System.out.println("activity onKeyDown");//打印消息
      return super.onKeyDown(keyCode, event);}
  class MyButton extends Button{//自己定义的 Button
      public MyButton(Context context) {//构造器
          super(context);}
      public boolean onKeyDown(int keyCode, KeyEvent event){//重写键盘按下监听
          System.out.println("MyView onKeyDown");//打印消息
          return false;
```

```
    }}
}
```

图 3-38　手机键盘被按下事件图

步骤三：运行该程序，查看运行效果，如图 3-38 所示。

打开 Eclipse 的 DDMS 中的 LogCat 窗口，单击右上侧的绿色加号创建一个 System.out，在 Log Tag 文本框中输入“System.out”，单击 OK 按钮过滤信息。当按钮控件获得焦点时单击手机键盘上的任意键，通过 LogCat 中看到的日志内容可知，此时先调用自定义的 Button 中的 onKeyDown 方法，再调用 Activity 中的 onKeyDown 方法，而当按钮控件没有获得焦点时，将只调用 Activity 中的 onKeyDown 方法。如果将 MyButton 类里面 onKeyDown 方法代码的 false 改成 true 再次运行，当按钮获得焦点时单击按键，只会调用自定义的 Button 中的 onKeyDown 方法，而不会再调用 Activity 中的该方法。

3.5.2　onKeyUp 方法简介

onKeyUp 方法的原理及使用方法与 onKeyDown 方法基本一样，只是该方法会在按键抬起时被调用。如果用户需要对按键抬起事件进行处理，通过重写该方法可以实现。该方法同样是接口 KeyEvent.Callback 中的一个抽象方法，并且所有的 View 同样全部实现了该接口并重写了该方法，onKeyUp 方法用来捕捉手机键盘按键抬起的事件，方法的声明格式如下所示：

```
public boolean onKeyUp (int keyCode, KeyEvent event)
```

参数 keyCode：同样为触发事件的按键码，需要注意的是，同一个按键在不同型号的手机中的按键码可能不同。参数 event：同样为事件封装类的对象，其含义与 onKeyDown 方法中的完全相同，在此不再赘述。返回值：该方法返回值表示的含义与 onKeyDown 方法相同，同样通知系统是否希望其他回调方法再次对该事件进行处理。

3.5.3　onTouchEvent 方法简介

onTouchEvent 事件方法是手机屏幕事件的处理方法。该方法在 View 类中的定义，并且所有的 View 子类全部重写了该方法，应用程序可以通过该方法处理手机屏幕的触摸事件。该方法的声明格式如下所示：

```
public boolean onTouchEvent (MotionEvent event)
```

参数 event：为手机屏幕触摸事件封装类的对象，其中封装了该事件的所有信息，例如触摸的位置、触摸的类型以及触摸的时间等。该对象会在用户触摸手机屏幕时被创建。返回值：该方法的返回值机理与键盘响应事件的相同，同样是当已经完整地处理了该事件且不希望其他回调方法再次处理时返回 true，否则返回 false。

该方法并不像之前介绍过的方法只处理一种事件，一般情况下以下三种情况的事件全部由 onTouchEvent 方法处理，只是三种情况中的动作值不同。

屏幕被按下事件：当屏幕被按下时，会自动调用该方法来处理事件，此时 MotionEvent.getAction()的值为 MotionEvent.ACTION_DOWN，如果在应用程序中需要处理屏幕被按下的事件，只需重新回调该方法，然后在方法中进行动作的判断即可。

屏幕被抬起事件：当触控笔离开屏幕时触发的事件，该事件同样需要 onTouchEvent 方法来捕捉，然后在方法中进行动作判断。当 MotionEvent.getAction()的值为 MotionEvent.ACTION_UP 时，表示是屏幕被抬起的事件。

在屏幕中拖动事件：该方法还负责处理触控笔在屏幕上滑动的事件，同样是调用 MotionEvent.getAction()方法来判断动作值是否为 MotionEvent.ACTION_MOVE 再进行处理。

接下来我们通过一个简单的案例来介绍该方法的使用。在该案例中，会在用户点击的位置绘制一个矩形，然后监测用户触控笔的状态，当用户在屏幕上移动触控笔时，使矩形随之移动，而当用户触控笔离开手机屏幕时，停止绘制矩形。该案例的开发步骤如下。

步骤一：创建一个名为 Example_onTouchEvent 的 Android 项目。

步骤二：打开 Example_onTouchEvent.java 文件，按如下所示编写代码。

```
public class Example_onTouchEvent extends Activity {
    MyView myView;//自定义 View 的引用
   public void onCreate(Bundle savedInstanceState) {//重写的 onCreate 方法
        super.onCreate(savedInstanceState);
        myView = new MyView(this);//初始化自定义的 View
        setContentView(myView); }//设置当前显示的用户界面
//重写的屏幕监听方法，在该方法中，根据事件动作的不同执行不同的操作
    @Override
    public boolean onTouchEvent(MotionEvent event) {
        switch(event.getAction()){
        case MotionEvent.ACTION_DOWN://按下
//通过调用 MotionEvent 的 getX()和 getY()方法得到事件发生的坐标，然后设置给自定义
View 的 x 与 y 成员变量。
            myView.x = (int) event.getX();//改变 x 坐标
            myView.y = (int) event.getY()-52;//改变 y 坐标
            myView.postInvalidate();//重绘
            break;
        case MotionEvent.ACTION_MOVE://移动
            myView.x = (int) event.getX();//改变 x 坐标
            myView.y = (int) event.getY()-52;//改变 y 坐标
            myView.postInvalidate();//重绘
            break;
        case MotionEvent.ACTION_UP://抬起
            myView.x = 50;//改变 x 坐标，使正方形绘制在初始的位置
            myView.y = 50;//改变 y 坐标
            myView.postInvalidate();//重绘
```

```
                break;}
            return super.onTouchEvent(event); }
        class MyView extends View{//自定义的View
            Paint paint;// 定义画笔
            int x = 50, y = 50;// 定义正方形的初始x、y坐标
            int w = 50; //定义所绘制矩形的边长
            public MyView(Context context) {//构造器
                super(context);
                paint = new Paint();}//初始化画笔
            @Override
            protected void onDraw(Canvas canvas) {//绘制方法
                canvas.drawColor(Color.GRAY);//绘制背景色
                canvas.drawRect(x, y, x+w, y+w, paint);//绘制矩形
                super.onDraw(canvas);
            } } }
```

步骤三：运行该案例，将看到如图3-39所示的效果。

当启动工程时，会在屏幕上绘出一个正方形；当点击屏幕时，正方形会移动到点击的位置；当触控笔在屏幕中滑动时，该正方形会随之移动；而当触控笔离开屏幕时，正方形便又回到原处。

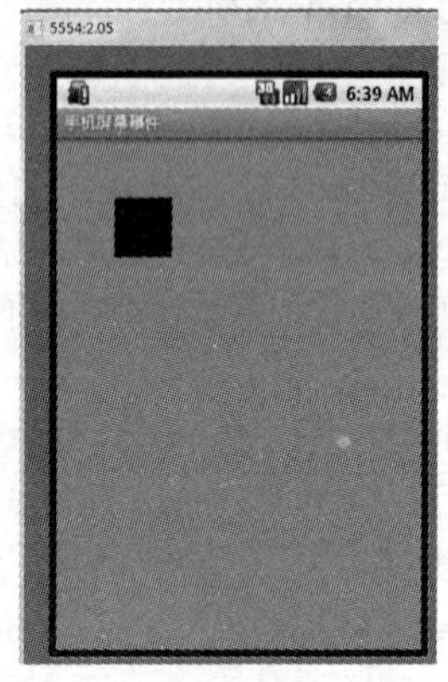

图3-39　手机屏幕事件图

3.5.4　onTrackBallEvent方法和onFocusChanged方法简介

onTrackBallEvent是手机中轨迹球的处理方法。所有的View同样全部实现了该方法。该方法的声明格式如下：

```
public boolean onTrackballEvent (MotionEvent event)
```

参数event：为手机轨迹球事件封装类的对象，其中封装了触发事件的详细信息，同样包括事件的类型、触发时间等，一般情况下，该对象会在用户操控轨迹球时被创建。返回值：该方法的返回值与前面介绍的各个回调方法的返回值机制完全相同，因本书篇幅有限，不再赘述。

轨迹球与手机键盘的区别如下所示。

（1）某些型号的手机设计出的轨迹球会比只有手机键盘时更美观，可增添用户对手机的整体印象。

（2）轨迹球使用更为简单，例如在某些游戏中使用轨迹球控制会更为合理。

（3）使用轨迹球会比键盘更为细化，即滚动轨迹球时，后台的表示状态的数值会变化得更细微、更精准。

onTrackBallEveni 使用方法与前面介绍过的各个回调方法基本相同，可以在 Activity 中重写该方法，也可以在各个 View 的实现类中重写。

提示：在模拟器运行状态下，通过 F6 键打开模拟器的轨迹球，然后便可以通过鼠标的移动来模拟轨迹球事件。

onFocusChanged 只能在 View 中重写，该方法是焦点改变的回调方法，当某个控件重写了该方法后，当焦点发生变化时，会自动调用该方法来处理焦点改变的事件。该方法的声明格式如下：

```
    protected void onFocusChanged (boolean gainFocus, int direction, Rect
previously FocusedRect)
```

参数 gainFocus：表示触发该事件的 View 是否获得了焦点，当该控件获得焦点时，gainFocus 等于 true，否则等于 false。参数 direction：表示焦点移动的方向，用数值表示，有兴趣的读者可以重写 View 中的该方法打印该参数进行观察。参数 previously FocusedRect：表示在触发事件的 View 的坐标系中，前一个获得焦点的矩形区域，即表示焦点是从哪里来的。如果不可用则为 null。表 3-22 为焦点有关的常用方法。

表 3-22 焦点有关的常用方法

方法名称	方法说明
setFocusable()	设置 View 是否可以拥有焦点
isFocusable()	监测此 View 是否可以拥有焦点
setNextFocusDownId() setNextFocusLeftId() setNextFocusRightId() setNextFocusUpId()	设置 View 的焦点向上、下、左、右移动后获得焦点 View 的 ID
hasFocus()	返回了 View 的父控件是否获得了焦点
requestFocus()	尝试让此 View 获得焦点
isFocusableInTouchMode()	设置 View 是否可以在触摸模式下获得焦点，在默认情况下是不可以获得的

对于手机轨迹球方法和焦点改变事件方法的具体应用，由于篇幅等原因，就不在此介绍了，读者可以去查相关资料。

3.5.5 OnClickListener 接口简介

对于一个 Android 应用程序来说，事件处理是必不可少的，用户与应用程序之间的交互便是通过事件处理来完成的。当用户与应用程序交互时，一定是通过触发某些事件来完成的，让事件来通知程序应该执行哪些操作，在这个繁杂的过程中主要涉及两个对象，事件源与事件监听器。事件源指的是事件所发生的控件，各个控件在不同情况下触发的事件不尽相同，而且产生的事件对象也可能不同。监听器则是用来处理事件的对象，实现了特定的接口，根据事件的不同重写不同的事件处理方法来处理事件。将事件源与事件监听器联系到一起，就需要为事件源注册监听，当事件发生时，系统才会自动通知事件监听器来处理相应的事件。接下来用图 3-40 来说明事件处理的整个流程。

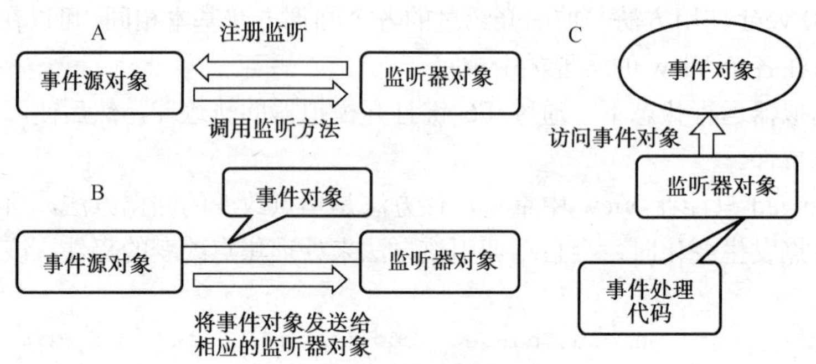

图 3-40　事件处理流程图

事件处理的过程一般分为三步，如下所示。

第一步，应该为事件源对象添加监听，这样当某个事件被触发时，系统才会知道通知谁来处理该事件，如事件处理流程图 3-40（A）所示。

第二步，当事件发生时，系统会将事件封装成相应类型的事件对象，并发送给注册到事件源的事件监听器，如事件处理流程图 3-40（B）所示。

第三步，当监听器对象接收到事件对象之后，系统会调用监听器中相应的事件处理方法来处理事件并给出响应，如事件处理流程图 3-40（C）所示。

OnClickListener 接口，是处理点击事件的。在触控模式下，是在某个 View 上按下并抬起的组合动作，而在键盘模式下，是某个 View 获得焦点后单击“确定”按钮或者按下轨迹球事件。该接口对应的回调方法签名如下：

```
    public void onClick(View v) ;    参数 v 便为事件发生的事件源。
```

接下来同样通过一个学生选择系部的简单案例来介绍该接口的使用方法，步骤如下。

步骤一：创建一个名为 Example_OnClickListener 的 Android 项目。设计其布局文件，使布局界面如图 3-41 所示。

图 3-41　选择系部初始图

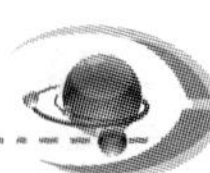

步骤二：修改主逻辑页面。该页面完成当用户单击其中任何一个按钮时，界面将显示用户所单击按钮的名称。或者当某个按钮获得焦点的时候，用户单击“确定”按钮同样显示该按钮的名称。其修改代码如下：

```
    public class Example_OnClickListener extends Activity implements
    OnClickListener{
        Button[] buttons =  new Button[3];
        TextView textView;
       public void onCreate(Bundle savedInstanceState) {
           super.onCreate(savedInstanceState);
           setContentView(R.layout.main);
    buttons[0]= (Button)this.findViewById(R.id.button01);
           buttons[1]= (Button)this.findViewById(R.id.button02);
           buttons[2]= (Button)this.findViewById(R.id.button03);
           textView= (TextView)this.findViewById(R.id.textView01);
           textView.setTextSize(18);
           for(Button button : buttons){
            button.setOnClickListener(this); } }//为每个按钮注册监听
        @Override
        public void onClick(View v) {//实现事件监听方法
        textView.setText("您选择了" + ((Button)v).getText()+", 欢迎你！");}}
```

接下来我们简单介绍 OnLongClickListener 接口，该接口与 OnClickListener 接口原理基本相同，只是该接口为 View 长按事件的捕捉接口，即当长时间按下某个 View 时触发的事件，该接口对应的回调方法为：

```
    public boolean onLongClick(View v)
```

参数 v：为事件源控件，当长时间按下此控件时才会触发该方法。返回值：该方法的返回值为一个 boolean 类型的变量，当返回 true 时，表示已经完整地处理了这个事件，并不希望其他的回调方法再次进行处理；当返回 false 时，表示并没有完全处理完该事件，更希望其他方法继续对其进行处理。

我们只需要修改 Example_OnClickListene 项目里面的三个地方，就可测试该接口的使用方法。其一为定义类的接口为 OnLongClickListener，其二在绑定监听器时修改为：button.setLongOnClickListener(**this**)；其三把实现监听方法该为：

```
    public boolean onLongClick(View v) {//实现事件监听方法
      textView.setText("您选择了" + ((Button)v).getText()+", 欢迎你！");
      return true;  }
```

这样，我们就可以测试 OnLongClickListener 接口的使用了。

3.5.6　OnFocusChangeListener 接口简介

OnFocusChangeListener 接口用来处理控件焦点发生改变的事件。如果注册了该接口，当某个控件失去焦点或者获得焦点时都会触发该接口中的回调方法，该接口对应的回调方法声明格式如下：

```
public void onFocusChange(View v, Boolean hasFocus)
```

参数 v：便为触发该事件的事件源。

参数 hasFocus：表示 v 的新状态，即 v 是否是获得焦点。

下面通过一个简单案例来介绍该接口的使用方法，步骤如下。

步骤一：创建一个名为 Example_OnFocusChangeListener 的 Android 项目。将程序需要的图片资源（见图 3-42），存放到 res\drawable-mdpi 目录下。

步骤二：编写项目的界面文件效果如图 3-43 所示。该界面布局文件在垂直的线性布局中插入了两个水平布局线性布局，然后在水平线性布局中添加两个 ImageButton 控件。

图 3-42　资源图片

图 3-43　焦点案例效果图

步骤三：编写主逻辑文件。在文件中实现 OnFocusChangeListener 接口，并重写相应的接口方法，从而达到实现效果的目的。修改代码如下：

```
public class Example_OnFocusChangeListener extends Activity implements
OnFocusChangeListener{
    TextView myTextView;
    ImageButton[] imageButtons = new ImageButton[4];
   @Override
   public void onCreate(Bundle savedInstanceState) {
      super.onCreate(savedInstanceState);
      setContentView(R.layout.main);
      myTextView = (TextView) this.findViewById(R.id.myTextView);//得到
      myTextView 的引用
      imageButtons[0] = (ImageButton) this.findViewById(R.id.button01);
      //得到 button01 的引用
      imageButtons[1] = (ImageButton) this.findViewById(R.id.button02);
      //得到 button02 的引用
      imageButtons[2] = (ImageButton) this.findViewById(R.id.button03);
      //得到 button03 的引用
      imageButtons[3] = (ImageButton) this.findViewById(R.id.button04);
      //得到 button04 的引用
      for(ImageButton imageButton : imageButtons){
```

```
        imageButton.setOnFocusChangeListener(this);//添加监听
    } }
    @Override
    public void onFocusChange(View v, boolean hasFocus) {//实现了接口中的方法，在方法中根据事件源的 ID 判断是哪个控件触发了该方法，然后实现相应的效果
        if(v.getId() == R.id.button01){//改变的是 button01 时
            myTextView.setText("您选中了羊！");
        }else if(v.getId() == R.id.button02){//改变的是 button02 时
            myTextView.setText("您选中了猪！");
        }else if(v.getId() == R.id.button03){//改变的是 button03 时
            myTextView.setText("您选中了牛！");
        }else if(v.getId() == R.id.button04){//改变的是 button04 时
            myTextView.setText("您选中了鼠！");
        }else{//其他情况
            myTextView.setText("");
        }}}
```

步骤四：运行调试，将看到我们想要的效果。

3.5.7 OnKeyListener 接口简介

OnKeyListener 是对手机键盘进行监听的接口，通过对某个 View 注册该监听，当 View 获得焦点并有键盘事件时，便会触发该接口中的回调方法。该接口中的抽象方法声明格式如下：

```
public boolean onKey(View v, int keyCode, KeyEvent event)
```

参数 v：为事件的事件源控件。

参数 keyCode：为手机键盘的键盘码。

参数 event：为键盘事件封装类的对象，其中包含了事件的详细信息，例如发生的事件、事件的类型等。

同样，我们也可以通过一个简单案例来实现该接口的使用，步骤如下：

步骤一：创建一个名为 Example_OnKeyListener 的 Android 项目。准备图片资源，将需要的图片存放到 res\drawable-mdpi 目录下，如图 3-44 所示。

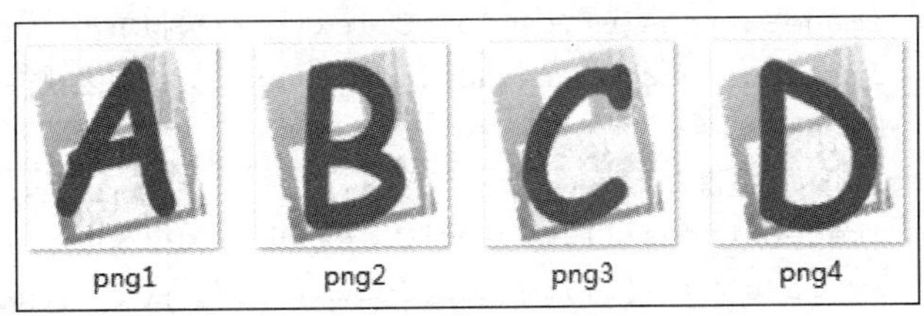

图 3-44

步骤二：开发主逻辑代码，编写 Example_OnKeyListener.java 类，代码如下：

```
public class Example_OnKeyListener extends Activity implements
OnKeyListener,OnClickListener{
```

```
    ImageButton[] imageButtons = new ImageButton[4];//声明按钮数组
    TextView myTextView;//声明 TextView 的引用
    @Override
    public void onCreate(Bundle savedInstanceState) {
        super.onCreate(savedInstanceState);
        this.setContentView(R.layout.main);
   myTextView = (TextView) this.findViewById(R.id.myTextView);//得到
   myTextView 的引用
 //获取每一个 button 的引用
 imageButtons[0] = (ImageButton) this.findViewById(R.id.button01);
 …//其他省略
    for(ImageButton imageButton : imageButtons){
     imageButton.setOnClickListener(this);//添加单击监听
        imageButton.setOnKeyListener(this); } }//添加键盘监听
    @Override
    public void onClick(View v) {//实现了接口中的方法
    if(v==imageButtons[0]){myTextView.setText("您单击了按钮 A! ");}
        if(v.getId() == R.id.button01){//改变的是 button01 时
            myTextView.setText("您单击了按钮 A! ");
        }else if(v.getId() == R.id.button02){//改变的是 button02 时
            myTextView.setText("您单击了按钮 B! ");
        }else if(v.getId() == R.id.button03){//改变的是 button03 时
            myTextView.setText("您单击了按钮 C! ");
        }else if(v.getId() == R.id.button04){//改变的是 button04 时
            myTextView.setText("您单击了按钮 D! ");
        }else{//其他情况
            myTextView.setText("");
        } }
    @Override
public boolean onKey(View v, int keyCode, KeyEvent event) {//键盘监听
        switch(keyCode){//判断键盘码
        case 29://按键 A
            imageButtons[0].performClick();//模拟单击
            imageButtons[0].requestFocus();//尝试使之获得焦点
            break;
        case 30://按键 B
            imageButtons[1].performClick();//模拟单击
            imageButtons[1].requestFocus();//尝试使之获得焦点
            break;
        case 31://按键 C
            imageButtons[2].performClick();//模拟单击
            imageButtons[2].requestFocus();//尝试使之获得焦点
            break;
        case 32://按键 D
```

```
            imageButtons[3].performClick();//模拟单击
            imageButtons[3].requestFocus();//尝试使之获得焦点
            break;}
        return false;  }  }
```

步骤三：此时运行该案例，观察运行效果如图 3-45 所示，当按手机键盘上的 A、B、C、D 键时，相当于单击 A、B、C、D 按钮。

图 3-45　手机键盘事件效果图

3.5.8　OnTouchListener 接口简介

OnTouchListener 接口是用来处理手机屏幕事件的监听接口，当在 View 的范围内发生触摸按下、抬起或滑动等动作时都会触发该事件。该接口中的监听方法声明格式如下：

```
    public boolean onTouch(View v, MotionEvent event)
```

参数 v：同样为事件源对象。

参数 event：为事件封装类的对象，其中封装了触发事件的详细信息，同样包括事件的类型、触发时间等信息。

接下来我们介绍一个通过监听接口的方式实现在屏幕上拖动按钮移动的案例。开发步骤如下：

步骤一：创建一个名为 Example_**OnTouchListener** 的 Android 项目。这里省略关于描述文件和字符串资源文件的编写。读者可以参考案例运行效果自己设计。

步骤二：接下来开始开发主要的逻辑代码。编写 Example_OnTouchListener.java 文件，其代码如下所示。

```
    public class Example_OnTouchListener extends Activity {
        final static int WRAP_CONTENT=-2;//表示 WRAP_CONTENT 的常量
        final static int X_MODIFY=4;//在非全屏模式下 X 坐标的修正值
        final static int Y_MODIFY=52;//在非全屏模式下 Y 坐标的修正值
        int xSpan;//在触控笔单击按钮的情况下相对于按钮自己坐标系的
        int ySpan;//X,Y 位置
      public void onCreate(Bundle savedInstanceState) {//重写的 onCreate 方法
          super.onCreate(savedInstanceState);
          setContentView(R.layout.main);//设置当前的用户界面
```

```
        Button bok=(Button)this.findViewById(R.id.Button01);
        bok.setOnTouchListener(new OnTouchListener(){
    public boolean onTouch(View view, MotionEvent event) {
        switch(event.getAction()){
        case MotionEvent.ACTION_DOWN://触控笔按下
        Button bok1=(Button)findViewById(R.id.Button01);
            bok1.setText("被按下");
             break;
        case MotionEvent.ACTION_UP:
        Button bok2=(Button)findViewById(R.id.Button01);
            bok2.setText("已抬起");
        case MotionEvent.ACTION_MOVE://触控笔移动
        Button bok=(Button)findViewById(R.id.Button01);
         //让按钮随着触控笔的移动一起移动
ViewGroup.LayoutParams lp=new AbsoluteLayout.LayoutParams(
WRAP_CONTENT,  WRAP_CONTENT, (int)event.getRawX()-xSpan-X_MODIFY,
(int)event.getRawY()-ySpan-Y_MODIFY  ) ;
        bok.setLayoutParams(lp);
            break;}
        return true;
            } } ); } }
```

步骤三：运行该案例，通过触控笔便可拖动屏幕中的按钮移动，如图 3-46 所示。

图 3-46　手机触屏事件监听接口运行案例

3.5.9　OnCreateContextMenuListener 接口简介

OnCreateContextMenuListener 接口是用来处理上下文菜单显示事件的监听接口。该方法是定义和注册上下文菜单的另一种方式。该接口中事件处理的回调方法签名如下所示。

```
public void onCreateContextMenu(ContextMenu menu, View v, ContextMenuInfo
info)
```

参数 menu：为事件的上下文菜单。

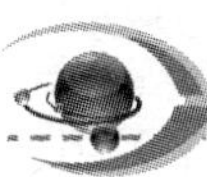

参数 v：为事件源 View，当该 View 获得焦点时才可能接收该方法的事件响应。

参数 info：info 对象中封装了有关上下文菜单额外的信息，这些信息取决于事件源 View。

该方法会在某个 View 中显示上下文菜单时被调用，开发人员可以通过实现该方法来处理上下文菜单显示时的一些操作。其使用方法与前面介绍的各个监听接口没有任何区别，因本书篇幅有限，不作详细介绍，读者可以参考之前的例子完成该接口的学习。

3.6 动画播放技术

本节将介绍 Android 平台下动画播放的技术，在进行用户界面的开发时，除了为空间设置合理的布局和外观外，让控件播放动画或许更加提高用户的体验。本节要介绍的动画播放技术主要有两种：帧动画和补间动画。帧动画是通过若干帧图片的轮流显示来实现的，而补间动画主要包括位置、角度、尺寸等属性的变化。

3.6.1 帧动画

帧动画（Frame Animation）是比较传统的动画方式，它将一系列的图片文件如放电影般依次进行播放，帧动画主要用到的类是 AnimationDrawable，每个帧动画都是一个 AnimationDrawable 对象。定义帧动画可以在 Java 代码中直接进行，也可以通过 XML 文件定义，定义帧动画的 XML 文件要存放在项目的 res\anim 目录下，我们建立的项目里面没有这个目录，读者可以自己创建一个。XML 文件中指定了图片帧出现的顺序及每个帧的持续时间。如表 3-23 所示为帧动画中的标记及其属性说明。

表 3-23 Frame Animation 中的标记及其属性说明

标记名称	属 性 值	说 明
<animation-list>	android:oneshot：如果设置为 true，则该动画只播放一次，然后停止在最后一帧；如果设置为 false，则该动画一直循环播放	Frame Animation 的根标记，可以包含若干<item>标记
<item>	android:drawable：图片帧的引用 android:duration：图片帧的停留时间 android:visible：图片帧是否可见	每个<item>标记定义了一个图片帧，其中包含图片资源的引用等属性

提示：AnimationDrawable 对象的 start 方法不可以在 Activity 的 onCreate 方法中调用，如果希望在程序一启动就播放动画，则应该在 OnWindowFocusChanged()方法中调用 start()方法。

接下来我们通过桌式足球项目欢迎界面的动画播放案例来介绍帧动画的具体使用，该案例的开发步骤如下。

步骤一：建立项目，准备资源。打开 Eclipse 工具，建立一个名为 Example_Animation 的项目，准备如图 3-47 所示图片存放在项目的 res\drawable-mdpi 目录下。

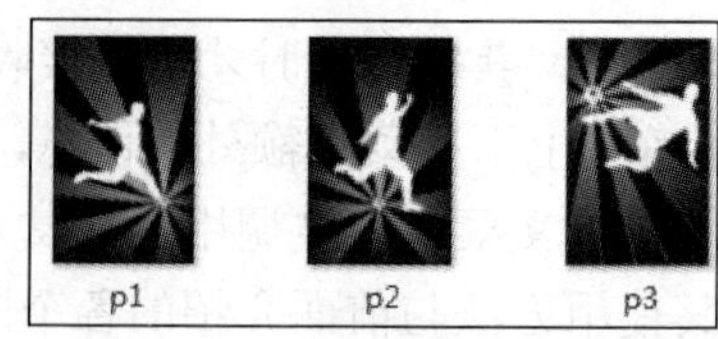

图 3-47

步骤二：在 res 目录下新建一个 anim 目录，在该目录下新建一个 XML 文件名称为 frame_ani.xml。在其中编写如下的代码：

```
<?xml version="1.0" encoding="utf-8"?>
 <animation-list xmlns:android="http://schemas.android.com/apk/res/android"
   android:oneshot="false">
    <item android:drawable="@drawable/p1" android:duration="1000" android:
    visible="true"/>
    <item android:drawable="@drawable/p2" android:duration="1000" android:
    visible="true"/>
    <item android:drawable="@drawable/p3" android:duration="1500" android:
    visible="true"/>
 </animation-list>
```

步骤三：修改布局文件。打开项目 res\layout 目录下的 main.xml 文件，在文件中添加一个 ImageView 控件和 Button 控件，添加代码如下：

```
<ImageView android:id="@+id/iv"
        android:background="@anim/frame_ani"
        android:layout_width="wrap_content"
        android:layout_height="wrap_content"
        android:layout_gravity="center_horizontal"
        />        <!-- 声明一个 ImageView 对象 -->
    <Button android:id="@+id/btn"
        android:text="@string/btn"
        android:layout_width="fill_parent"
        android:layout_height="wrap_content"
        android:layout_gravity="center_horizontal"<!-设置在父控件中水平位置
        居中-->
        />
```

步骤四：主逻辑页面开发。编写项目自动生成的 Example_Animation.java 类文件，编写代码如下：

```
public class Example_Animation extends Activity {
    @Override
    public void onCreate(Bundle savedInstanceState) {    //重写onCreate方法
        super.onCreate(savedInstanceState);
        setContentView(R.layout.main);
        Button btn = (Button)findViewById(R.id.btn);
//其下为单击按钮则播放动画的代码
```

```
btn.setOnClickListener(new OnClickListener() {//为按钮设置监听器
        //@Override
    public void onClick(View v) {          //重写 onClick 方法
      ImageView iv = (ImageView)findViewById(R.id.iv);
      iv.setBackgroundResource(R.anim.frame_ani);
      AnimationDrawable ad = (AnimationDrawable)iv.getBackground();
          ad.start();      //启动 AnimationDrawable
        } }); }
  //其下为启动程序就播放动画的代码
  @Override
  public void onWindowFocusChanged(boolean hasFocus) {
  // TODO Auto-generated method stub
super.onWindowFocusChanged(hasFocus);
  ImageView iv = (ImageView)findViewById(R.id.iv);
    iv.setBackgroundResource(R.anim.frame_ani);
    AnimationDrawable ad = (AnimationDrawable)iv.getBackground();
    ad.start();}
}
```

步骤五：调试程序，查看案例效果，如图 3-48 所示。

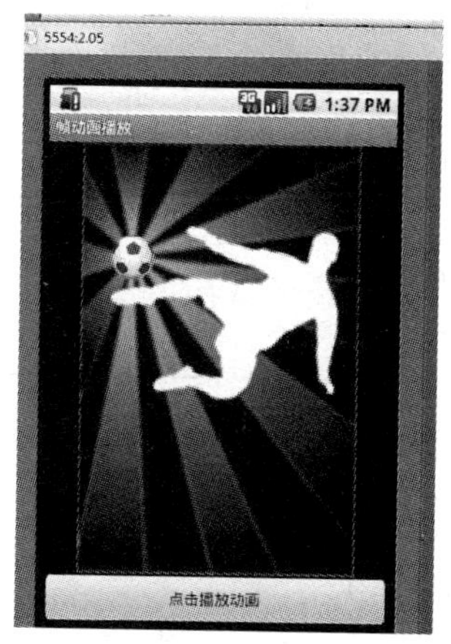

图 3-48 帧动画效果图

3.6.2 补间动画

补间动画（Tween Animation）作用于 View 对象，主要包括对 View 对象的位置、尺寸、旋转角度和透明度的变换。补间动画涉及的类主要有 Animation、AnimationSet 等，这些类位于 android.view.animation 包下。补间动画通过一系列的指令来定义，和布局管理器一样，补间动画既可以在 XML 文件中声明，也可以在 Java 代码中动态定义。我们一般推荐使用 XML 文件定义动画，因为 XML 文件可读性和可用性高，而且便于替换。

补间动画的XML文件位于程序的res\anim目录下，在XML文件中可以指定进行何种变换、何时进行变换以及持续多长时间。当需要在XML文件中定义多个变换时，需要多个变换包含在一组<set></set>标记中。表3-24所示为补间动画中的标记及属性值说明，表3-25为属性值说明。

表3-24　Tween Animation中的标记及属性值说明

标记名称	属 性 值	说　明
<set>	shareInterpolator：是否在子元素中共享插入器	可以包含其他动画变换的容器，同时也可以包含<set>标记
<alpha>	fromAlpha：变换的起始透明度 toAlpha：变换的终止透明度，取值为0.0~1.0，其中0.0代表全透明	实现透明度变换效果
<scale>	fromXScale：起始的X方向上的尺寸 toXScale：终止的X方向上的尺寸 fromYScale：起始的Y方向上的尺寸 toYScale：终止的Y方向上的尺寸，其中1.0代表原始大小 pivotX：进行尺寸变换的中心X坐标 pivotY：进行尺寸变换的中心Y坐标	实现尺寸变换效果，可以指定一个变换中心，例如指定pivotX和pivotY为(0,0)，则尺寸的拉伸或收缩均从左上角的位置开始
<translate>	fromXDelta：起始X坐标 toXDelta：终止X坐标 fromYDelta：起始Y坐标 toYDelta：终止X坐标	实现水平或竖直方向上的移动效果。如果属性值以"%"结尾，代表相对于自身的比例；如果以"%p"结尾，代表相对于父控件的比例；如果不以任何后缀结尾，代表绝对的值
<rotate>	fromDegrees：开始旋转位置 toDegrees：结束旋转位置，以角度为单位 pivotX：旋转中心点的X坐标 pivotY：旋转中心点的Y坐标	实现旋转效果，可以指定旋转定位点
<interpolator tag>	无	插入器，描述变换的速度曲线。如先慢后快、先快后慢等

表3-25　Tween Animation中标记共有属性值说明

属 性 值	说　明
duration	变换持续的时间，以毫秒为单位
startOffset	变换开始的时间，以毫秒为单位
repeatCount	定义该动画重复的次数
interpolator	为每个子标记变换设置插入器，系统已经设置好一些插入器，可以在R.anim包下找到

提示：startAnimation 方法不可以在 Activity 的 onCreate 方法中调用，如果希望在程序一启动就播放动画，则应该在 OnWindowFocusChanged() 方法中调用 startAnimation ()方法。

接下来我们通过一个补间动画案例的介绍，来具体学习步间动画的使用方法。其案例的开发步骤如下：

步骤一：建立项目，准备资源。打开 Eclipse 工具，建立一个名为 Example_Tween Animation 的项目，准备如图 3-49 所示图片存放在项目的 res/drawable-mdpi 目录下。

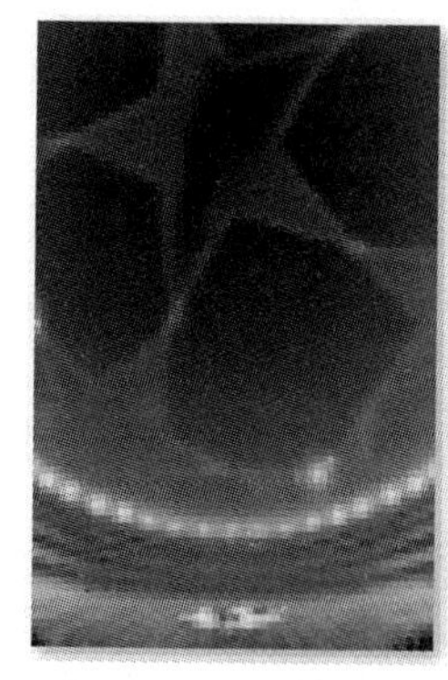

图 3-49

步骤二：在 res 目录下新建一个 anim 目录，在该目录下新建一个 XML 文件名称为 tween_ani.xml。编写如下代码：

```
<?xml version="1.0" encoding="utf-8"?><!-- XML 的版本以及编码方式 -->
<set xmlns:android="http://schemas.android.com/apk/res/android">
  <alpha android:fromAlpha="0.0" android:toAlpha="1.0"
    android:duration="6000"     android:repeatCount="5"
  /> <!-- 透明度的变换 -->
  <scale android:interpolator= "@android:anim/accelerate_decelerate_
  interpolator"
    android:fromXScale="0.0"    android:toXScale="1.0"
    android:fromYScale="0.0"    android:toYScale="1.0"
    android:pivotX="50%"        android:pivotY="50%"
    android:fillAfter="true"    android:duration="6000"
    android:repeatCount="5"/> <!-- 尺寸的变换 -->
  <translate  android:fromXDelta="30"
    android:toXDelta="0"   android:fromYDelta="30"
    android:toYDelta="0"   android:duration="6000"
    android:repeatCount="5" /> <!-- 尺位置的变换 -->
  <rotate android:interpolator="@android:anim/accelerate_decelerate_
  interpolator"
    android:fromDegrees="0"    android:toDegrees="+360"
    android:pivotX="50%"      android:pivotY="50%"
    android:duration="6000"    android:repeatCount="5"
  /> <!-- 旋转变换 --> </set>
```

步骤三：修改布局文件。打开项目 res/layout 目录下的 main.xml 文件，在文件中添加一个 ImageView 控件和 Button 控件，添加代码如下：

```
<ImageView  android:id="@+id/iv"
android:src="@drawable/welcome" android:layout_width="wrap_content"
    android:layout_height="wrap_content"
    android:layout_gravity="center_horizontal"
    />  <!-- 声明一个 ImageView 控件 -->
<Button   android:id="@+id/btn"
    android:text="Click"    android:layout_width="fill_parent"
```

```
        android:layout_height="wrap_content"/>
```

步骤四：主逻辑页面开发。编写项目自动生成的 Example_TweenAnimation.java 类文件，编写代码如下：

```
public class Example_TweenAnimation extends Activity {
    @Override
    public void onCreate(Bundle savedInstanceState) {   //重写onCreate方法
        super.onCreate(savedInstanceState);
        setContentView(R.layout.main);                //设置屏幕
        Button btn = (Button)findViewById(R.id.btn);//获取 Button 对象
        btn.setOnClickListener(new OnClickListener() {  //为 Button 对象添加
        OnClickListener 监听器
            @Override
public void onClick(View v) {                   //重写 onClick 方法
ImageView iv = (ImageView)findViewById(R.id.iv);
 Animation animation = AnimationUtils.loadAnimation(Example_
 TweenAnimation.this, R.anim.tween_ani);
    iv.startAnimation(animation);               //启动动画
            }});
 }
    @Override//启动程序自动播放补间动画
    public void onWindowFocusChanged(boolean hasFocus) {
     // TODO Auto-generated method stub
     super.onWindowFocusChanged(hasFocus);
     ImageView iv = (ImageView)findViewById(R.id.iv);
        Animation animation = AnimationUtils.loadAnimation(Example_
        TweenAnimation.this, R.anim.tween_ani);
        iv.startAnimation(animation); }
}
```

步骤五：调试程序，查看案例效果，如图 3-50 所示。足球赛场图片会循环 5 次，旋转显示。

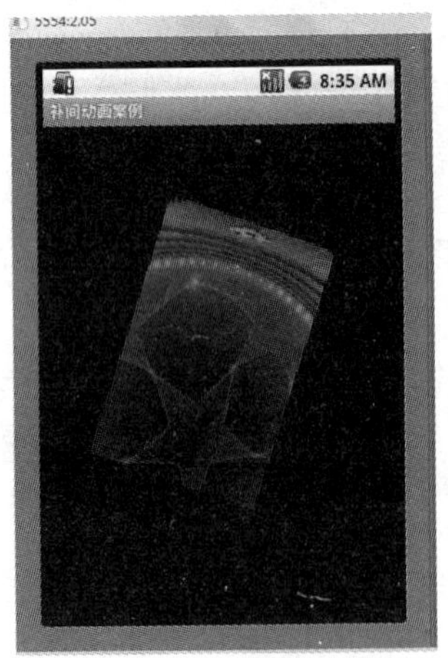

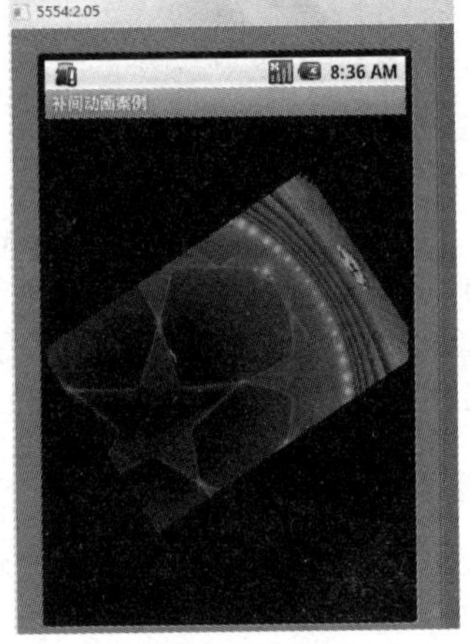

图 3-50

第 4 章

Android 常用高级控件

上一章我们学习了 Android 的一些基本控件的使用，但这些基本控件并不能完全满足我们程序开发设计的应用，本章将继续学习 Android 高级控件的使用。

4.1 自动完成文本框

本节将介绍自动完成文本框 AutoCompleteTextView 控件的使用方法。所谓“自动完成”就是在文本框中输入文字信息时，会显示与之相似的关键字让你来选择。

AutoCompleteTextView 类继承自 EditView 类，位于 android.widget 包下。自动完成文本框控件的外观与图片文本框几乎相同，只是当用户输入某些文字信息时，会自动出现下拉菜单显示与用户输入文字相关的信息，用户直接点击需要的文字便可自动填写到文本控件中。在 XML 文件中，使用属性进行设置可以自动完成文本框，也可以在 Java 代码中通过方法进行设置。下面给出常用属性与方法使用的对照表，见表 4-1。

表 4-1 自动完成文本的属性与方法对照表

属性名称	对应方法	属性说明
android:completionThreshold	setThreshold(int)	定义需要用户输入的字符数
android:dropDownHeight	setDropDownHeight(int)	设置下拉菜单高度
android:dropDownWidth	setDropDownWidth(int)	设置下拉菜单宽度
android:popupBackground	setDropDownBackgroundResource(int)	设置下拉菜单背景

ListAdapter 继承与 Adapter，它是 ListView 和其里边数据的适配器。ArrayAdapter 是 ListAdapter 的一个直接子类，可以翻译成数组适配器，它是一个数组和 ListView 之间的桥梁，可以将数组里面定义的数据一一对应显示在 ListView 里面。ArrayAdapter 是由 3 个参数进行构造的，第一个参数表示上下文的应用，为当前应用实例；第二个参数为一个在 R 文件里面定义的 Layout，可以通过 R.layout.XX 访问（XX 为资源的名称），也可以通过 Android.R.layout.XX 来进行对 Android 系统的默认布局进行访问，Android 的默认布局有很多种，我们常见的有以下几个。

android.R.layout.simple_list_item_single_choice：每一项只有一个 TextView，但这一项可以被选择。

android.R.layout.simple _list_item_1：每一项只有一个 TextView。

android.R.layout.simple_ list_item_2：每一项有两个 TextView。

第三个参数为字符串数组。

接下来我们通过一个简单的案例来具体介绍 AutoCompleteTextView 的使用方法，开发步骤如下。

步骤一：创建项目，准备资源。在 Eclipse 工具中创建一个名称为 Example_4_1 的 Android 项目。在项目的 res\drawable-mdpi 目录下，添加两张名字分别为 a.jpg 和 d.jpg 的图片，作为文本框的背景图片应用。

步骤二：修改布局文件。打开项目 res\layout 目录下的 main.xml 文件，修改里面的代码如下：

```
<?xml version="1.0" encoding="utf-8"?><!—XML 文件的版本与编码方式 -->
<LinearLayout
xmlns:android="http://schemas.android.com/apk/res/android"
    android:orientation="vertical"
    android:layout_width="fill_parent"
    android:layout_height="fill_parent"
    ><!—添加一个垂直的线性布局 -->
    <AutoCompleteTextView
    android:id="@+id/myAutoCompleteTextView"<!—自动文本框 ID 编号 -->
    android:layout_width="fill_parent" <!-- 自动文本框宽度 -->
    android:layout_height="wrap_content" <!-- 自动文本框高度 -->
    android:dropDownHeight="350px" <!—下拉菜单高度 -->
    android:dropDownWidth="100px" <!--下拉菜单宽度 -->
    android:completionThreshold="1" <!—定义需要用户输入的字符数，弹出下拉菜单 -->
    android:background="@drawable/d" <!--自动文本框背景图片 -->
    android:popupBackground="@drawable/a"<!--下拉菜单背景图片 -->
```

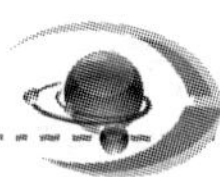

```
        /><!-添加一个自动完成文本框 -->
</LinearLayout>
```

步骤三：主逻辑页面开发。编写项目自动生成的 Example_4_1.java 类文件，修改代码如下：

```
    public class Example_4_1 extends Activity {
        private static final String[] myStr = new String [] {//常量数组
        "aaa", "bbb", "ccc", "bab", "aac", "aad", "abd", "abc"};
        public void onCreate(Bundle savedInstanceState) {//重写的onCreate方法
            super.onCreate(savedInstanceState);
            setContentView(R.layout.main);//设置当前显示的用户界面
            ArrayAdapter<String> aa = new ArrayAdapter<String> (//创建适配器
                    this, //Context
        android.R.layout.simple_dropdown_item_1line,//Android系统自带的布局格式
                myStr);//资源数组
          //得到控件的引用
            AutoCompleteTextView myAutoCompleteTextView =
        (AutoCompleteTextView)
    findViewById(R.id.myAutoCompleteTextView);
            myAutoCompleteTextView.setAdapter(aa);//设置适配器
        //  myAutoCompleteTextView.setThreshold(1);//定义需要用户输入的字符数
        } }
```

步骤四：运行项目。运行该项目，用户可以看到添加到文本框里面的背景图片，当用户在文本框中输入字母“a”或“b”或“c”时，界面将自动弹出一个下拉菜单，该下拉菜单的高度、宽度与背景图片都是我们所定义的。其中的方法 setThreshold(int)里面的参数就是要弹出对话框让用户来输入字符的长度。

4.2　滚动视图和列表视图

4.2.1　滚动视图

滚动视图 ScrollView 类继承自 FrameLayout 类，因此，实际上它是一个帧布局，同样位于 android.widget 包下。ScrollView 控件是当需要显示的信息在一个屏幕内显示不下时，在屏幕上会自动生成一个滚动条，以达到用户可以对其进行滚动，显示更多信息的目的。ScrollView 控件的使用与普通布局没有太大的区别，可以在 XML 文件中进行配置，也可以通过 Java 代码进行设置。在 ScrollView 控件中可以添加任意满足条件的控件，当一个屏幕显示不下其中所包含的信息时，便会自动添加滚动功能。

需要注意的是：ScrollView 中同一时刻只能包含一个 View。

下面我们就用一个简单的案例，从 XML 配置和 Java 代码设置两个方面对滚动视图 ScrollView 的应用做一个详细介绍。先从 XML 配置方面使用 ScrollView 控件，开发项目步骤如下。

步骤一：项目搭建。通过 Eclipse 开发工具建立项目名称为 Example_4_2 的项目。

步骤二：修改布局文件。打开项目 res\layout 目录下的 main.xml 文件，修改里面的代码如下：

```
<?xml version="1.0" encoding="utf-8"?>
<ScrollView xmlns:android="http://schemas.android.com/apk/res/android"
   android:layout_width="fill_parent"
   android:layout_height="fill_parent">
  <EditText
  android:id="@+id/textview"
   android:layout_width="fill_parent"
   android:layout_height="wrap_content"
   />
</ScrollView>
```

步骤三：主逻辑页面开发。编写项目自动生成的 Example_4_2.java 类文件，修改代码如下：

```
public class Example_4_2  extends Activity {
    String msg = "世间自有公道，付出总有回报，说到不如做到，要做就做最好！";
    String str = "";
   public void onCreate(Bundle savedInstanceState) {
       super.onCreate(savedInstanceState);
       setContentView(R.layout.main);
       EditText tv =(EditText)findViewById(R.id.textview);
    for(int i=0 ;i<12; i++){
        str = str + msg;     }//循环生成较长的文本字符内容
      tv.setTextSize(24);//设置文本视图中文字的大小
       tv.setText(str);}}//设置文本控件的内容
```

步骤四：调试程序。运行该案例，我们则可以看到如图 4-1 所示的效果。

接下来，我们再从 Java 代码设置方面，来学习使用 ScrollView 控件，开发项目步骤如下。

步骤一：与前面的步骤一完全一样。

步骤二：直接修改 Example_4_2.java 文件的代码如下：

```
public class Sample_5_2 extends Activity {
    ScrollView scrollView; //滚动视图的引用
 String msg = "世间自有公道，付出总有回报，说到不如做到，要做就做最好！";
    String str = "";
   public void onCreate(Bundle savedInstanceState) {
       super.onCreate(savedInstanceState);
        scrollView = new ScrollView(this);//初始化滚动视图
        EditText tv =new EditText(this);    //初始化文本视图
        for(int i=0 ;i<12; i++){
        str = str + msg;     }
       tv.setTextSize(24);
       tv.setText(str);
        scrollView.addView(tv);
     setContentView(scrollView);
     } }
```

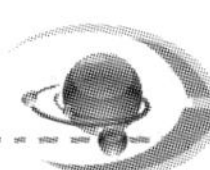

步骤三：调试程序。运行该案例，可以看到与前面相同的效果，如图 4-1 所示。

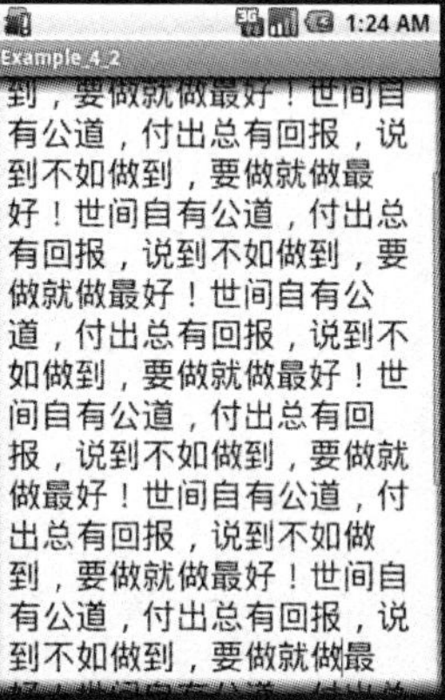

图 4-1　滚动视图效果图

4.2.2　列表视图

ListView 类位于 android.widget 包下，是一种列表视图，将 ListAdapter 所提供的各个控件显示在一个垂直且可滚动的列表中。该类的使用方法非常简单，只需先初始化所需要的数据，然后创建适配器并将其设置给 ListView，ListView 将信息以列表的形式显示到页面中。

我们通过实现一个简单的名人录，其中包括各个名人的照片及描述的案例，来具体介绍 ListView 控件的使用。具体开发步骤如下。

步骤一：创建项目，准备资源。打开 Eclipse 开发工具，建立一个名称为 Example_4_3 的 Android 项目，将如图 4-2 所示的名人图片资源图存放到 res\drawable-mdpi 目录下。

图 4-2　名人照片图

步骤二：修改 strings.xml 文件。打开 res\values 目录下的 strings.xml 文件，修改里面的代码如下：

```
<?xml version="1.0" encoding="utf-8"?>
<resources>
<string name="hello">您选择了</string>
<!-定义名人姓名和说明的字符串 -->
    <string name="app_name">Example_4_3</string>
    <string name="andy">Andy Rubin \nAndroid 的创造者</string>
    <string name="bill">Bill Joy \nJava 创造者之一</string>
    <string name="edgar">Edgar F. Codd \n 关系数据库之父</string>
    <string name="torvalds">Linus Torvalds \nLinux 之父</string>
    <string name="turing">Turing Alan    \nIT 的祖师爷</string>
    <string name="ys">您选择了</string>
</resources>
```

步骤三：创建颜色资源文件。在 res\values 目录下创建一个名称为 colors.xml 文件，并编写其代码如下：

```
<?xml version="1.0" encoding="utf-8"?>
<resources>
<!-定义颜色的名称值对-->
<color name="red">#fd8d8d</color>
<color name="green">#9cfda3</color>
<color name="blue">#8d9dfd</color>
<color name="white">#FFFFFF</color>
<color name="black">#000000</color>
<color name="gray">#050505</color>
</resources>
```

步骤四：修改布局文件。打开项目 res\layout 目录下的 main.xml 文件，修改里面的代码如下：

```
<?xml version="1.0" encoding="utf-8"?>
<LinearLayout
xmlns:android="http://schemas.android.com/apk/res/android"
    android:orientation="vertical"
    android:layout_width="fill_parent"
    android:layout_height="fill_parent">
   <TextView
    android:id="@+id/TextView01"
    android:layout_width="fill_parent"
    android:layout_height="wrap_content"
    android:textSize="24dip"          <!-定义字体大小-->
    android:textColor="@color/white"  <!-定义字体颜色-->
    android:text="@string/hello"/>
   <ListView
      android:id="@+id/ListView01"
      android:layout_width="fill_parent"
      android:layout_height="wrap_content"
      android:choiceMode="singleChoice"/><!-定义选择模式为单一 -->
</LinearLayout>
```

步骤五：主逻辑页面开发。编写项目自动生成的 Example_4_3.java 类文件，编写代码如下：

```
public class Example_4_3 extends Activity {
    //所有资源图片（andy、bill、edgar、torvalds、turing）id 的数组
    int[] drawableIds=
    {R.drawable.andy,R.drawable.bill,R.drawable.edgar,R.drawable.torval
ds,R.drawable.turing};
    //所有资源字符串（andy、bill、edgar、torvalds、turing）id 的数组
    int[]
msgIds={R.string.andy,R.string.bill,R.string.edgar,R.string.tor
valds,R.string.turing};
 public void onCreate(Bundle savedInstanceState) {
```

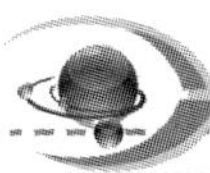

```
    super.onCreate(savedInstanceState);
    setContentView(R.layout.main);
    ListView lv=(ListView)this.findViewById(R.id.ListView01);//初始化 ListView
   BaseAdapter ba=new BaseAdapter(){//为 ListView 准备内容适配器
  public int getCount() {return 5;}//总共 5 个选项
  public Object getItem(int arg0) { return null; }
  public long getItemId(int arg0) { return 0; }
  public View getView(int arg0, View arg1, ViewGroup arg2) {
      //动态生成每个下拉项对应的 View，每个下拉项 View 由 LinearLayout
      //中包含一个 ImageView 和一个 TextView 构件
  LinearLayout linearLayout=new LinearLayout(Example_4_3.this);//初始化
  LinearLayout
  linearLayout.setOrientation(LinearLayout.HORIZONTAL);//设置朝向
      linearLayout.setPadding(5,5,5,5);//设置四周留空白
      ImageView  imageView=new ImageView(Example_4_3.this);//初始化
      ImageView
  imageView.setImageDrawable(getResources().getDrawable(drawableIds[arg0]
  ));//设置图片
      imageView.setScaleType(ImageView.ScaleType.FIT_XY);
      imageView.setLayoutParams(new
  Gallery.LayoutParams(100,98));
      linearLayout.addView(imageView);//添加到 LinearLayout 中
      TextView tv=new TextView(Example_4_3.this);//初始化 TextView
      tv.setText(getResources().getText(msgIds[arg0]));//设置内容
      tv.setTextSize(24);//设置字体大小
  tv.setTextColor(Exanple_4_3.this.getResources().getColor(R.color.white)
);//设置字体颜色
      tv.setPadding(5,5,5,5);//设置四周留空白
      tv.setGravity(Gravity.LEFT);
      linearLayout.addView(tv);//添加到 LinearLayout 中

      return linearLayout;}  };
      lv.setAdapter(ba);//为 ListView 设置内容适配器
      lv.setOnItemSelectedListener(//设置选项选中的监听器
      new OnItemSelectedListener(){
      public void onItemSelected(AdapterView<?> arg0, View arg1,
                   int arg2, long arg3) {//重写选项被选中事件的处理方法
  TextView tv=(TextView)findViewById(R.id.TextView01);//获取主界面 TextView
  LinearLayout ll=(LinearLayout)arg1;//获取当前选中选项对应的 LinearLayout
      TextView tvn=(TextView)ll.getChildAt(1);//获取其中的 TextView
      StringBuilder sb=new StringBuilder();//用 StringBuilder 动态生成信息
      sb.append(getResources().getText(R.string.ys));
      sb.append(":");
      sb.append(tvn.getText());
```

```
        String stemp=sb.toString();
        tv.setText(stemp.split("\\n")[0]); }//信息设置进主界面TextView

        public void onNothingSelected(AdapterView<?> arg0){} } );
            lv.setOnItemClickListener(//设置选项被单击的监听器
              new OnItemClickListener(){
        public void onItemClick(AdapterView<?> arg0, View arg1, int arg2,long
    arg3) {//重写选项被单击事件的处理方法
    TextView tv=(TextView)findViewById(R.id.TextView01);//获取主界面TextView
    LinearLayout ll=(LinearLayout)arg1;//获取当前选中选项对应的LinearLayout
    TextView tvn=(TextView)ll.getChildAt(1);//获取其中的TextView
    StringBuilder sb=new StringBuilder();//用StringBuilder动态生成信息
            sb.append(getResources().getText(R.string.ys));
            sb.append(":");
            sb.append(tvn.getText());
            String stemp=sb.toString();
            tv.setText(stemp.split("\\n")[0]);//信息设置进主界面TextView
              }    }    ); } }
```

步骤六：调试程序。在模拟器里运行我们开发的案例，将看到我们所需要的效果，如图 4-3 所示。

图 4-3　列表视图效果图

SimpleAdapter 和我们前面所说的 ArrayAdapter 一样，也是 ListAdapter 的直接子类。前面学习的 ListView 与 ArrayAdapter 绑定的列表有一定的局限性，在这种绑定中，ListView 的每一项只有一个 TextView，并且 TextView 里边的内容都是调用了数组里面每一个对象的 toString()方法生成的字符串。而 SimpleAdapter 与 ListView 的绑定生成的列表就会有很大的用户可定制性。通常将 ListView 中某项的布局信息写在一个 XML 的布局文件中，这个布局文件通过 R.layout.XX（XX 为文件的名称）获得。

ArrayAdapter 的作用是数组和 ListView 间的桥梁，而 SimpleAdapter 的作用是 ArrayList 和 ListView 的桥梁。需要注意的是，这个 ArrayList 里边的每一项都是一个 Map<String,?>类型，ArrayList 当中的每一项 Map 对象都和 ListView 当中的一项进行数据绑定和一一对应。SimpleAdapter 类的构造方法结构如下：

```
    Public  SimpleAdapter(Context  context,List<?  Extends  Map<String,?>>
data,int resource,String[] from,int[] to);
```

Context 参数：负责上下文应用的传递。

Data 参数：基于 Map 的 List，Data 里面的每一项都和 ListView 里面的每一项对应，Data 里面的每一项是一个 Map 类型，这个 Map 类里边包含了 ListView 每一行需要的数据。比较常用的用法是：data=new

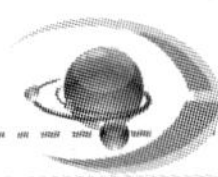

```
    ArrayList<Map<String,Object>();
```

Resource 参数：这个 Resource 就是一个 Layout，这个 Layout 最起码要包含在 to 中出现的那些 View。一般用系统提供的就可以了，当然也可以自己定义。

Form 参数：是一个名字的数组，每一个名字都是为了在 ArrayList 中的每一个 item 中索引 Map<String,Object>的 Object 用的。

To 参数：是一个 TextView 的数组，这些 TextView 是以 id 的形式来表示的，如 Android.R.id.text1，这个 text1 在 layout 当中是可以索引到的。

接下来我们就 SimpleAdapter 和 ListView 的绑定使用方法，通过一个案例来详细说明，本案例的开发步骤如下。

步骤一：创建项目。打开 Eclipse 开发工具，建立一个名称为 Example_ListView2 的 Android 项目。

步骤二：自定义布局文件。打开 res\layout 目录，并建立一个自定义的 XML 文件的名称为 list_item.xml。其代码如下：

```
<?xml version="1.0" encoding="utf-8"?>
<LinearLayout
    xmlns:android="http://schemas.android.com/apk/res/android"
    android:orientation="horizontal"
   android:layout_width="fill_parent"
    android:layout_height="wrap_content">
    <TextView android:id="@+id/mview1"
android:layout_width="100px"
        android:layout_height="wrap_content" />
    <TextView
    android:id="@+id/mview2" android:layout_width="wrap_content"
      android:layout_height="wrap_content" />
</LinearLayout>
```

步骤三：编写主逻辑文件。其代码如下：

```
public class Example_ListView2 extends Activity {
    private List<Map<String, Object>> data;
    private ListView listView = null;
    @Override
    public void onCreate(Bundle savedInstanceState) {
        super.onCreate(savedInstanceState);
        PrepareData();//准备了一个 Map 类型的 ArrayList 对象
        listView = new ListView(Example_ListView2.this);
        // 利用系统的 layout 显示一项
    /* SimpleAdapter adapter = new SimpleAdapter(this, data,
    android.R.layout.simple_list_item_1, new String[] { "姓名" },
        new int[] { android.R.id.text1 });*/
        // 利用系统的 layout 显示两项
    //  SimpleAdapter adapter = new SimpleAdapter(this, data,
    //  android.R.layout.simple_list_item_2, new String[] { "姓名","系别
```

```
" },new int[] { android.R.id.text1 , android.R.id.text2});
/* 利用自己的 layout 来进行显示两项，负责生成一个 SimpleAdapter 的实例
  R.layout.list_item 关联在 list_item.xml 文件当中的 Layout
  String[]{"姓名","系别"},这是一个 String 的数组，里边每一个字符串用于访问得到
  Map 对象当中的值。
  int[] {R.id.mview1, R.id.mview2 }，里面的值就是在 list_item.xml 里面
  定义的两个 TextView 的 Id
*/
   SimpleAdapter adapter = new SimpleAdapter(this,
data,R.layout.list_item, new String[] { "姓名", "系别" }, new
int[] {R.id.mview1, R.id.mview2 });
    listView.setAdapter(adapter);
    setContentView(listView);
/*增加一个单击事件，最常用的方法就是采用监听器的方法，Button 采用的是注
册 OnClickListener 的方式，对应的处理 List 的单击事件是把一个
OnItemClickListener 注册到 ListView 当中。
*/
OnItemClickListener listener = new OnItemClickListener() {
public void onItemClick(AdapterView<?> parent, View view, int
 position,long id) {
//getItemAtPosition()方法可以通过 position 获得和这一行绑定的数据
 setTitle(parent.getItemAtPosition(position).toString());}};
listView.setOnItemClickListener(listener);}//注册监听器
private void PrepareData() {
   data = new ArrayList<Map<String, Object>>();
   Map<String, Object> item;
   item = new HashMap<String, Object>();
   item.put("姓名", "张三");   item.put("系别", "游戏学院");
   data.add(item);
   item = new HashMap<String, Object>();
   item.put("姓名", "王五");   item.put("系别", "网络技术系");
   data.add(item);
   item = new HashMap<String, Object>();
   item.put("姓名", "小李");  item.put("系别", "艺术设计系");
   data.add(item);
   item = new HashMap<String, Object>();
   item.put("姓名", "陈飞");  item.put("系别", "软件技术系");
   data.add(item); } }
```

步骤四：运行程序，查看结果，如图 4-4 所示。

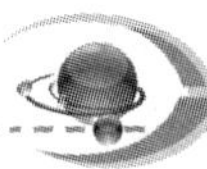

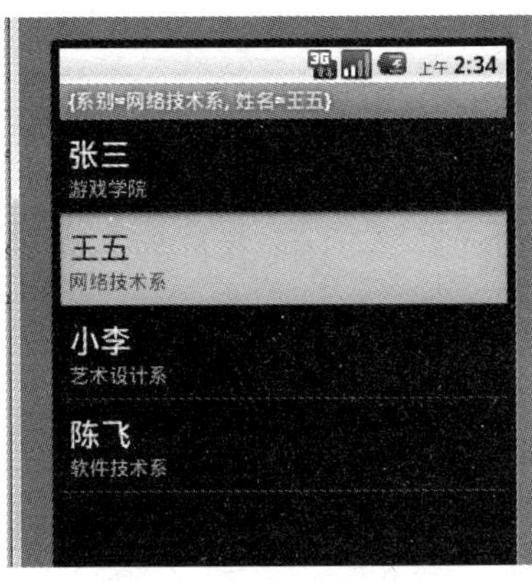

图 4-4　SimpleAdapter 和 ListView 的绑定使用案例图

4.3　滑块与进度条

滑块类似于声音控制条，主要完成与用户的简单交互，而进度条则是需要长时间加载某些资源时用户显示加载进度的控件。

ProgressBar 类同样位于 android.widget 包下，但其继承自 View，主要用于显示一些操作的进度。应用程序可以修改其长度表示当前后台操作的完成情况。因为进度条会移动，所以长时间加载某些资源或者执行某些耗时的操作时，不会使用户界面失去响应。ProgressBar 类的使用非常简单，只需将其显示到前台，然后启动一个后台线程定时更改表示进度的数值即可。SeekBar 继承自 ProgressBar，是用来接收用户输入的控件。SeekBar 类似于拖动条，可以直观地显示用户需要的数据，常用于声音调节等场合。SeekBar 不但可以直观地显示数值的大小，还可以为其设置标度，类似于显示在屏幕中的一把尺子。RatingBar 是另一种滑块控件，一般用于星级评分的场合，其位于 android.widget 包下，外观是 5 个星星，可以通过拖动来改变进度，除图片形式外，还有较小的以及较大的两种表现形式。

接下来我们通过一个案例来介绍 ProgressBar 和 SeekBar 两个控件的使用方法。该案例的开发步骤如下。创建一个名为 SeekBarExample 的项目，打开 res\layout 目录下的 main.xml 文件，编写其布局文件，代码如下。

```
<?xml version="1.0" encoding="utf-8"?>
<LinearLayout
 xmlns:android="http://schemas.android.com/apk/res/android"
  android:orientation="vertical"
  android:layout_width="fill_parent"
  android:layout_height="fill_parent">
<ProgressBar   android:id="@+id/ProgressBar01"
   android:layout_width="fill_parent"
   android:layout_height="wrap_content"
   android:max="100"
   android:progress="20"
```

```
        style="@android:style/Widget.ProgressBar.Horizontal"/>
    <SeekBar   android:id="@+id/SeekBar01"
        android:layout_width="fill_parent"
        android:layout_height="wrap_content"
        android:max="100"
        android:progress="20"/>
    </LinearLayout>
```

然后编写 Activity 的逻辑代码，打开 SeekBarExample.java 文件，编写代码如下。

```
    public class SeekBarExample extends Activity {
    final static double MAX=100;//SeekBar、ProgressBar 的最大值
    SeekBar sb;
    public void onCreate(Bundle savedInstanceState) {
        super.onCreate(savedInstanceState);
        setContentView(R.layout.main);
        //普通拖拉条被拉动的处理代码
         sb=(SeekBar)this.findViewById(R.id.SeekBar01);
        sb.setOnSeekBarChangeListener(
         new SeekBar.OnSeekBarChangeListener(){
         public void onProgressChanged(SeekBar seekBar, int progress,
                    boolean fromUser) {
         ProgressBar
 pb=(ProgressBar)findViewById(R.id.ProgressBar01);
                //SeekBar
 sb=(SeekBar)findViewById(R.id.SeekBar01);
                pb.setProgress(sb.getProgress()); }
                public void onStartTrackingTouch(SeekBar seekBar){ }
                public void onStopTrackingTouch(SeekBar seekBar){ }
                } ); }
        }
```

最后运行程序，查看代码运行效果，如图 4-5 所示。

图 4-5　滑块与进度条案例效果图

前面我们对滑块和进度条的使用方法进行了介绍，这里我们在对星级滑块的使用通过同样的方法进行介绍说明。建立一个名为 RatingBarExample 的 Android 项目。打开 res\layout 目录下的 main.xml 文件，编写其布局文件，代码如下。

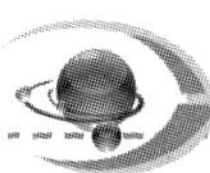

```
<?xml version="1.0" encoding="utf-8"?>
<LinearLayout
xmlns:android="http://schemas.android.com/apk/res/android"
  android:orientation="vertical"
  android:layout_width="fill_parent"
  android:layout_height="fill_parent" >
<ProgressBar  android:id="@+id/ProgressBar01"
  android:layout_width="fill_parent"
  android:layout_height="wrap_content"
  android:max="100"
  android:progress="20"
  style="@android:style/Widget.ProgressBar.Horizontal" />
<RatingBar  android:id="@+id/RatingBar01"
  android:layout_width="wrap_content"
  android:layout_height="wrap_content"
  android:max="5"
  android:rating="1"  />
</LinearLayout>
```

然后编写 Activity 的逻辑代码，打开 RatingBarExample.java 文件，编写代码如下。

```
public class RatingBarExample extends Activity {
   final static double MAX=100;//ProgressBar 的最大值
   final static double MAX_STAR=5;//RatingBar 的最大星星数
   public void onCreate(Bundle savedInstanceState) {
      super.onCreate(savedInstanceState);
      setContentView(R.layout.main);
      //星型拖拉条被拉动的处理代码
      RatingBar rb=(RatingBar)findViewById(R.id.RatingBar01);
      rb.setOnRatingBarChangeListener(
         new RatingBar.OnRatingBarChangeListener(){
               @Override
   public void onRatingChanged(RatingBar ratingBar, float rating,
               boolean fromUser) {
   ProgressBar
pb=(ProgressBar)findViewById(R.id.ProgressBar01);
   RatingBar rb=(RatingBar)findViewById(R.id.RatingBar01);
                float rate=rb.getRating();
     //将 0～5 星星数折算成 0～100 的进度值
   pb.setProgress((int) (rate/MAX_STAR*MAX));
      }} );    }
   }
```

最后运行程序，查看代码运行效果，如图 4-6 所示。

图 4-6　星级滑块进度条案例效果图

4.4　画廊控件与消息提示

4.4.1　画廊控件

画廊控件（Gallery）是 Android 中一种较为常见的高级控件，其效果酷炫且使用方式简单，是设计相册或者图片选择器的首选控件。Gallery 是一种水平滚动的列表，一般情况下用来显示图片等资源，可以使图片在屏幕上滑来滑去。Gallery 所显示的图片资源同样来自适配器。

Gallery 是 View 的子类，Gallery 控件可以在 XML 布局文件中配置，也可以通过 Java 代码直接操控。Gallery 的常用属性和方法见表 4-2 所示。

表 4-2　Gallery 的常用属性和方法表

属性名称	对应方法	说　明
android:animationDuration	setAnimationDuration(int)	设置动画过渡时间
android: gravity	set Gravity(int)	在父控件中的对齐方式
android:unselectedAlpha	setUnselectedAlphafloat()	设置选中图片的透明度
android:spacing	setSpacing(int)	图片之间的空白大小

我们通过介绍一个画廊的案例来具体学习 Gallery 控件的使用方法。该案例的开发步骤如下。

步骤一：建立项目，准备资源。打开 Eclipse 工具，建立一个名为 Example_Gallery 的项目，准备相应的图片，如图 4-7 所示，存放在项目的 res\drawable-mdpi 目录下。

图　4-7　资料图片

步骤二：编写布局文件，添加 Gallery 控件。参考代码如下：

```
<?xml version="1.0" encoding="utf-8"?>
<LinearLayout xmlns:android="http://schemas.android.com/apk/res/android"
```

```
   android:orientation="vertical"
   android:layout_width="fill_parent"
   android:layout_height="wrap_content"
   android:gravity="center_vertical">
  <Gallery
    android:id="@+id/Gallery01"
    android:layout_width="fill_parent"
    android:layout_height="wrap_content"
    android:spacing="10dip"<!–设置图片之间的空白 -->
    android:unselectedAlpha="1" <!–设置选中图片的透明度 -->
    android:animationDuration="2"<!--  设置动画过度时间-->
    android:gravity="top" /> <!–设置图片在父控件中的对齐方式 -->
  </LinearLayout>
```

步骤三：开发主逻辑代码。参考代码如下：

```
public class Example_Gallery extends Activity {
    int[] imageIDs={  //定义图片的数组
    R.drawable.club_1,R.drawable.club_2,R.drawable.club_3,
    R.drawable.club_4,R.drawable.club_5,R.drawable.club_6,R.drawable.
    club_7,R.drawable.club_8};
    @Override
    public void onCreate(Bundle savedInstanceState) {
      super.onCreate(savedInstanceState);
      setContentView(R.layout.main);//设置当前用户界面
      Gallery gl=(Gallery)this.findViewById(R.id.Gallery01);//得到Gallery
      的引用
      BaseAdapter ba=new BaseAdapter(){//初始化适配器
            @Override
        public int getCount() {//重写getCount方法
                return imageIDs.length;}
            @Override
        public Object getItem(int arg0) {   //重写getItem方法
                return null;}
            @Override
        public long getItemId(int arg0) {//重写getItemId方法
                return 0;}
            @Override  //重写getView方法
        public View getView(int arg0, View arg1, ViewGroup arg2) {
        ImageView iv = new ImageView(Example_Gallery.this);//初始化ImageView
        iv.setImageResource(imageIDs[arg0]);//设置图片资源
        iv.setScaleType(ImageView.ScaleType.FIT_XY);
        iv.setLayoutParams(new Gallery.LayoutParams(100,100));
        return iv; } };
     gl.setAdapter(ba);//设置适配器
   gl.setOnItemClickListener(new OnItemClickListener(){//添加监听
```

```
                    @Override
    public void onItemClick(AdapterView<?> arg0, View arg1,int arg2, long arg3) {
        Gallery gl=(Gallery)findViewById(R.id.Gallery01);
        gl.setSelection(arg2);//设置选中项
                    } } ); } }
```

步骤四：调试程序，便可以观察到效果图，如图 4-8 所示。当用户点击左边或者右边的图片时，所点击的图片将自动居中，其他图片依次左右滑动。当最左边或者最右边的图片居中时，就只能单方向的点击图片，实现画廊功能。

图 4-8　画廊控件效果图

4.4.2　Toast 的使用

Toast 向用户提供比较快速的即时消息，当 Toast 被显示时，虽然其悬浮于应用程序的最上方，但是 Toast 从不获取焦点。Toast 对象的创建是通过 Toast 类的静态方法 makeText 来实现的，该方法有两个重载实现，主要的不同是一个接收字符串，而另一个接收字符串的资源标识符作为参数。Toast 对象创建好之后，调用 show()方法即可将其消息提示显示在屏幕上。一般来讲，Toast 只显示比较简短的文本信息，但也可以显示图片。下面我们就将一个个人爱好例子作为介绍。本案例开发步骤如下：

图 4-9

步骤一：建立项目，准备资源。打开 Eclipse 工具，建立一个名为 Example_Toast 的项目，准备相应的图片，如图 4-9 所示，存放在项目的 res\drawable-mdpi 目录下。

步骤二：修改字符串资源描述文件。打开项目 res\values 目录下的 strings.xml 文件，在<resources>和</resources>标记之间加入如下的代码：

```
<string name="like">个人爱好</string>
<string name="message">你的爱好是篮球</string>
```

步骤三：修改布局文件。打开项目 res\layout 目录下的 main.xml 文件，修改代码如下：

```
<?xml version="1.0" encoding="utf-8"?>
<LinearLayout xmlns:android="http://schemas.android.com/apk/res/android"
    android:orientation="vertical"
    android:layout_width="fill_parent"
    android:layout_height="fill_parent"
    >    <!-- 声明一个线性布局 -->
    <Button
        android:id="@+id/btn"
```

```
        android:layout_width="fill_parent"
        android:layout_height="wrap_content"
        android:text="@string/like"
        />   <!-- 声明一个 Button 控件 -->
</LinearLayout>
```

步骤四：开发界面文件。打开项目中的 Example_Toast.java 文件，编写代码如下：

```
public class Example_Toast extends Activity {
    int mess=R.string.message;
    @Override
    public void onCreate(Bundle savedInstanceState) {   //重写onCreate方法
      super.onCreate(savedInstanceState);
      setContentView(R.layout.main);                   //设置当前屏幕
      Button btn = (Button)findViewById(R.id.btn);
      btn.setOnClickListener(new View.OnClickListener(){//为按钮添加监听器
    @Override
    public void onClick(View v) {
    ImageView iv = new ImageView(Example_Toast.this);  //创建 ImageView
    iv.setImageResource(R.drawable.basketball);//设置 ImageView 的显示内容
    Toast toast1 = Toast.makeText(Example_Toast.this, "你的爱好是足球",
    Toast.LENGTH_LONG);
    Toasttoast2 = Toast.makeText(Example_Toast.this,getResources().
    getString(mess), Toast.LENGTH_LONG);
    LinearLayout ll = new LinearLayout(Example_Toast.this);
                                                //创建一个线性布局
    toast2.setGravity(Gravity.CENTER, 0, 0);
    View toastView = toast2.getView(); //获得 Toast 的 View
    ll.setOrientation(LinearLayout.HORIZONTAL);//设置线性布局的排列方式
        ll.addView(iv); //将 ImageView 添加到线性布局
        ll.addView(toastView); //将 Toast 的 View 添加到线性布局
        toast2.setView(ll);
        toast1.show();       //显示 Toast
        toast2.show();}}); }}
```

步骤五：调试程序，查看效果，如图 4-10 和图 4-11 所示。

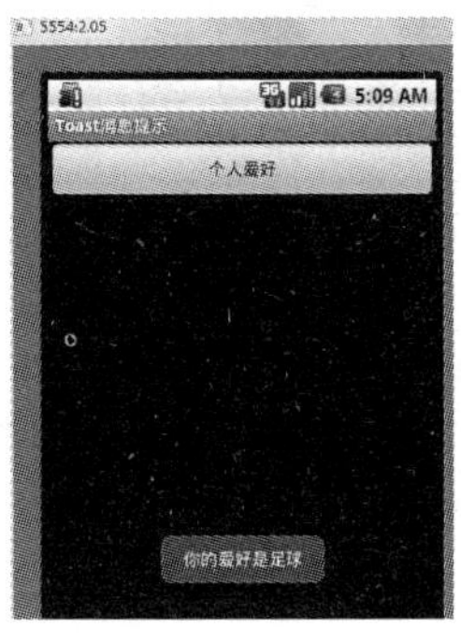

图 4-10　不带图片的效果图

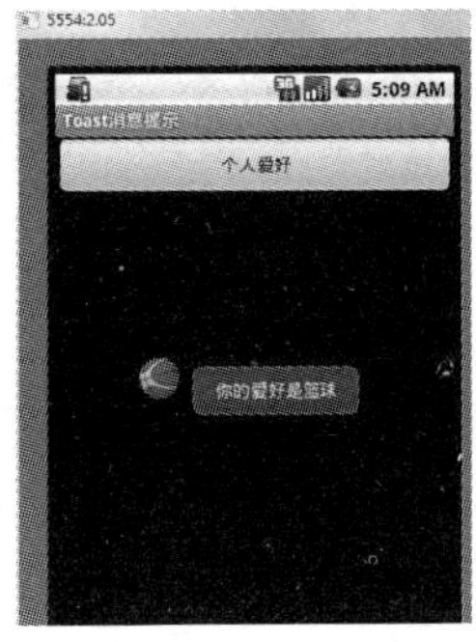

图 4-11　带图片的效果图

4.4.3 Notification 的使用

Notification 是另外一种消息提示的方式。Notification 位于手机的状态栏（Status Bar），状态栏位于手机屏幕的最上层，通常显示电池电量、信号强度等信息，在 Android 手机中，用手指按下状态栏并往下拉可以打开状态栏查看系统的提示消息。在应用程序中可以开发自己的 Notification 并将其添加到系统的状态栏中，下面通过一个案例来具体理解 Notification 的使用方法。其案例开发步骤如下。

步骤一：建立项目，准备资源。打开 Eclipse 工具，建立一个名为 Example_Notification 的项目，准备如图 4-9 所示图片，存放在项目的 res\drawable-mdpi 目录下。该图片将作为图标显示到手机屏幕的状态栏。

步骤二：修改字符串资源描述文件。打开项目 res\values 目录下的 string.xml 文件，在 <resources>和</resources>标记之间加入如下的代码：

```
<string name="btn">点击添加 Notification</string>
<string name="tv">Notification 启动</string>
<string name="notification">你的爱好是篮球</string>
```

步骤三：修改布局文件。打开项目 res\layout 目录下的 main.xml 文件，修改代码如下：

```
<?xml version="1.0" encoding="utf-8"?>
<LinearLayout xmlns:android="http://schemas.android.com/apk/res/android"
    android:orientation="vertical"
    android:layout_width="fill_parent"
    android:layout_height="fill_parent"
    >              <!-- 声明一个线性布局 -->
    <Button  android:id="@+id/btn"
        android:layout_width="fill_parent"
        android:layout_height="wrap_content"
        android:text="@string/btn"
        />             <!-- 声明一个 Button 对象 -->
</LinearLayout>
```

步骤四：开发界面文件。打开项目中的 Example_Notification.java 文件，编写代码如下：

```
public class Example_Notification extends Activity {
    @Override
  public void onCreate(Bundle savedInstanceState) {//重写 onCreate 方法
      super.onCreate(savedInstanceState);
      setContentView(R.layout.main);//设置当前屏幕
      Button btn = (Button)findViewById(R.id.btn);  //获取 Button 对象
    btn.setOnClickListener(new View.OnClickListener(){//为按钮设置监听器
        @Override
```

/*重写 onClick 方法，在该方法中首先创建一个 Notification 对象，并设置该对象的图标、提示信息等属性，其中最重要的是设置点下状态栏中该 Notification 时发送的 Intent 对象*/

```
        public void onClick(View v) {
```

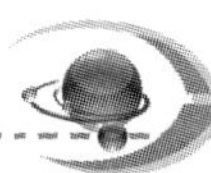

```
Intent i = new Intent(Example_Notification.this, NotifiedActivity.class);
PendingIntent pi = PendingIntent.getActivity(Example_Notification.this, 0, i, 0);
    Notification myNotification = new Notification();    //创建一个
    Notification 对象
    myNotification.icon=R.drawable.basketball;//设置 Notification 显示的图
    标 id 为篮球
myNotification.tickerText=getResources().getString(R.string.notification);
//设置 Notification 提示的文本信息
    myNotification.defaults=Notification.DEFAULT_SOUND;//表示当前的这个
    Notification 显示出来的时候，手机会伴随着音乐
/*
myNotification.defaults=Notification.DEFAULT_VIBRATE;//表示当前的这个
Notification 显示出来的时候，手机会伴随着震动，需要设置权限
myNotification.defaults=Notification.DEFAULT_ALL;//表示当前的这个
Notification 显示出来的时候，手机会伴随着音乐和震动，需要设置权限
*/
/*通过 setLatestEventInfo 方法设置两方面的内容：一方面当展开 Notification 列表的时
候，如何呈现 Notification；另外一方面设置当单击 Notification 的时候，如何处理单击*/
    myNotification.setLatestEventInfo(Example_Notification.this,
    "Notification 示例", "欢迎查看", pi);
/*所有的 Notification 都是有 NotificationManager 来管理，必须得到一个
NotificationManager 的实例*/
    NotificationManager notificationManager = (NotificationManager)
    getSystemService(NOTIFICATION_SERVICE);
//发送 Notification,每个 Notification 都有一个唯一的 id，这个 id 是由应用开发者来指
定的，这里指定为 0
notificationManager.notify(0, myNotification);
            }   });  }}
```

步骤五：编写另一个布局文件。打开项目 res\layout 目录，创建一个名称为 notified.xml 的 XML 文件。其参考代码如下：

```
<?xml version="1.0" encoding="utf-8"?>
<LinearLayout
  xmlns:android="http://schemas.android.com/apk/res/android"
  android:layout_width="fill_parent"
  android:layout_height="wrap_content"><!-- 声明线性布局 -->
<EditText android:text="@string/tv"
        android:layout_width="fill_parent"
        android:layout_height="wrap_content"
        android:editable="false"
        android:cursorVisible="false"
        android:layout_gravity="center_horizontal"
        />          <!-- 声明 EditText 控件 -->
</LinearLayout>
```

步骤六：创建另一个 Activity，名称为 NotifiedActivity.java，作为 Intent 对象发送的目的 Activity 页面。其代码如下：

```
public class NotifiedActivity extends Activity {
    @Override
    public void onCreate(Bundle savedInstanceState) {   //重写onCreate方法
        super.onCreate(savedInstanceState);
        setContentView(R.layout.notified);   }}//设置当前屏幕
```

步骤七：修改应用程序系统控制文件 AndroidManifest.xml 文件，使我们新创建的 Activity 能够正常运行。在控制文件中添加代码如下：

```
<activity android:name=".NotifiedActivity"
 android:label="@string/app_name"></activity>
```

步骤八：调试程序，查看效果。启动模拟器成功，运行该案例，其效果如图 4-12 所示，点击按钮时，则添加 Notification 信息到状态栏中，如图 4-13 所示，鼠标左键点住篮球标记，往下拉动，则会显示如图 4-14 所示的效果。当用户点击图 4-14 标记，则发送 Intent 对象，显示图 4-15 所示的效果。

图 4-12　实例效果

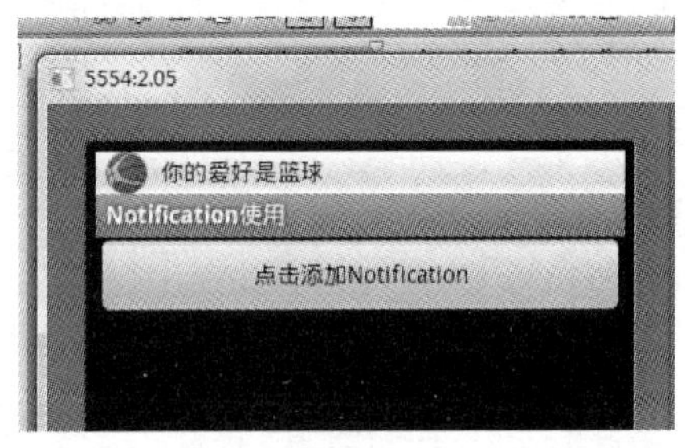

图 4-13　添加 Notification 信息到状态栏

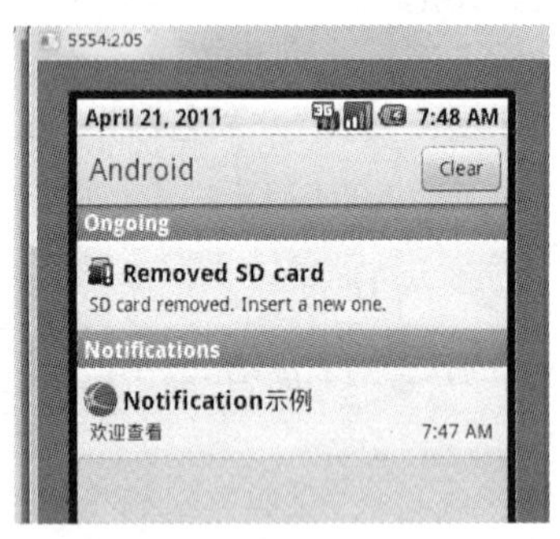

图 4-14　拉动篮球标记效果

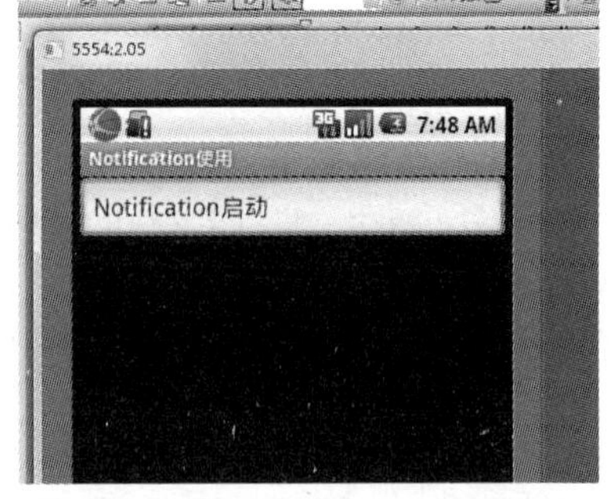

图 4-15　发送 Internet 对象效果

4.5　下拉列表控件与选项卡

下拉列表控件是最常用的控件之一，一般用来从多个选项中选择一个需要时，例如所在城市的选择、爱好的选择等。选项卡主要用来构建属于自己的 Android 用户界面。

4.5.1　下拉列表控件

下拉列表控件（Spinner）是最常用的高级控件之一，一般用来从多个选项中选择一个需要的，例如出生日期的选择、居住城市的选择等。Spinner 控件位于 android.widget 包下，是 View 类的一个子类。每次只显示用户选中的元素，当用户再次点击时，会出现选择列表供用户选择，而选择列表中的元素同样来自适配器。

接下来我们通过一个个人爱好选择的案例来具体介绍 Spinner 控件的使用方法，需要注意的是，Android 中的下拉列表并不像其他系统中直接下拉显示选项，而是相当于弹出菜单供用户选择。该项目的开发过程如下：

步骤一：建立项目，准备资源。打开 Eclipse 工具，建立一个名为 Example_ Spinner 的项目，准备如图 4-16 所示的图片，存放在项目的 res\drawable-mdpi 目录下。

图 4-16

步骤二：修改字符串资源描述文件。打开项目 res\values 目录下的 string.xml 文件，在<resources>和</resources>标记之间加入如下的代码：

```
<string name="ys">您的爱好</string>
<string name="lq">篮  球</string>
<string name="zq">足  球</string>
<string name="pq">排  球</string>
```

步骤三：修改颜色资源文件。打开项目 res\values 目录下的 colors.xml 文件，在<resources>和</resources>标记之间加入如下的代码：

```
<color name="red">#fd8d8d</color>
    <color name="green">#9cfda3</color>
    <color name="blue">#8d9dfd</color>
    <color name="white">#FFFFFF</color>
    <color name="black">#000000</color>
```

步骤四：修改布局文件。打开项目 res\layout 目录下的 main.xml 文件，修改代码如下：

```
<?xml version="1.0" encoding="utf-8"?>
<LinearLayout android:id="@+id/LinearLayout01"
  android:layout_width="fill_parent"
  android:layout_height="fill_parent"
  android:orientation="vertical"
  xmlns:android="http://schemas.android.com/apk/res/android">
  <TextView android:text="@string/ys"
    android:id="@+id/TextView01"
    android:layout_width="fill_parent"
```

```
    android:layout_height="wrap_content"
    android:textSize="20dip"/> <!-添加一个 TextView 控件 -->
  <Spinner  android:id="@+id/Spinner01"
    android:layout_width="fill_parent"
    android:layout_height="wrap_content"/> <!-添加一个下拉列表控件  -->
</LinearLayout>
```

步骤五：开发界面文件。打开项目中的 Example_Spinner.java 文件，编写代码如下：

```
public class Example_Spinner extends Activity {
    //所有资源图片（足球、篮球、排球）id 的数组
int[] drawableIds={R.drawable.football,R.drawable.basketball,
R.drawable.volleyball};
    //所有资源字符串（足球、篮球、排球）id 的数组
int[] msgIds={R.string.zq,R.string.lq,R.string.pq};
    @Override
  public void onCreate(Bundle savedInstanceState) {
      super.onCreate(savedInstanceState);
      setContentView(R.layout.main);
 Spinner sp=(Spinner)this.findViewById(R.id.Spinner01);//初始化 Spinner
     BaseAdapter ba=new BaseAdapter(){//为 Spinner 准备内容适配器
            @Override
    public int getCount() {return 3;}//总共三个选项
            @Override
    public Object getItem(int arg0) { return null; }
            @Override
    public long getItemId(int arg0) { return 0; }
            @Override
public View getView(int arg0, View arg1, ViewGroup arg2) {
                // 动态生成每个下拉项对应的 View，每个下拉项 View 由 LinearLayout
                //中包含一个 ImageView 及一个 TextView 构成
                //初始化 LinearLayout
LinearLayout ll=new LinearLayout(Example_Spinner.this);
ll.setOrientation(LinearLayout.HORIZONTAL);          //设置朝向
                //初始化 ImageView
ImageView  ii=new ImageView(Example_Spinner.this);
ii.setImageDrawable(getResources().getDrawable(drawableIds[arg0]));//设置图片
    ll.addView(ii);//添加到 LinearLayout 中
                //初始化 TextView
    TextView tv=new TextView(Example_Spinner.this);
 tv.setText(" "+getResources().getText(msgIds[arg0]));//设置内容
                tv.setTextSize(24);//设置字体大小
                tv.setTextColor(R.color.red);//设置字体颜色
                ll.addView(tv);//添加到 LinearLayout 中
                return ll;}  };
    sp.setAdapter(ba);//为 Spinner 设置内容适配器
```

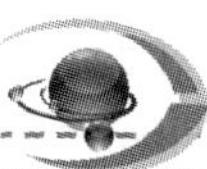

```
//设置 Spinner 控件的监听器，当有选项被选中时会自动调用 onItemSelected 方法
 sp.setOnItemSelectedListener( new OnItemSelectedListener(){
            @Override
    public void onItemSelected(AdapterView<?> arg0, View arg1,int arg2, long
    arg3) {//重写选项被选中事件的处理方法
    TextView tv=(TextView)findViewById(R.id.TextView01);//获取主界面
    TextView
    LinearLayout ll=(LinearLayout)arg1;//获取当前选中选项对应的 LinearLayout
    TextView tvn=(TextView)ll.getChildAt(1);//获取其中的 TextView
StringBuilder sb=new StringBuilder();//用 StringBuilder 动态生成信息
    sb.append(getResources().getText(R.string.ys));
    sb.append(":");
    sb.append(tvn.getText());
    tv.setText(sb.toString()); }//信息设置进主界面 TextView
    @Override
    public void onNothingSelected(AdapterView<?> arg0){}});}}
```

步骤六：运行案例，查看效果，如图 4-17 和图 4-18 所示。

图 4-17　未被打开的个人爱好选择

图 4-18　打开时的个人爱好选择

4.5.2　选项卡

选项卡（TabHost）类位于 android.widget 包下，是选项卡的封装类，用于创建选项卡窗口。TabHost 类继承自 FrameLayout，是帧布局的一种，其中可以包含多个布局，用户可以根据自己的爱好选择显示不同的界面。

表 4-3　选项卡（TabHost）常用方法表

方法名称	方法说明
addTab(TabHost.TabSpec tabSpec)	添加一项 Tab 页
clearAllTabs()	清除所有与之相关联的 Tab 页
getCurrentTab()	返回当前 Tab 页
getTabContentView()	返回包含内容的 FrameLayout
newTabSpec(String tag)	返回一个与之关联的新的 TabSpec

接下来我们通过介绍一个点菜单的案例来具体学习选项卡（TabHost）类的使用方法。其案例的具体开发步骤如下。

步骤一：建立项目，准备资源。打开 Eclipse 工具，建立一个名为 Example_TabHost 的项目，准备如图 4-19 所示的图片，存放在项目的 res\drawable-mdpi 目录下。

图 4-19　资料图片

步骤二：修改字符串资源描述文件。打开项目 res\values 目录下的 string.xml 文件，在<resources>和</resources>标记之间加入如下的代码：

```
<string name="app_name">点菜单</string>
<string name="image1">清脆酥饼</string>
<string name="image2">香辣可乐鸡翅</string>
<string name="image3">清蒸泉水鱼</string>
```

步骤三：修改布局文件，需要使用 FrameLayout。打开项目 res\layout 目录下的 main.xml 文件，修改代码如下：

```
<?xml version="1.0" encoding="utf-8"?>
<FrameLayout xmlns:android="http://schemas.android.com/apk/res/android"
  android:layout_width="fill_parent"
  android:layout_height="fill_parent">
<LinearLayout android:id="@+id/linearLayout01"
  android:layout_width="fill_parent"
  android:layout_height="fill_parent"
  android:gravity="center_horizontal"
  android:orientation="vertical">
<ImageView android:id="@+id/ImageView01"
  android:scaleType="fitXY"
  android:layout_gravity="center"
  android:layout_width="wrap_content"
  android:layout_height="wrap_content"
  android:src="@drawable/image1"/>
<TextView android:id="@+id/TextView01"
  android:layout_width="wrap_content"
  android:layout_height="wrap_content"
  android:textSize="24dip"
  android:text="@string/image1"/>
</LinearLayout>  <!—菜单一选项卡的内容-->
<LinearLayout android:id="@+id/linearLayout02"
```

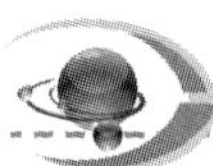

```
    android:layout_width="fill_parent"
    android:layout_height="fill_parent"
    android:gravity="center_horizontal"
    android:orientation="vertical" >
  <ImageView android:id="@+id/ImageView02"
    android:scaleType="fitXY"
    android:layout_gravity="center"
    android:layout_width="wrap_content"
    android:layout_height="wrap_content"
    android:src="@drawable/image2"/>
  <TextView  android:id="@+id/TextView02"
    android:layout_width="wrap_content"
    android:layout_height="wrap_content"
    android:textSize="24dip"
    android:text="@string/image2"/>
  </LinearLayout> <!—菜单二选项卡的内容-->
  <LinearLayout android:id="@+id/linearLayout03"
   android:layout_width="fill_parent"
   android:layout_height="fill_parent"
   android:gravity="center_horizontal"  android:orientation="vertical">
  <ImageView  android:id="@+id/ImageView03"
    android:scaleType="fitXY"  android:layout_gravity="center"
    android:layout_width="wrap_content"
    android:layout_height="wrap_content"
    android:src="@drawable/image3"/>
  <TextView  android:id="@+id/TextView03"
     android:layout_width="wrap_content"
    android:layout_height="wrap_content"
    android:textSize="24dip"
    android:text="@string/image3"/>
  </LinearLayout> <!—菜单三选项卡的内容-->
</FrameLayout>
```

步骤四：开发界面文件，继承 TabActivity。打开项目中的 Example_ TabHost.java 文件，编写代码如下：

```
public class Example_ TabHost extends TabActivity {
 private TabHost myTabhost;
 public void onCreate(Bundle savedInstanceState) {
     super.onCreate(savedInstanceState);
     //从 TabActivity 上面获取放置 Tab 的 TabHost
     myTabhost=this.getTabHost();/*设置布局，from(this)是从 TabActivity 获取
     LayoutInflater，R.layout.main 表示存放着 Tab 布局的文件；第二个参数表示通过
     TabHost 获得存放 Tab 标签页内容的 FrameLayout；而第三个参数表示是否将 inflate
     拴系到根布局元素上*/
```

```
LayoutInflater.from(this).inflate(R.layout.main,
myTabhost.getTabContentView(), true);
  myTabhost.addTab(//添加 Tab 页
myTabhost.newTabSpec("菜单一")//创建一个 newTabSpec
.setIndicator("菜单一", //设置选项标题
getResources().getDrawable(R.drawable.png1))//设置选项图标
.setContent(R.id.linearLayout01) ); //设置选项卡的布局文件
myTabhost.addTab(
    myTabhost.newTabSpec("菜单二").setIndicator("菜单二",
    getResources().getDrawable(R.drawable.png2))
    .setContent(R.id.linearLayout02));
myTabhost.addTab(
    myTabhost.newTabSpec("菜单三") .setIndicator("菜单三",
    getResources().getDrawable(R.drawable.png3))
    .setContent(R.id.linearLayout03));  }}
```

步骤五：运行案例，查看效果，如图 4-20 所示。

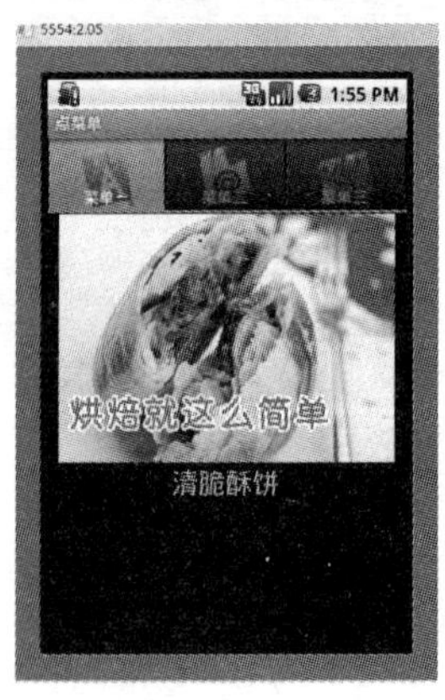

图 4-20　点菜单案例效果图

第5章

Android 游戏应用程序开发

Android 平台下的应用开发，一般来说主要分为商业应用和游戏应用两种，在开发商业应用中，界面主要用到我们前面曾经介绍过的控件、菜单和对话框等，而游戏应用开发中只能自己定义所需要的控件和界面了。本章将重点介绍 Android 平台下使用 View 和 SurfaceView 类开发游戏中所需要的界面。

5.1 自定义 View 的使用

本节将要介绍的是如何使用自定义的 View 来开发自己的游戏或应用。View 类是 Android 的一个超类，这个类几乎包含了所有的屏幕类型。每一个 View 都有一个用于绘图的画布，这个画布可以进行任意扩展。在游戏开发中也可以自定义视图（View），这个画布的功能更能满足我们在游戏开发中的需要。在 Android 中，任何一个 View 类都只需重写 onDraw 方法来实现界面显示，自定义的视图可以是复杂的 3D 实现，也可以是非常简单的文本形式等。使用自定义的 View 来开发程序需要开发继承来自 View 的子类，其主要的工作包括如下两个方面。

（1）绘制屏幕：由于使用自定义的 View，程序的画面需要从头到尾进行设计，这些工作都需要放在绘制屏幕模块中进行。

（2）刷新屏幕：在程序的运行中，如果后台数据发生了变化，需要开发人员自己刷新屏幕以将最新的数据显示给用户，这个功能通常用多线程来完成。

下面我们通过案例来说明如何开发继承自 View 的子类来实现自己的程序功能，本案例包括 MyView、MyThread 和 Example_5_1 三个类，下面就此案例的开发步骤进行详细介绍。

步骤一：项目建立，资源准备。本案例需要建立一个名称为 Example_5_1 的 Android 项目，在案例中用到的图片资源主要包括代表英雄的图片 man.png（40×40）和游戏场景中不断循环轮换的照片 a.jpg（320×200），b.jpg，c.jpg，d.jpg，如图 5-1 所示，并且把这些图片复制到项目的\res\drawable-mdpi 文件夹里面。

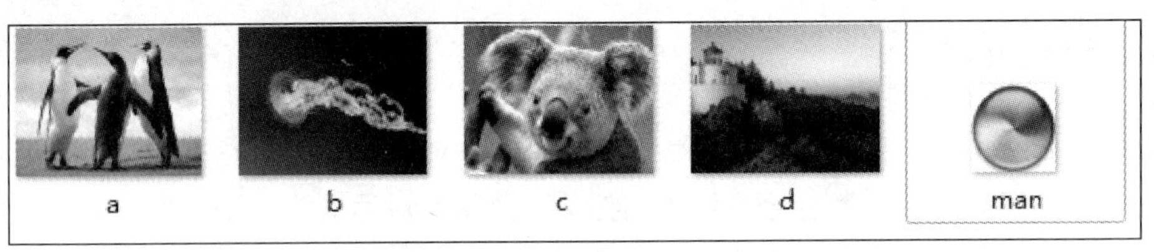

图 5-1　资料图片

步骤二：MyView 类的创建。MyView 类是 View 类的子类，其代码主要由三部分构成：游戏场景所用到的常量与变量的声明，初始化游戏所用图片资源的构造方法和重写绘制游戏需要元素的绘制方法 onDraw()，其主要代码如下：

```
public class MyView extends View{
    //游戏场景所用到的常量与变量的声明
    static final int ANGLE_MAX=50; //张嘴最大角度
    static final int SPEED=3;         //移动的速度
    static final int SCREEN_WIDTH = 480;//设置屏幕的宽度
    static final int SCREEN_HEIGHT = 320;//设置屏幕的高度
    static final int LEFT=2;     //英雄移动四个方向常数  左
    static final int RIGHT=0;   //右
    static final int UP=3;       //上
    static final int DOWN=1;      //下
    int angle=30;                   //当前张嘴总角度
    int angleChange=3;                  //角度变化值
    int radius=20;                      //半径
    int centerX=radius;                 //当前位置
    int centerY=radius;                 //当前位置
    long timeStamp=System.currentTimeMillis();//记录换照片的时间
    int currPhoto;                  //当前照片数组中的索引
    int direction = RIGHT;          //记录精灵方向
    Bitmap bmpMan;                  //精灵的图片
    Bitmap [] bmpPhotos;                //照片数组
    int[] imgIds = {
```

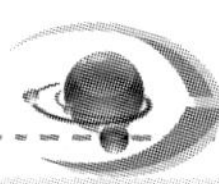

```
    R.drawable.a,R.drawable.b,R.drawable.c,R.drawable.d};//图片 id 数组
    //构造方法，对所用到的图片进行初始化
    public MyView(Context context) {
        super(context);
        //初始化精灵的图片
    bmpMan =
BitmapFactory.decodeResource(getResources(),R.drawable.man);
        //初始化照片数组
        bmpPhotos = new Bitmap[imgIds.length];
        //初始化每张照片
    for(int i=0;i<bmpPhotos.length;i++){bmpPhotos[i] =
BitmapFactory.decodeResource(getResources(), imgIds[i]);
        }
    }
    //重写绘制游戏需要元素的绘制方法 onDraw()
    protected void onDraw(Canvas canvas) {
        Paint paint = new Paint();          //创建画笔对象
        canvas.drawColor(Color.BLACK);      //绘制背景为黑色
        canvas.drawBitmap(bmpMan, centerX-radius, centerY-radius, null);
        //绘制吃豆小人
        paint.setColor(Color.BLACK);                //设置画笔颜色
        paint.setAntiAlias(true);            //设置抗据齿
        RectF oval = new RectF(centerX-radius,centerY-radius,
                centerX+radius,centerY+radius);
    //绘制扇形实现张嘴闭嘴
    canvas.drawArc(oval, 360-angle+90*direction, 2*angle, true, paint);
    if(System.currentTimeMillis()-timeStamp > 5000){//如果时间间隔超过 5 秒
          timeStamp= System.currentTimeMillis(); //重新记录时间
          currPhoto = (currPhoto+1)%bmpPhotos.length;//更换照片
        }
    canvas.drawBitmap(bmpPhotos[currPhoto], 85, 50, null);//绘制照片
        super.onDraw(canvas);
    }
}
```

步骤三：MyThread 类的创建。MyThread 类是 Thead 类的子类，其功能主要是负责修改 MyView 类绘制的数据，从而实现游戏场景动画播放。该类主要功能由三部分构成：控制英雄张嘴闭嘴的功能，控制英雄进行移动的功能和检测英雄是否与屏幕边界碰撞的功能。其主要代码如下：

```
public class MyThread extends Thread{
    int sleepSpan = 30;        //休眠时间
    MyView myView;             //MyView 对象引用
    public MyThread(MyView myView){ //MyThread 类的构造器
        this.myView = myView;   }
```

```
    public void run(){          //线程执行方法
        while(true){  //用死循环控制英雄的嘴不断地一张一合
            //动态张闭嘴=====================begin=============
            myView.angle=myView.angle+myView.angleChange;
        if(myView.angle>MyView.ANGLE_MAX){//超过最大值则进入闭嘴环节
                myView.angleChange=-3; }
            else if(myView.angle<0){//小于 0 则进入张嘴环节
                myView.angleChange=+3; }
            //动态张闭嘴=====================end===============

            //移动英雄=====================begin=============
            switch(myView.direction){
               case MyView.RIGHT://向右
                 myView.centerX=myView.centerX+MyView.SPEED;
               break;
               case MyView.UP://向上
                 myView.centerY=myView.centerY-MyView.SPEED;
               break;
               case MyView.LEFT ://向左
                 myView.centerX=myView.centerX-MyView.SPEED;
               break;
               case MyView.DOWN://向下
                 myView.centerY=myView.centerY+MyView.SPEED;
               break;
            }
   //判断英雄是否与屏幕边界发生碰撞，并改变英雄的移动方向
   if(myView.centerY+myView.radius>MyView.SCREEN_HEIGHT){//碰到下边界
            myView.centerY=myView.centerY-MyView.SPEED;
            myView.direction=MyView.LEFT;  }
   if(myView.centerY-myView.radius<0){// 碰到上边界
            myView.centerY=myView.centerY+MyView.SPEED;
            myView.direction=MyView.RIGHT;  }
   if(myView.centerX-myView.radius<0){// 碰到左边界
            myView.centerX=myView.radius;
            myView.direction=MyView.UP;  }
   if(myView.centerX+myView.radius>MyView.SCREEN_WIDTH){// 碰到右边界
            myView.centerX=myView.centerX-MyView.SPEED;
            myView.direction=MyView.DOWN;  }
        //移动英雄=====================end=============
        myView.postInvalidate();  //重绘屏幕
        try{
            Thread.sleep(sleepSpan);  //线程休眠
        }
        catch(Exception e){
```

```
                e.printStackTrace();
            }   }  }  }
```

步骤四：Activity 类的开发。Activity 类是用户与模拟器交互的视图类，该类中需要把自定义视图和动画播放效果显示在屏幕上。其主要代码如下：

```
public class Example_5_1 extends Activity {
    MyView myView;          //MyView 对象引用
    MyThread mt;            //MyThread 对象引用
    @Override
    public void onCreate(Bundle savedInstanceState) {   //重写onCreate方法
        super.onCreate(savedInstanceState);
        requestWindowFeature(Window.FEATURE_NO_TITLE);   //设置不显示应用程序
标题栏
      getWindow().setFlags(WindowManager.LayoutParams.FLAG_FULLSCREEN,
        WindowManager.LayoutParams.FLAG_FULLSCREEN); //设置全屏
        myView = new MyView(this);              //创建 MyView 对象
        setContentView(myView);                 //设置当前屏幕
        mt = new MyThread(myView);              //创建 MyThread 对象
        mt.start();                             //启动 MyThread
}   }
```

到此为止，我们所说的自定义 View 实例开发结束，下面开始调试我们开发的项目。由于该项目需要在宽屏幕模式下运行，所以在调试前需要将模拟器屏幕设置为横屏模式，可以通过同时按 Ctrl+F12 键来实现。

5.2 SurfaceView 的使用

在前面的章节中，我们学习了开发继承自 View 的子类来实现自定义 View，本节我们将继续学习在游戏开发中经常用到的 SurfaceView 类。

在一般的情况下，应用程序的 View 都是在相同的 GUI 线程中绘制的。这个主应用程序线程同时也用来处理所有的用户交互（例如，按钮单击或者文本输入）。SurfaceView 封装了一个 Surface 对象，而不是 Canvas。当需要快速地更新 View 的 UI，或者当渲染代码阻塞 GUI 线程的时间过长的时候，SurfaceView 就是解决上述问题的最佳选择。这一点很重要，因为 Surface 可以使用后台线程绘制。对于那些资源敏感的操作，或者那些要求快速更新或者高帧速率的地方，例如，使用 3D 图形，创建游戏，或者实时预览摄像头，这一点特别有用。

SurfaceView 使用的方式与任何 View 所派生的类都是完全相同的。可以像其他 View 那样应用动画，并把它们放到布局中。SurfaceView 封装的 Surface 支持所有的标准 Canvas 方法进行绘图，同时也支持完全的 OpenGL ES 库。要创建一个新的 SurfaceView，需要创建一个新的扩展了 SurfaceView 的类，并实现 SurfaceHolder.Callback 接口。SurfaceHolder 回调可以在底层的 Surface 被创建和销毁的时候通知 View，并传递给它对 SurfaceHolder 对象的引用，其中包含了当前有效的 Surface。一个典型的 SurfaceView 设计模型包括一个由 Thread 所派生的类，它可以接收对当前的 SurfaceHolder 的引用，并独立地更新它。

下面我们通过一个弹性小球的案例来具体说明 SurfaceView 类的使用，本案例中涉及到 Example_5_2、GameView、BallGoThread 和 TimeThread 类，下面我们对该项目的开发步骤进行比较详细的介绍。

步骤一：项目的创建和图片资源的准备。在 Eclipse 中新建一个 Android 项目，名称为 Example_5_2，并在其 res\drawable-mdpi 目录下复制粘贴入项目开发需要应用到的图片资源，资源内容如图 5-2 所示。

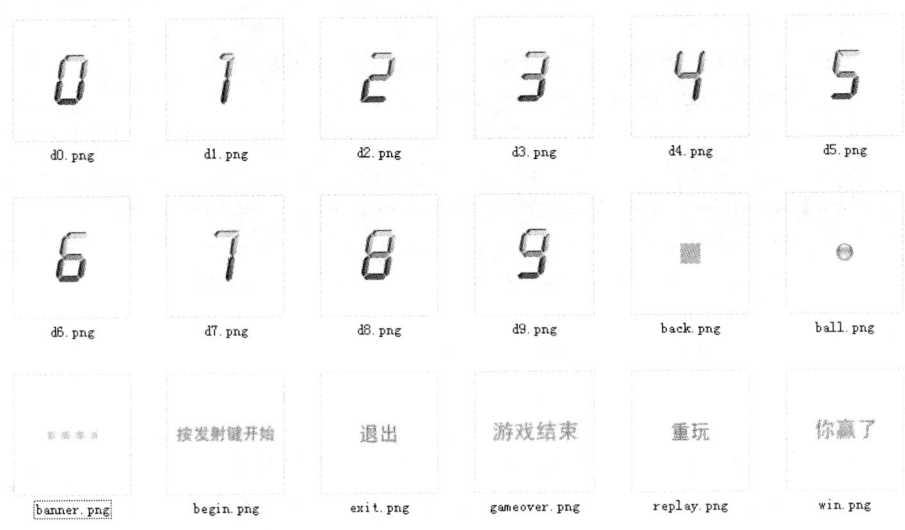

图 5-2　资料图片

步骤二：GameView 类的创建。由于 GameView 类是 SurfaceView 类的子类，如果要实现 SurfaceView 类，还需要添加一个 SurfaceHolder.Callback 接口的实现，该接口需要重写 surface Changed、surfaceCreated、surfaceDestroyed 三个方法。该类的功能主要是声明该案例开发过程中所需要的变量，对图片资源的初始化，清除上次游戏的存留数据，处理用户单击屏幕事件，重写绘制游戏需要元素的绘制方法以及重写 SurfaceHolder.Callback 接口的相应方法。主要参考代码如下：

```
public class GameView extends SurfaceView implements
SurfaceHolder.Callback{
//声明该项目开发过程中所需要的变量
    Example_5_2Activity activity ; //activity的引用
    TimeThread tt;              //TimeThread引用
    BallGoThread bgt;           //BallGoThread引用
    int backSize=16;//背景尺寸，游戏中的背景图片是通过多个相同图片平铺绘制而成
    int screenWidth = 320;//屏幕宽度
    int screenHeight = 480;//屏幕高度
    int bannerWidth=40;//负责接收小球挡板的宽度
    int bannerHeight=6;//负责接收小球挡板的高度
    int bottomSpance=16;//底下空白
    int bannerSpan=5;//板步进
```

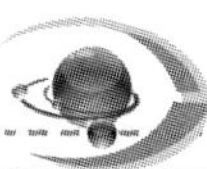

```
    int ballSpan=8;//球步进
    int ballSize=16;//小球尺寸
    int hintWidth=100;//提示宽度
    int hintHeight=20;//提示高度
//记录游戏状态值的成员变量：0-等待开始、1-进行中、2-游戏结束、3-游戏胜利
    int status=0;
    int score=0;//得分
    int ballx;//小球 x 坐标
    int bally;//小球 y 坐标
    int direction=0;//初始化小球方向 0 为右上、1 为右下、2 为左下、3 为左上
    int bannerX;// 负责接收小球挡板的 x 坐标
    int bannerY;// 负责接收小球挡板的 y 坐标
    int scoreWidth = 32;  //数字图片的宽度

    Bitmap iback;//背景图片声明
    Bitmap[] iscore=new Bitmap[10];//得分图片组声明
    Bitmap iball;//球图片声明
    Bitmap ibanner;//板图片声明
    Bitmap ibegin;//开始图片声明
    Bitmap igameover;//游戏结束图片声明
    Bitmap iwin;//游戏胜利图片声明
    Bitmap iexit;//退出游戏图片声明
    Bitmap ireplay;//重玩游戏图片声明
    GameViewDrawThread gameViewGrawThread;//构造方法，对图片资源初始化
public GameView(Example_5_2 activity) {
    super(activity);
    getHolder().addCallback(this);//注册接口
    this.activity = activity;
    initBitmap();  //调用初始化常用方法
gameViewGrawThread = new GameViewDrawThread(this);//启动后台刷屏线程
}
public void initBitmap(){//初始化图片资源方法
iback = BitmapFactory.decodeResource(getResources(), R.drawable.back);
//背景图片初始化
     ... //此处省略图片资源的初始化代码
    initBallAndBanner();//初始化小球位置及板坐标的方法
}
public void initBallAndBanner(){
bally=screenHeight-bottomSpance-bannerHeight-ballSize; //计算小球 y 坐标
    ballx=screenWidth/2-ballSize/2; //计算小球 x 坐标
    bannerX=screenWidth/2-bannerWidth/2; //计算挡板 x 坐标
    bannerY=screenHeight-bottomSpance-bannerHeight;     //计算挡板 y 坐标
}
//replay()方法功能是清除上一轮游戏中留下的数据，以便开始新的游戏
```

```
        public void replay(){
            if(status==2||status==3){
                initBallAndBanner(); //初始化小球位置及板坐标
                score=0;      //清零得分
                status=0;     //设置新的初始状态
                direction=3; //初始化一个方向
            } }
        //绘制游戏界面的方法
        public void doDraw(Canvas canvas) {
        //------绘制背景图片开始-------
        int
cols=(screenWidth%backSize==0)?(screenWidth/backSize):(screenWidth/backSize+1);
//背景小图片排列的列数
        int
rows=(screenHeight%backSize==0)?(screenHeight/backSize):(screenHeight/backSize+
1);// 背景小图片排列的行数
        //通过二重循环对背景图片排列，形成游戏的背景整体图片
        for(int i=0;i<rows;i++){
            for(int j=0;j<cols;j++){
                canvas.drawBitmap(iback, 16*j,16*i, null);
            }   }
        //------绘制背景图片结束-------
        //绘制游戏得分的整体图片
        String scoreStr=score+"";
        int loop=3-scoreStr.length();
        //通过循环在分值的前面添加 0，构成三位数得分图片组
        for(int i=0;i<loop;i++){
            scoreStr="0"+scoreStr;  }
        //计算得分图片组绘制的开始 x 坐标
         int startX=screenWidth-scoreWidth*3-10;
         for(int i=0;i<3;i++){//图片组绘制
           //通过 ASCII 码计算，得出绘制的图片
        int tempScore=scoreStr.charAt(i)-'0';
        canvas.drawBitmap(iscore[tempScore], startX+i*scoreWidth,5, null);}
        canvas.drawBitmap(iball,ballx,bally, null);    //绘制小球
        //绘制接收小球的挡板
        canvas.drawBitmap(ibanner,bannerX,bannerY, null);
        //绘制开始提示图片按钮
        if(status==0){
            canvas.drawBitmap( ibegin,screenWidth/2-hintWidth/2,
               screenHeight/2-hintHeight/2, null); }
            //绘制失败提示图片按钮
        if(status==2){
            canvas.drawBitmap(igameover,screenWidth/2-hintWidth/2,
```

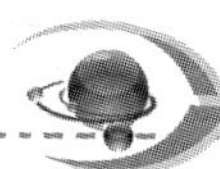

```
        screenHeight/2-hintHeight/2, null);}
    //绘制胜利图片按钮
    if(status==3){
        canvas.drawBitmap(iwin,screenWidth/2-hintWidth/2,
            screenHeight/2-hintHeight/2, null);}
    //绘制退出选项图片按钮
        canvas.drawBitmap(iexit,
            screenWidth-32,screenHeight-16,null );
    //绘制重玩选项图片按钮
    if(status==2||status==3){
        canvas.drawBitmap(ireplay ,0, screenHeight-16,null);
    }    }
    /*处理用户点击屏幕事件*/
    public boolean onTouchEvent(MotionEvent event) {
        int x = (int) event.getX(); //获取被点击处的 x 坐标
        int y = (int) event.getY(); //获取被点击处的 y 坐标
        if(x<screenWidth&&x>screenWidth-32
                &&y<screenHeight&&y>screenHeight-16){//按下退出选项
            System.exit(0);  //退出程序
        }
        if(status == 0){      //等待状态，按下任意键开始游戏
            status=1;    //设置状态值
            tt=new TimeThread(this);      //创建新的计时线程
            bgt=new BallGoThread(this); //创建新的 BallThread
            tt.start(); //启动相关线程
            bgt.start();      }
        else if(status == 1){//如果是游戏状态下
            bannerX = x; //移动挡板的位置
        }
        else if(status==2||status==3){//如果是游戏失败或胜利
        if(x<32&&x>0&&y<screenHeight&&y>screenHeight-16){//按下重玩选项
             replay();  //调用方法重新开始游戏
            }  }
        return super.onTouchEvent(event); }
    //主要处理当 Surface 的尺寸发生变化时执行的操作
    public void surfaceChanged(SurfaceHolder holder, int format, int width, int
    height) {}
    //当 Surface 被创建之后，将需要执行的代码放置在该处
    public void surfaceCreated(SurfaceHolder holder) {//创建时启动相应进程
       this.gameViewGrawThread.flag = true;
       gameViewGrawThread.start();}
    //当 Surface 被移除当前显示屏幕时回调的方法
    public void surfaceDestroyed(SurfaceHolder holder) {//摧毁时释放相应进程
       boolean retry = true;
```

```
        this.gameViewGrawThread.flag =  false;
        while (retry) {
            try {
            gameViewGrawThread.join();
              retry = false; }
          catch (InterruptedException e) {}//不断地循环，直到刷帧线程结束
        } } }
```

步骤三：BallGoThread 类的开发。该类继承自 Thread 类，主要功能是根据小球当前的方向不断移动小球，并不断检测小球是否与屏幕边沿或挡板发生碰撞。开发代码如下：

```
    public class BallGoThread extends Thread{
        GameView father;//创建 GameView 类的对象
    boolean flag=true;//设置控制线程是否执行的变量，如果值为 false 则停止线程执行
        //构造方法，实例化 GameView 类的对象
        public BallGoThread(GameView father){
            this.father=father;}
        //线程自动运行的 run 方法
        public void run(){
            while(flag){    //判断当前线程是否需要执行
                switch(father.direction){ //获取小球当前位置
                    case 0:
                      //小球向右上方向移动相应的坐标
                      father.ballx=father.ballx+father.ballSpan;
                      father.bally=father.bally-father.ballSpan;
            //判断小球是否碰到右壁，如果碰到则改变小球移动方向
            if(father.ballx>=father.screenWidth-father.ballSize){
                        father.direction=3;  }
            //判断小球是否碰到上壁，如果碰到则改变小球移动方向
            else if(father.bally<=0){
                        father.direction=1; }
                      break;
                    case 3://小球向左上方向移动相应的坐标
                      father.ballx=father.ballx-father.ballSpan;
                      father.bally=father.bally-father.ballSpan;
            //判断小球是否碰到左壁，如果碰到则改变小球移动方向
                      if(father.ballx<=0){
                        father.direction=0; }
            //判断小球是否碰到上壁，如果碰到则改变小球移动方向
                      else if(father.bally<=0){
                        father.direction=2; }
                      break;
                //以下代码功能与上面类似，在此不作详述
                  case 1:
                      father.ballx=father.ballx+father.ballSpan;
```

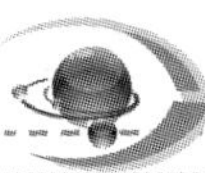

```
            father.bally=father.bally+father.ballSpan;
        //检测小球与挡板是否发生碰撞
            if(father.bally>=
            father.screenHeight-father.bannerHeight
            -father.bottomSpance-father.ballSize){
               checkCollision(1);}
                else if(father.ballx>=
          father.screenWidth-father.ballSize){
              father.direction=2;}
            break;
           case 2:
            father.ballx=father.ballx-father.ballSpan;
            father.bally=father.bally+father.ballSpan;
            if(father.bally>=
                  father.screenHeight-father.bannerHeight
                  -father.bottomSpance-father.ballSize){
             checkCollision(2); }
           else if(father.ballx<=0){father.direction=1;}
              break;  }
        try{Thread.sleep(100);}  //线程休眠时间
        catch(Exception e){e.printStackTrace();} } }
    //检测小球是否与挡板发生碰撞的方法
      public void checkCollision(int direction){
       if(father.ballx>=father.bannerX-father.ballSize&&
          father.ballx<=
            father.bannerX+father.bannerWidth){//碰到板
          switch(direction){
              case 1:father.direction=0;break;
              case 2: father.direction=3; break;} }
        else{//没有碰到板，游戏失败，相关状态重新赋值
          father.tt.flag=false;
          father.bgt.flag=false;
          father.status=2; } }    }
```

步骤四：TimeThread 类的开发。该类继承自 Thread，该类的主要功能是对游戏进行计时，不断修改玩家的得分（即游戏持续的时间），并根据玩家的得分决定游戏是否胜利。其主要代码如下：

```
public class TimeThread extends Thread{
int highest=60;        //胜利值，游戏的最高得分
GameView gameView;   //GameView 引用
boolean flag=true;   //线程是否执行标志位
public TimeThread(GameView gameView){   //构造器
    this.gameView=gameView;}
public void run(){
```

```
    while(flag){
        gameView.score++;              //增加得分
        if(gameView.score==highest){  //判断时间是否达到胜利时间
            gameView.status=3;         //游戏胜利
            gameView.tt.flag=false;  //停止 TimeThread 的执行
            gameView.bgt.flag=false;    //停止 BallGoThread 的执行
        }
        try{ Thread.sleep(1000);    }    //休眠 1 秒钟
        catch(Exception e){e.printStackTrace(); } } } }
```

步骤五：GameViewDrawThread 类的开发。该类继承自 Thread，该类的主要功能是定时刷新 GameView 的屏幕，在这里对 SurfaceView 进行绘制时，首先要通过 SurfaceHolder 对象将 SurfaceView 的画布加锁，然后对获得的画布进行绘制，最后解锁画布并将其传回。其主要代码如下：

```
public class GameViewDrawThread extends Thread{
boolean flag = true;
int sleepSpan = 100;
GameView gameView;
SurfaceHolder surfaceHolder;
public GameViewDrawThread(GameView gameView){
    this.gameView = gameView;
    this.surfaceHolder = gameView.getHolder();}
public void run(){
    Canvas c;
   while (this.flag) {
       c = null;
       try {
     // 锁定整个画布，在内存要求比较高的情况下，建议参数不要为null
          c = this.surfaceHolder.lockCanvas(null);
          synchronized (this.surfaceHolder) {
           gameView.doDraw(c); }//绘制
         } finally { if (c != null) {
            //解锁并更新屏幕显示内容
          this.surfaceHolder.unlockCanvasAndPost(c);
          } }
       try{Thread.sleep(sleepSpan); }//睡眠指定毫秒数
       catch(Exception e){ e.printStackTrace();} //打印堆栈信息
        } } }
```

步骤六：Example_5_2Activity 类的开发。前面完成了案例所有功能模块的开发，最后开发 Activity 类，该类就是创建 GameView 对象并将程序的屏幕设置为 GameView 对象。其主要代码如下：

```
public class EXample_5_2Activity extends Activity {
    GameView gameView;    //GameView 对象引用
    @Override
```

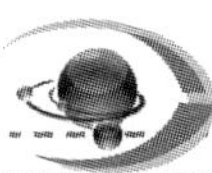

```
        public void onCreate(Bundle savedInstanceState) {   //重写onCreate方法
            super.onCreate(savedInstanceState);
             //设置全屏
             requestWindowFeature(Window.FEATURE_NO_TITLE);
        getWindow().setFlags(WindowManager.LayoutParams.FLAG_FULLSCREEN ,
                        WindowManager.LayoutParams.FLAG_FULLSCREEN);
            gameView = new GameView(this);    //创建GameView
            setContentView(gameView);       //设置当前屏幕
        } }
```

到此为止，我们所说的通过SurfaceView类开发游戏，其代码基本完成，然后启动模拟器，就可以看到我们所需要的效果，如图5-3和图5-4所示。

图5-3 按发射键开始时界面

图5-4 游戏结束时界面

5.3 游戏碰撞与检测技术

碰撞检测是游戏中必不可少的一个功能，无论是角色扮演游戏还是其他类型的游戏，二维游戏还是三维游戏，都会涉及到碰撞检测模块的开发。如玩家打中怪物、足球碰撞反弹、炮弹击中飞机、玩家捡到宝箱等，这些都用到了碰撞检测技术。碰撞检测技术的实现也是基于数学和物理方面的知识来进行计算和判断的。在游戏开发中不同的场合有不同的碰撞检测方式，本节将主要讲解碰撞检测在游戏中的使用方式以及使用碰撞检测时的基本原则。

5.3.1 碰撞检测技术基础

一般情况下，碰撞检测只会发生在游戏中的实体对象的位置发生了变化之后，如怪物走动、炮弹沿轨道移动、玩家跳起等。不同的碰撞检测其目的也不尽相同，如怪物走动后检测是否遇到玩家，炮弹移动后检测是否打中目标等。

对于移动之后进行碰撞检测的场合，程序中通常按照以下的流程来应用碰撞检测：

（1）更新实体对象的位置。

（2）进行碰撞检测。

（3）如果碰撞到了，进行相应处理。

上述流程是针对单个实体对象来说的，即每个实体在自己位置更新了之后就进行碰撞检测。还有一种方法是在一类实体的位置全部更新完毕之后，再逐个进行碰撞检测。如果屏幕中所有炮弹全部移动位置，然后判断每个炮弹是否碰到目标，这种方法不能适用于所有的场合。步骤三是对碰撞检测进行处理，最简单的一种处理就是让实体对象后退回到原来的位置，除去这种简单的情况，通常碰撞检测环节和处理碰撞环节是结合在一起的。实现碰撞检测所涉及的内容主要有如下 3 个方面。

（1）确定检测对象。游戏在运行中会产生很多实体对象，在进行碰撞检测时并不需要对所有的实体对象都检测一遍，如玩家没有必要检测是否和自己发出的子弹碰撞，静止的宝箱也没有必要检测是否和另外的宝箱碰撞。所以在开始碰撞检测之前，首先要确定检测的对象是什么。

（2）检测是否碰撞。这是碰撞检测的核心环节，在这个环节需要综合考虑游戏本身需求和运行平台的性能问题，合理地选择碰撞检测的算法。

（3）处理碰撞。当检测到有碰撞发生时，就需要根据碰撞的类型进行相应的处理，如炮弹会在碰到目标后爆炸并给目标造成伤害。

考虑到游戏的最佳用户体验问题，通常在发生开发游戏的碰撞检测模块需要遵循以下几个原则。

（1）尽量避免碰撞检测，如果无法避免就尽量减少检测的次数。由于进行碰撞检测基本上都会进行一些数学计算，所以应该尽量少用，减少检测的机会。

（2）采用不影响游戏性能的算法尽快检测。对于碰撞检测来说，精确和速度就像是算法设计中的空间复杂度和时间复杂度，二者很难达到两全。所以在游戏开发时多做衡量，选择一种对于精确程度和执行速度来说都比较适中的算法。

（3）碰撞处理要柔和，不要让玩家感到不适应。碰撞检测的处理不能过于生硬，如物体从地面抛向远处，在飞行的时候如果检测到与地面发生了碰撞，不应该立刻让其停止运动，而是就着惯性向前滑动一段距离。

5.3.2 游戏中实体对象之间的碰撞检测

具体来讲，碰撞检测主要分为游戏实体对象（如玩家控制的英雄、怪物、发射的炮弹等）之间的碰撞检测以及游戏实体对象与环境（如游戏场景中的墙、台阶、树等）之间的碰撞检测。游戏中实体对象与环境之间的碰撞检测无法偷工减料（否则将会出现游戏实体对象穿墙而过的奇怪现象），但是实体间的碰撞检测可以稍加优化。所以在研究实体间碰撞的算法前，需要考虑如何减少待检测实体的个数，一般有如下几种可考虑的方案。

（1）静止的实体不负责碰撞检测。如游戏中静止的宝箱不应该定时检测玩家控制的英雄有没有与自己发生碰撞，这项工作应该交给二者中进行移动的一方即玩家自己来负责。

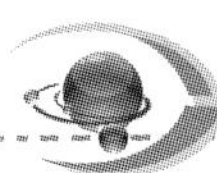

（2）只进行单向碰撞检测。如射击游戏中，不应该出现这样的检测算法：敌人射出的子弹会在移动中检测是否遇到了玩家控制的英雄，而玩家控制的英雄在移动中也会检测是否有子弹打中自己。这种算法首先是多此一举，降低游戏的性能；其次还会出现玩家被一颗子弹打中，受到双倍伤害（进行了两次碰撞处理）的奇怪现象。一般来说，碰撞检测应该由两个实体对象中主动的一方来进行。如对于子弹和英雄，是子弹击中英雄而不是英雄迎接子弹，所以应该有子弹负责二者的碰撞检测，而不是“被碰撞”的英雄。

（3）距离远的实体对象不进行碰撞检测。在游戏中，如果某两个实体之间的距离太远，在检测碰撞时会将较远的实体忽略，这样会对游戏的执行速度提高不少，实现这种策略主要有如下两种方式。

① 划分网格法。

将游戏地图划分为若干个格子，每个实体隶属于格子单元（如果有个横跨多个格子的实体，其隶属格子也只能有一个，如中心点所在的格子）。这样进行碰撞检测时，实体对象只需要检测隶属格子及相邻格子中的实体对象是否与自己发生碰撞即可，如图 5-5 所示。

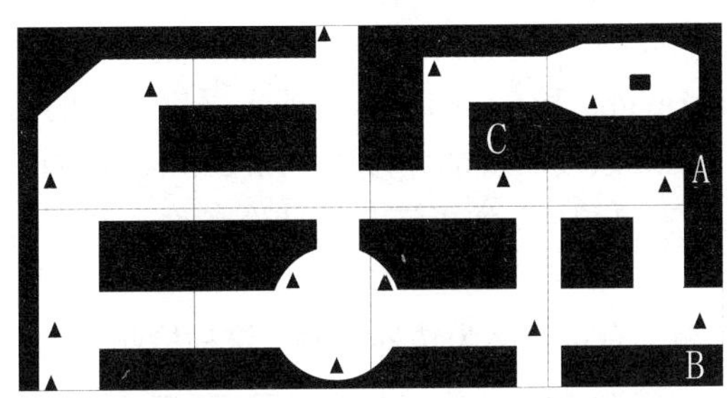

图 5-5　划分成若干格子的游戏地图

在图 5-5 中所示的游戏地图中，方块代表需要进行碰撞检测的实体，三角形表示被检测的实体，采用网格划分法对游戏地图进行划分后。方块所在的格子为 A，与其相邻的格子为 B 和 C，那么方块在进行碰撞检测的时候只需要与格子 A、B 和 C 中的三角形进行碰撞检测即可。

② 实体排序法。

这种方法的实现途径首先是保持被检测实体是有序的，如按照实体某个方向轴的坐标大小排列，这样在对实体序列中某一个实体进行碰撞检测时，只需要检测与该实体在实体序列中相邻的实体即可。不过这种方式需要消耗一定的性能来维护实体的有序性。

在尽可能减少了碰撞检测的次数后，下面讲解必须要进行的碰撞检测，它是通常采用的检测算法，主要有矩（圆柱）形检测和圆（球）形检测两种。

（1）矩（圆柱）形检测。

这种检测算法是给实体外层套上矩形（二维游戏中采用）或圆柱形（三维游戏中用），下面以二维游戏为例说明矩形检测的用法。首先为实体套上一个外接矩形框，在进行实体间碰撞检测时，只需检测两个实体的外接矩形是否发生了碰撞，如图 5-6 和图 5-7 所示。

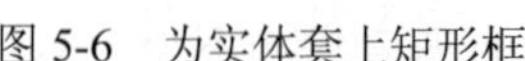

图 5-6　为实体套上矩形框

图 5-7　对实体进行矩形检测

具体检测的算法可描述为：取两个实体的左上角坐标（$x1, y1$）和（$x2, y2$）以及实体宽度 w 和高度 h；声明 4 个变量 max$x$、min$x$、max$y$ 和 min$y$，并将其分别赋值为两个实体中 x 坐标的最大值、x 坐标的最小值、y 坐标最大值和 y 坐标的最小值；判断是否 $maxx<minx+w$，且 $maxy<miny+h$，如果满足这两个条件，则说明两个实体发生碰撞，进行碰撞处理。

上述算法比较简单，但是前提条件是进行检测的两个实体的宽度和高度必须相同，有一些游戏中并不满足这些前提条件。那么就应该在上述第二个步骤中多记录几个变量 maxWidth、minWidth、maxHeight 和 minHeight。或者采用下面的第二种算法：取两个实体的左上角坐标（$x1, y1$）和（$x2, y2$）以及二者的宽度 $W1$、$W2$ 和高度 $H1$、$H2$；判断是否 $x1<x2+W2$，且 $x2<x1+W1$，$y1<y2+H2$，$y2<y1+H1$。如果满足这四个条件，则说明两个实体发生了碰撞，进行碰撞处理。

对于二维游戏中矩形检测，有可以称为边界检测，这种检测在大多数情况能够很好地满足游戏的需要。但是有些情况下两个实体的边界发生碰撞后，实体并没有发生碰撞，如图 5-8 所示。

解决图中出现的误差问题的一个办法就是在发生碰撞后计算重合面积，即当检测到实体的矩形框碰撞时，计算两个矩形框的重合面积，只有重合比例（即重合部分面积占整个矩形框面积的百分比）达到一定数值时才认定两个实体发生了碰撞。以上面介绍的第一矩形框碰撞检测算法为例，检测到矩形框碰撞后重合的面积，如图 5-9 所示。计算可以按照如下的步骤来进行：

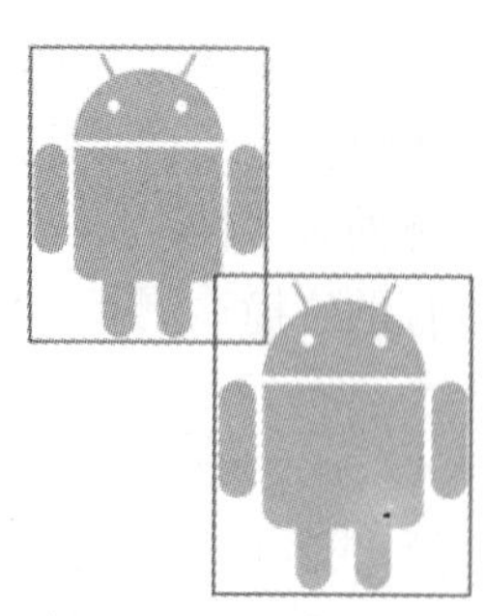

图 5-8　矩形碰撞产生误差示意图

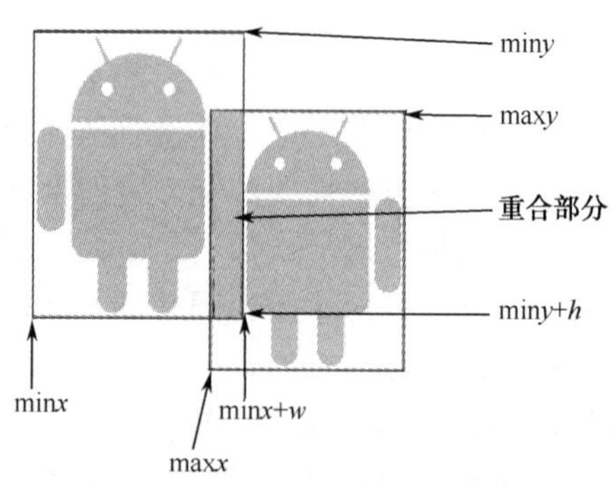

图 5-9　计算重合面积示意图

① 计算重合部分的宽度 W，其值为 $minx+W-maxx$。

② 计算重合部分的高度 H，其值为 $miny+H-maxy$。

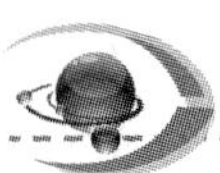

③ 计算重合部分的面积 $S=W\times H$。

④ 计算重合比例，如果超过某个值，则说明两个实体之间发生碰撞，进行碰撞处理。

（2）圆（球）形检测。

圆（球）形碰撞检测的算法是给实体对象套上圆形（二维游戏中采用）或球形（三维游戏中采用）的边框，如图 5-10 所示。下面以二维游戏为例说明圆形检测的用法。圆形检测就是在需要进行检测的实体上套上一个外接圆，进行碰撞检测时，只需要对相关实体的圆形框进行检测，如果两个圆相交，则这两个实体就发生碰撞，如图 5-11 所示。

图 5-10 为实体套上圆形框

图 5-11 对实体间进行圆形检测

圆形检测的碰撞检测算法可描述如下：取两个实体的中心点坐标（x_1, y_1）和（x_2, y_2）；取两个实体的半径，求其和为 R；计算两个实体中心点间的距离 D，使用 $\sqrt{(x_2-x_2)_2+(y_1-y_2)_2}$ 公式。将 D 与 R 对比，如果 D 小于 R，则二者发生了碰撞，进行碰撞处理。在实际开发中，为了提高性能，往往比较 D 和 R 的平方，这样免去了开发的计算。圆形检测实现起来比较简单，速度也够快，但是不够准确。例如两个实体的圆框已经碰上，但是实体之间并没有碰撞，这样就产生了误差。这种误差可能会给玩家比较糟糕的体验，如游戏中玩家还没有碰到怪物可能就已经受到伤害了。解决这种误差带来的问题可以在使用圆形检测到碰撞后进行更深入、全面的碰撞检测。

5.3.3 游戏实体对象与环境之间的碰撞检测

前面介绍了游戏中常见的实体对象之间的碰撞检测技术，游戏中实体对象不但要检测是否与其他实体对象发生碰撞，还需要检测是否与游戏环境发生碰撞。在此，我们就来简单介绍一些常用的游戏实体与环境的碰撞检测技术。

很多时候游戏的地图都是由图元（Tile）组成，而游戏中的实体与环境的检测就是与所遇到的地图中的图元进行判断。一般的做法是为实体对象设定一个定位点（如定位点在实体中心位置），在与地图进行碰撞检测时，先计算并获得定位点所在的位置对应地图中的那个格子的距离，然后判断该格子是否属于不可通过的图元对象，如果是，则发生碰撞，进行碰撞处理。这种算法有一个缺点，就是碰撞检测的可靠程度完全取决于定位点的选取，例如将定位点选在实体的中心位置，那么可能会出现如图 5-12 所示的情况。实体明明都已经陷入墙里面了，但是由于定位点隶属的地图图元仍然还是可以通过，碰撞检测的结果将会是没有发生碰撞。

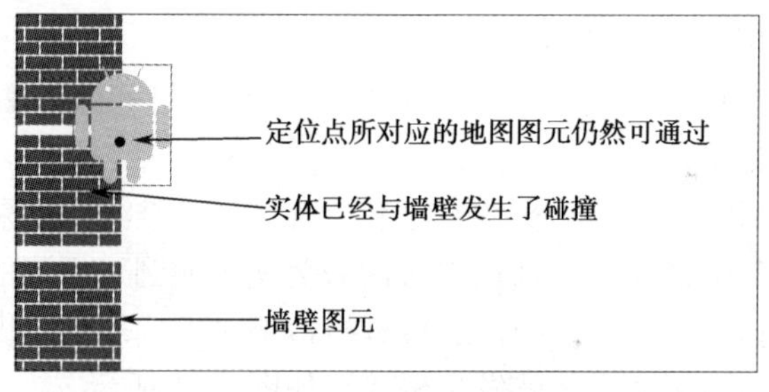

图 5-12　定位点在中心位置时的碰撞

如果将实体对象的定位点设置为左上角，则其在判断左上方向上的墙壁会比较灵敏，而在其他方向就变得不可靠了。因此这种碰撞检测方式不适合所有的场合，如果希望实体与环境之间的碰撞检测足够事实，就得采用如下的解决方案了。

（1）移动实体对象的位置。

（2）求出实体对象的左上、左下、右上、右下 4 个角的坐标。

（3）分别检测这 4 个点所对应的地图图元是否可以通过。只要有一个点检测到了不可通过的地图图元，则发生碰撞，进行碰撞处理。

上述的碰撞检测算法不仅适合于地图由图元构成的场合，对于其他场合也是适用的。有些情况下是不必对四角上的点进行检测的（如向左上运动），所以该算法还可以进行优化。

一般情况下，实体对象与环境发生碰撞后，实体对象的处理方式通常是退回到移动前的位置，这种算法并不是十分完美。比如实体对象当前的位置靠在墙边，运动方向为左下方运动，那么其在移动单位距离时必然会与墙壁发生碰撞，如图 5-13 所示。

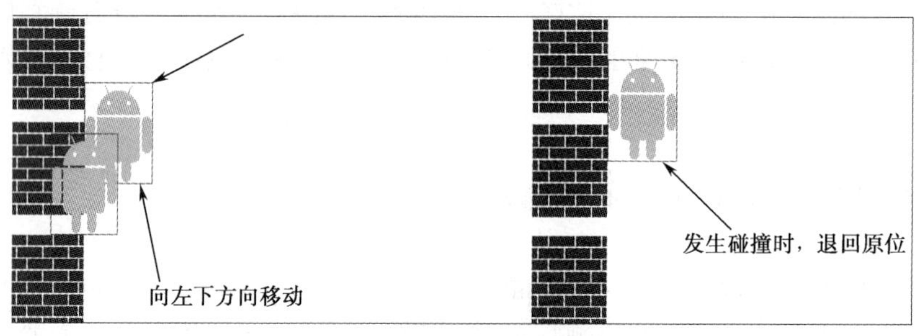

图 5-13　靠墙边运动时的碰撞

这时如果让实体退回到移动前的位置显得太不合逻辑了，正确的处理方式是让实体沿着墙壁向下移动一段距离。要想实现这种处理方式，在碰撞检测前就必须将 x、y 方向的判断单独进行，其检测步骤如下所示。

（1）根据指令在 x 方向上移动指定的距离，进行碰撞检测，如果发生碰撞，退回原位，x 方向上的移动无效。

（2）根据指令在 y 方向上移动指定的回距离，进行碰撞检测，如果发生碰撞，退回原位，y 方向上的移动无效。

采用上述算法来处理图 5-13 中的情况，实体在检测到 x 方向上无法向左移动后，并不影响其在 y 方向上的移动，实体会沿着墙壁向下滑动，而不是留在原位。对于实体不紧靠墙壁但是

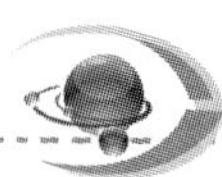

仍然会在一个单位移动距离内与墙壁发生碰撞的情况，其处理方式与实体紧靠墙壁时比较类似，只是需要对 *x* 方向的位置进行修正，使其在本次移动过程结束后 *x* 方向上紧靠墙壁。

下面我们通过一个物理小球在游戏中的应用实例，来具体理解和感受碰撞与检测的应用。本例中的小球为一个可移动物体 Movable 对象，该类中除了包含小球图片对象之外，还包括了如位置坐标、水平坐标、竖直速度等一系列用于模拟小球运动的成员变量和一些方法，这些方法将涉及到物理力学方面的知识，我们在项目开发中将做详细的介绍。

步骤一：项目创建，资源准备。在 Eclipse 中新建一个 Android 项目，名称为 Example_5_3，并在其 res\drawable-mdpi 目录下复制粘贴入项目开发需要应用到的图片资源，资源内容如图 5-14 所示。

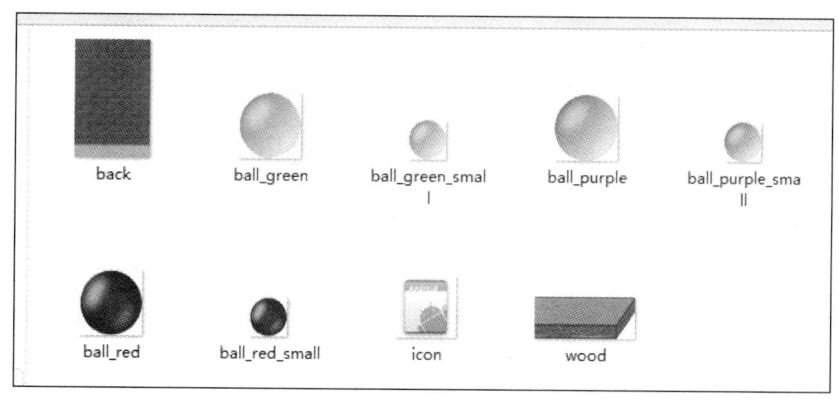

图 5-14　图片资源

步骤二：创建物理小球类。在 src 目录下创建文件 Movable.java，并在其中声明成员变量。由于小球在运动过程中涉及到物理计算，则类中声明了一些跟物理方面相关的成员变量，以及 Movable 类的构造方法和成员方法，在 Movable 类的构造方法中，将会对其部分成员变量进行初始化，并启动物理引擎 BallThread 类。Movable 类中包含一个成员方法 drawself()，该方法负责绘制小球到屏幕上。该类的主要参考代码如下：

```
public class Movable{
//声明记录小球每一个阶段(如从最高点下落到最低点)开始的初始 x,y 坐标，在以后小球的实时计算中，小球的实时 x,y 坐标会是初始坐标加上这段时间的位移。
    int startX=0,startY=0;
    int x,y;      //实时 X, Y 坐标
//声明小球在每一个阶段初始时刻在水平方向和竖直方向的速度，将用于计算小球的实时速度
    float startVX=0f, startVY=0f;//初始竖直和水平方向的速度
    float v_x=0f,   v_y=0f;      //实时水平和竖直方向速度
    int r;                  //可移动物体半径
    double timeX;               //X 方向上的运动时间
    double timeY;               //Y 方向上的运动时间
    Bitmap bitmap=null; //可移动物体图片
    BallThread bt=null; //负责小球移动的线程
    boolean bFall=false;//小球是否已经从木板上下落
    float impactFactor = 0.25f; //小球撞地后速度的损失系数
```

```
        //构造器
        public Movable(int x,int y,int r,Bitmap bitmap){
            this.startX = x;    this.x = x;
            this.startY = y;    this.y = y;
            this.r = r; this.bitmap = bitmap;
            timeX=System.nanoTime();     //获取系统时间初始化
    //初始化小球水平方向速度，这里用到的两个常量分别代表小球水平方向速度的最大值和最小值。
    及速度在最大值和最小值之间随机生成的一个数
            this.v_x = BallView.V_MIN + (int)((BallView.V_MAX-BallView.V_MIN)
*Math.random());
            bt = new BallThread(this);//创建并启动BallThread
            bt.start();}
        //方法：绘制自己到屏幕上
        public void drawSelf(Canvas canvas){
        canvas.drawBitmap(this.bitmap,x, y, null);} }
```

步骤三：物理引擎类开发。在src目录下建立BallThread类，该类继承Thread类，其功能是改变物理小球的运动轨迹。这里运用到了相关的物理知识，速度、加速度和位移的计算。速度是表示物体运动快慢的物理量，它是有方向的，该方向代表物体运动的方向；而加速度则是表示速度变化快慢的物理量，也是有方向的，代表物体速度是增加还是减少。速度与时间的乘积形成了位移。其计算位移公式：$s = v\times t + a\times t\times t/2$，速度公式：$v= v0 + a\times t$。该类主要代码如下：

```
    //继承自Thread的线程类，负责修改球的位置坐标
    public class BallThread extends Thread{
        Movable father;     //Movable对象引用
        boolean flag = false;   //线程执行标志位
        int sleepSpan = 30;     //休眠时间
        float g = 200;      //球下落的加速度
        double current; //记录当前时间
        //构造器：初始化Movable对象引用及线程执行标志位
        public BallThread(Movable father){
            this.father = father;
            this.flag = true;       //设置线程执行的标志位为true
        }
        //方法：负责根据物理公式修改小球位置
        public void run(){
            while(flag){
                current = System.nanoTime();//获取当前时间，单位为纳秒
                double timeSpanX =
                (double)((current-father.timeX)/1000/1000/1000);
                //获取从玩家开始到现在水平方向走过的时间
                //处理水平方向上的运动
        father.x = (int)(father.startX + father.v_x * timeSpanX);
                //处理竖直方向上的运动
        if(father.bFall){//判断球是否已经移出挡板
```

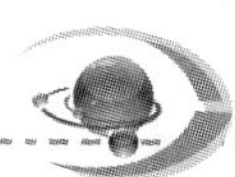

```
        double timeSpanY = (double)((current - father.timeY)/1000/1000/1000);
father.y = (int)(father.startY + father.startVY * timeSpanY +
timeSpanY*timeSpanY*g/2);
        father.v_y = (float)(father.startVY + g*timeSpanY);
        //判断小球是否到达最高点
    if(father.startVY < 0 && Math.abs(father.v_y) <= BallView.UP_ZERO){
    father.timeY = System.nanoTime();  //设置新的运动阶段竖直方向上的开始时间
    father.v_y = 0;       //设置新的运动阶段竖直方向上的实时速度
    father.startVY = 0; //设置新的运动阶段竖直方向上的初始速度
    father.startY = father.y; }//设置新的运动阶段竖直方向上的初始位置
            //判断小球是否撞地
    if(father.y + father.r*2 >= BallView.GROUND_LING && father.v_y >0){
    //改变水平方向的速度，衰减水平方向上的速度
    father.v_x = father.v_x * (1-father.impactFactor);
        //改变竖直方向的速度，衰减竖直方向上的速度并改变方向
    father.v_y = 0 - father.v_y * (1-father.impactFactor);
        //判断撞地后的速度，太小就停止
    if(Math.abs(father.v_y) < BallView.DOWN_ZERO){
    this.flag = false;}
            else{   //撞地后的速度还可以弹起继续下一阶段的运动
                                  //撞地之后水平方向的变化
        father.startX = father.x;   //设置新的运动阶段的水平方向的起始位置
    father.timeX = System.nanoTime();  //设置新的运动阶段的水平方向的开始时间
                                  //撞地之后竖直方向的变化
        father.startY = father.y;       //设置新的运动阶段竖直方向上的起始位置
    father.timeY = System.nanoTime();  //设置新的运动阶段竖直方向开始运动的时间
    father.startVY = father.v_y;    //设置新的运动阶段竖直方向上的初速度
                    }   }       }
    else if(father.x+father.r/2>= BallView.WOOD_EDGE){//判断球是否移出了挡板
    father.timeY = System.nanoTime();       //记录球竖直方向上的开始运动时间
            father.bFall = true;     }    //设置表示是否开始下落标志位
            try{Thread.sleep(sleepSpan);    }    //休眠一段时间
                catch(Exception e){e.printStackTrace();}}}}
```

步骤四：画面渲染视图类开发。在src目录下新建文件BallView类，在该类中声明了一些物理计算时要使用的静态常量，同时还声明了程序中要绘制的图片资源以及要绘制的小球对象列表，其主要代码如下：

```
    public class BallView extends SurfaceView implements SurfaceHolder.
    Callback{
        public static final int V_MAX=35;   //小球水平速度的最大值
        public static final int V_MIN=15;   //小球竖直速度的最大值
        public static final int WOOD_EDGE = 60;//木板的右边沿的x坐标
        public static final int GROUND_LING = 450;//游戏中代表地面的y坐标，小球
        下落到此会弹起
```

```
public static final int UP_ZERO = 30;  //小球在上升过程中，如果速度小于该
值就算为 0
public static final int DOWN_ZERO = 60;//小球在撞击地面后，如果速度小于该
值就算为 0
Bitmap [] bitmapArray = new Bitmap[6]; //各种颜色形状的小球图片引用
Bitmap bmpBack;     //背景图片对象
Bitmap bmpWood;     //木板图片对象
String fps="FPS:N/A";        //用于显示帧速率的字符串，调试使用
int ballNumber =8; //小球数目
ArrayList<Movable> alMovable = new ArrayList<Movable>();//小球对象数组
DrawThread dt; //后台屏幕绘制线程

public BallView(Context activity){
    super(activity);                 //调用父类构造器
    getHolder().addCallback(this);
    initBitmaps(getResources());     //初始化图片
    initMovables();                  //初始化小球
    dt = new DrawThread(this,getHolder()); }   //初始化重绘线程
//方法：初始化图片
public void initBitmaps(Resources r){
    bitmapArray[0] = BitmapFactory.decodeResource(r, R.drawable.ball_
    red_small); //红色较小球
    . . . 此处省略其他图片初始化代码
bmpBack = BitmapFactory.decodeResource(r, R.drawable.back);
bmpWood = BitmapFactory.decodeResource(r, R.drawable.wood);    }
//方法：初始化小球
public void initMovables(){
Random r = new Random();     //创建一个 Random 对象
for(int i=0;i<ballNumber;i++){
int index = r.nextInt(32);        //产生随机数
Bitmap tempBitmap=null;           //声明一个 Bitmap 图片引用
if(i<ballNumber/2){
tempBitmap = bitmapArray[3+index%3];//如果是初始化前一半球，就从大球中随机
找一个
        }else{
tempBitmap = bitmapArray[index%3]; }//如果是初始化后一半球，就从小球中随机
找一个
Movable m = new Movable(0,70-tempBitmap.getHeight(),tempBitmap
.getWidth()/2, tempBitmap); //创建 Movable 对象
alMovable.add(m);   //将新建的 Movable 对象添加到 ArrayList 列表中
    }}
//方法：绘制程序中所需要的图片等信息
public void doDraw(Canvas canvas) {
    canvas.drawBitmap(bmpBack, 0, 0, null);//绘制背景图片
```

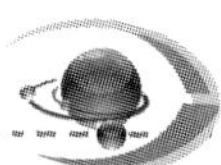

```
        canvas.drawBitmap(bmpWood, 0, 60, null);//绘制木板图片
        for(Movable m:alMovable){   //遍历 Movable 列表，绘制每个 Movable 对象
            m.drawSelf(canvas);}
        Paint p = new Paint();  //创建画笔对象
        p.setColor(Color.BLUE); //为画笔设置颜色
        p.setTextSize(18);      //为画笔设置字体大小
        p.setAntiAlias(true);   //设置抗锯齿
        canvas.drawText(fps, 30, 30, p); } //画出帧速率字符串
    @Override
    public void surfaceChanged(SurfaceHolder holder, int format, int width,
    int height) {}//重写 surfaceChanged 方法
    @Override
    public void surfaceCreated(SurfaceHolder holder) {//从写 surfaceCreated
    方法
        if(!dt.isAlive()){  //如果 DrawThread 没有启动，就启动这个线程
            dt.start();}}
    @Override
    public void surfaceDestroyed(SurfaceHolder holder) {//重写 surface
    Destroyed 方法
        dt.flag = false;    //停止线程的执行
        dt = null;          //将 dt 指向的对象声明为垃圾
    }   }
```

步骤五：屏幕绘制和帧速率计算类。在 src 目录下建立 DrawThread 类，该类继承自 Thread 类，主要功能负责重绘屏幕和计算帧速率。其主要代码如下：

```
public class DrawThread extends Thread{
    BallView bv;                    //BallView 对象引用
    SurfaceHolder surfaceHolder;//SurfaceHolder 对象引用
    boolean flag=false;             //线程执行标志位
    int sleepSpan = 30;             //休眠时间
long start = System.nanoTime();//记录起始时间，该变量用于计算帧速率
    int count=0;            //记录帧数，该变量用于计算帧速率
    //构造器
    public DrawThread(BallView bv,SurfaceHolder surfaceHolder){
    this.bv = bv;           //为 BallView 对象应用赋值
    this.surfaceHolder = surfaceHolder; //为 SurfaceHolder 对象应用赋值
        this.flag = true;   }   //设置标志位
    //方法：线程的执行方法，用于绘制屏幕和计算帧速率
    public void run(){
        Canvas canvas = null;//声明一个 Canvas 对象
        while(flag){
            try{
    canvas = surfaceHolder.lockCanvas(null);//获取 BallView 的画布
                synchronized(surfaceHolder){
```

```
            bv.doDraw(canvas); }}//调用 BallView 的 doDraw 方法进行绘制
            catch(Exception e){e.printStackTrace();}//捕获并打印异常
            finally{
            if(canvas != null){      //如果 canvas 不为空
        surfaceHolder.unlockCanvasAndPost(canvas); }}  //surfaceHolder 解锁并
        将画布对象传回
            this.count++;
            if(count == 20){     //如果计满 20 帧
            count = 0;       //清空计数器
            long tempStamp = System.nanoTime();//获取当前时间
            long span = tempStamp - start;       //获取时间间隔
            start = tempStamp;                   //为 start 重新赋值
        double fps = Math.round(100000000000.0/span*20)/100.0;//计算帧速率
    bv.fps = "FPS:"+fps;//将计算出的帧速率设置到 BallView 的相应字符串对象中
            }
                try{Thread.sleep(sleepSpan); }         //线程休眠一段时间
    catch(Exception e){ e.printStackTrace();}}}}     //捕获并打印异常
```

步骤六：Activity 代码开发。前面几个步骤把该实例的主要功能全部完成，最后需要开发 Activity 代码，检测实例运行的效果。该代码非常简单，和前面实例的代码大同小异，这里不做详细介绍，其代码复制如下：

```
    public class Example_5_3 extends Activity {
     BallView bv;         //BallView 对象引用
     @Override
     public void onCreate(Bundle savedInstanceState) {
        super.onCreate(savedInstanceState);
    requestWindowFeature(Window.FEATURE_NO_TITLE);//设置不显示标题
        getWindow().setFlags(              //设置为全屏模式
            WindowManager.LayoutParams.FLAG_FULLSCREEN,
            WindowManager.LayoutParams.FLAG_FULLSCREEN);
        bv = new BallView(this);      //创建 BallView 对象
        setContentView(bv); //将屏幕设置为 BallView 对象
    } }
```

到此为止，项目实例开发结束，让我们启动模拟器，看看实际的效果，如图 5-15 所示。

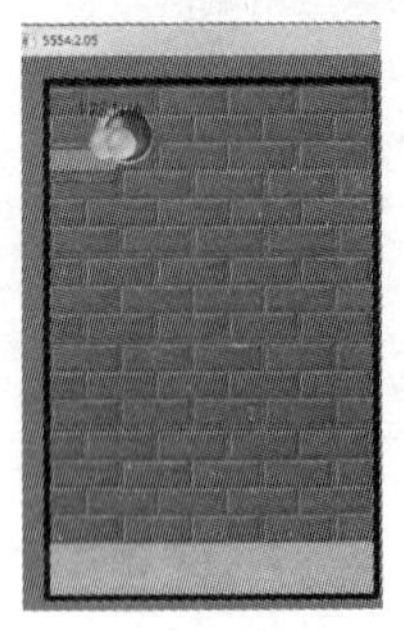
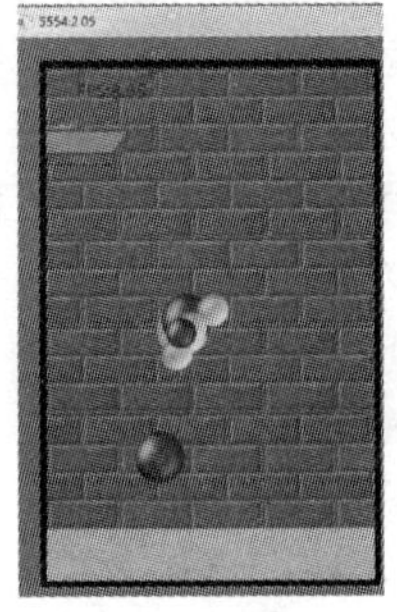

图 5-15 物理小球碰撞效果图

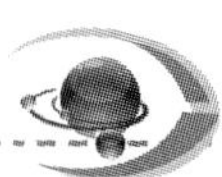

5.4　扫雷游戏实例

前面我们对 Android 游戏应用开发的一些基础进行了介绍，接下来介绍一个 Android 做的扫雷小游戏。该游戏的开发步骤如下。

步骤一：建立项目，准备资源。打开 Eclipse 开发工具，建立一个名称为 Main 的 Android 项目。准备相应的图片资源，如图 5-16 所示，存放在 res\drawable-mdpi 目录下。

i_cry	i_flag	i_flag_down	i_happy	i00
i01	i02	i03	i04	i05
i06	i07	i08	i09	i10
i11	i12	i13	i14	icon

图 5-16　图片资源

步骤二：编写游戏运行的界面类。打开 Main.java 文件，编写代码如下：

```
public class Main extends Activity {
  GameView gameView;
    @Override
    public void onCreate(Bundle icicle) {
        super.onCreate(icicle);
        gameView = new GameView(this);
        setContentView(gameView);}}
```

步骤三：编写游戏的主逻辑界面，新建 GameView.java 类，编写代码如下：

```
public class GameView extends View {
private static final String TAG = "GameView";
private RefreshHandler mRedrawHandler = new RefreshHandler();
class RefreshHandler extends Handler {
    @Override
    public void handleMessage(Message msg) {
        GameView.this.updateView();
        GameView.this.invalidate();}
public void sleep(long delayMillis) {
```

```
                this.removeMessages(0);
                sendMessageDelayed(obtainMessage(0), delayMillis);
            }};
        // 游戏状态 开始 胜利 失败 暂停
        public static final int STATE_PLAYING = 0;//声明游戏开始的状态常量
        public static final int STATE_WIN = 1;//声明游戏胜利的状态常量
        public static final int STATE_LOST = 2;//声明游戏失败的状态常量
        public static final int STATE_PAUSE = 3;//声明游戏暂停的状态常量
        public int gameState;   //声明表示游戏状态的变量
        private static final Random random = new Random();
        private Paint paint;//画笔
        private String message;
        private static final int tileWidth = 24;//声明地雷背景图片的宽度
        private static final int tileHeight = 24;//声明地雷背景图片的高度
        private static final int tilesCount = 19;//图片数组的元素个数
        private static final int margin = 16;
        private static final int titleHeight = 30;//声明标题栏的高度
        private Bitmap[] tiles;
        private int[][] mapSky;
        private int[][] mapGround;
        private int tileCountX, tileCountY;//背景地图的列数和行数
        private int offsetX, offsetY;
        int mineCount;//地雷数量
        int safeCount;
        private int mapX, mapY;
        private boolean altKeyDown = false;
        private static final int MILLIS_PER_TICK = 500;
        long startTime, lastTime;
        long time, remain;
        public GameView(Context context) {//构造方法
            super(context);
            //设置画笔的属性
            paint = new Paint();
            paint.setARGB(255, 60, 60, 200);//设置画笔颜色
            paint.setTextSize(12);//设置字体大小
            tiles = new Bitmap[tilesCount];//tilesCount = 19;
            loadTiles();//初始化图片信息
            setFocusable(true); }//获得焦点
    /* 将资源图片的ID放入tiles数组中*/
        private void loadTiles() {
            Resources r = this.getContext().getResources();
            for (int i = 0; i < tilesCount; i++) {//初始存下19个不同的图片
            tiles[i] = BitmapFactory.decodeResource(r, R.drawable.i00 + i);
                //十六进制自动编号
```

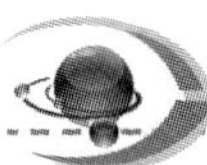

```
            }}
/* 初始化第一层 */
    @Override
    protected void onLayout(boolean changed, int left, int top, int right,
            int bottom) {
        // TODO Auto-generated method stub
        super.onLayout(changed, left, top, right, bottom);
        // if(Main.gameState == Main.STATE_START)
        init(right - left, bottom - top);
        startTime = System.currentTimeMillis();//获取游戏开始时间
        updateView();}
/* 初始化地图，根据屏幕的长宽计算地图中格子的数量和边缘的宽度
    * @param 宽
    * @param 高*/
    private void init(int w, int h) {
        mapX = -1;   mapY = -1;
    tileCountX = (int) Math.floor((w - margin * 2) / tileWidth);//计算列数
    tileCountY = (int) Math.floor((h - margin * 2 - titleHeight) /
    tileHeight);//计算行数
    offsetX = (w - (tileWidth * tileCountX)) / 2;//计算地雷前的空
    offsetY = (h - (tileHeight * tileCountY) + titleHeight) / 2;//计算地雷
    上面的空
    mineCount = (int) Math.sqrt(tileCountX * tileCountY) * tileCountX
            * tileCountY / 100;//计算地雷个数
        reset();}
/*  初始化两层地图，并放雷*/
    private void reset() {
        int x, y;
        mapSky = new int[tileCountX][tileCountY];
        mapGround = new int[tileCountX][tileCountY];
        for (int i = 0; i < mineCount; i++) {
            // 随机生产雷放置在二维数组中的位置
        do {
            x = random.nextInt(tileCountX);
            y = random.nextInt(tileCountY);
            System.out.println(x+"="+y);
            System.out.println("="+mapGround[x][y]);
        } while (mapGround[x][y] == 12);//避免在重复的位置布雷，设置循环
            mapGround[x][y] = 12;//12 指的是有雷的那幅图片
          //周围 8 个相邻的方格子
            increase(x - 1, y - 1);  increase(x - 1, y);
            increase(x - 1, y + 1);   increase(x, y + 1);
            increase(x + 1, y + 1);   increase(x + 1, y);
            increase(x + 1, y - 1);   increase(x, y - 1);}
```

```
        for (x = 0; x < tileCountX; x++) {
            for (y = 0; y < tileCountY; y++) {
                if (mapGround[x][y] == 0)
                    mapGround[x][y] = 9;}}
        safeCount = tileCountX * tileCountY - mineCount;//安全的方格数
        time = 0;
        remain = mineCount;//剩余雷数
        altKeyDown = false;
        // shuffle();
            }
    private void increase(int x, int y) {
        if (x > -1 && x < tileCountX && y > -1 && y < tileCountY) {
            if (mapGround[x][y] != 12)
                mapGround[x][y]++;}}
    @Override
    public void onDraw(Canvas canvas) {
        super.onDraw(canvas);
        // Log.v(TAG, "onDraw");
        if (altKeyDown) {//红旗的图标是否未按下样式
            canvas.drawARGB(255, 255, 0, 0);
            canvas.drawBitmap(tiles[17], 30, 0, paint);
        } else {
            canvas.drawBitmap(tiles[16], 30, 0, paint);}
        if (gameState != STATE_LOST) {//判断笑脸与哭脸
            canvas.drawBitmap(tiles[18], 0, 0, paint);
        } else {
            canvas.drawBitmap(tiles[15], 0, 0, paint);}
        message = "剩余: " + remain + "  耗时" + time + "秒";
        canvas.drawText(message, 0, message.length(), 80, 15, paint);
        for (int x = 0; x < tileCountX; x += 1) {
            for (int y = 0; y < tileCountY; y += 1) {
            Rect rDst = new Rect(offsetX + x * tileWidth, offsetY + y
                * tileHeight, offsetX + (x + 1) * tileWidth, offsetY
                + (y + 1) * tileHeight);
            //根据数组中的数字对应画出相应的图片
        canvas.drawBitmap(tiles[mapGround[x][y]], null, rDst, paint);
        if (gameState != STATE_LOST) {
            if (mapSky[x][y] > -1) {
        canvas.drawBitmap(tiles[mapSky[x][y]], null, rDst,paint); }
                } else {
            if (mapGround[x][y] != 12 && mapSky[x][y] == 10) {
                mapSky[x][y] = 14; }//游戏失败后，指明插红旗的雷
            if (mapSky[x][y] > -1 && mapGround[x][y] != 12
                || mapSky[x][y] == 13 || mapSky[x][y] == 10) {
```

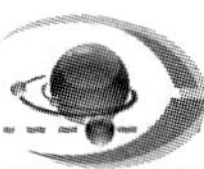

```
canvas.drawBitmap(tiles[mapSky[x][y]], null, rDst, paint); }
    }}}}//失败后保持状态的格子
private void updateView() {
    if (gameState == STATE_PLAYING) {
        time = (System.currentTimeMillis() - startTime) / 1000;
        mRedrawHandler.sleep(MILLIS_PER_TICK);}}
//手机屏幕事件处理方法
@Override
public boolean onTouchEvent(MotionEvent event) {
    int action = event.getAction();
    int x = (int) event.getX();
    int y = (int) event.getY();
//屏幕被按下
    if (action == MotionEvent.ACTION_DOWN) {
    mapX = screenX2mapX(x);//得到事件所点击方块在数组中的二维下标
        mapY = screenY2mapY(y);
        if (gameState != STATE_LOST && mapX > -1 && mapY > -1) {
            if (gameState == STATE_PAUSE) {
                gameState = STATE_PLAYING;
                startTime = System.currentTimeMillis();
                updateView();}
            if (altKeyDown) {//当Alt按下时 插旗
                if (mapSky[mapX][mapY] == 0) {
                    mapSky[mapX][mapY] = 10;
                    remain--;
                } else if (mapSky[mapX][mapY] == 10) {
                    remain++;
                    mapSky[mapX][mapY] = 11;
                } else if (mapSky[mapX][mapY] == 11) {
        mapSky[mapX][mapY] = 0;
        } else if (mapSky[mapX][mapY] == -1) {//模拟扫雷左右键同时按下，
        sky层没有被初始化的时候为-1
    int flags = flag(mapX - 1, mapY - 1)+ flag(mapX - 1, mapY)
            + flag(mapX - 1, mapY + 1)+ flag(mapX, mapY + 1)
            + flag(mapX + 1, mapY + 1)+ flag(mapX + 1, mapY)
            + flag(mapX + 1, mapY - 1)+ flag(mapX, mapY - 1);
        if (flags == mapGround[mapX][mapY]) {
            open(mapX - 1, mapY - 1); open(mapX - 1, mapY);
            open(mapX - 1, mapY + 1); open(mapX, mapY + 1);
            open(mapX + 1, mapY + 1); open(mapX + 1, mapY);
            open(mapX + 1, mapY - 1); open(mapX, mapY - 1);}}
        } else {
            open(mapX, mapY);}
            invalidate();
```

```
            } else {
                if (x < 26 && y < 26) {
                    gameState = STATE_PAUSE;
                    reset();
                    invalidate();}
                if (x > 30 && x < 56 && y > 0 && y < 26) {
                    altKeyDown = !altKeyDown;
                    invalidate();}  }}
        return true;}
    private int flag(int x, int y) {
        if (x > -1 && x < tileCountX && y > -1 && y < tileCountY) {
            if (mapSky[x][y] == 10)
                return 1;}
        return 0;}
    public void open(int x, int y) {
        if (x > -1 && x < tileCountX && y > -1 && y < tileCountY) {
            if (mapSky[x][y] == -1)
                return;
            if (mapSky[x][y] == 0 || mapSky[x][y] == 11) {
                if (mapGround[x][y] == 12) {
                    mapSky[x][y] = 13;
                    gameState = STATE_LOST;
                } else {
                mapSky[x][y] = -1;
                safeCount--;
            if (safeCount == 0) {
                gameState = STATE_WIN;}
            if (mapGround[x][y] == 9) {
                open(x - 1, y - 1);   open(x - 1, y);
                open(x - 1, y + 1);   open(x, y + 1);
                open(x + 1, y + 1);   open(x + 1, y);
                open(x + 1, y - 1);   open(x, y - 1);}}}}}}
    private int screenX2mapX(int c) {//得到横轴所在的格子
        if (c - offsetX < 0)
            return -1;
        int rtn = (c - offsetX) / tileWidth;
        if (rtn >= tileCountX)
            return -1;
        return rtn;}
    private int screenY2mapY(int c) {//得到纵轴所在的格子
        if (c - offsetY < 0)
            return -1;
        int rtn = (c - offsetY) / tileHeight;
        if (rtn >= tileCountY)
```

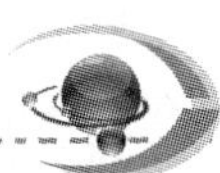

```
            return -1;
        return rtn;}}
```

步骤四：运行程序，查看结果。当我们运行程序时，出现图 5-17 的效果，并开始计时，当用户点击界面方格，没有点到地雷，就出现如图 5-18 所示有空格的现象，如果用户确定某一个方格为地雷，可以点击第一行的红旗图标，就可以为方格插上地雷图标。当用户不小心点到地雷了，则出现如图 5-19 效果。

图 5-17　运行程序

图 5-18　未碰撞到地雷

图 5-19　点到地雷后

第 6 章

Android 消息与广播

Intent 是轻量级的进程间通信机制，用于跨进程的组件通信和发送系统级的广播。通过本章的学习可以让读者基本了解 Android 系统的组件通信原理，掌握如何利用组件通信启动其他组件的方法，以及利用组件通信信息和发送广播消息的方法。

6.1 Intent

Intent 是一个动作的完整描述，包含了动作的产生组件、接收组件和传递的数据信息。当然，也可以将 Intent 视为一个在不同组件之间传递的消息，这个消息在到达接收组件后，接收组件会执行相关的动作。

由于 Intent 的存在，使得 Android 系统中相互独立的应用程序组件成为了一个可以互相通信的组件集合。因此，无论这些组件是否在同一个应用程序中，Intent 都可以将一个组件的数据和动作传递给另一个组件。

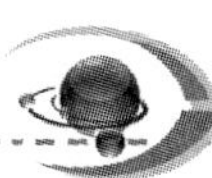

Intent 为 Activity、Service 和 BroadcastReceiver 等组件提供交互能力。Intent 的一个最常见的用途就是启动 Activity 和 Service，另一个用途是在 Android 系统上发布广播消息。广播消息可以是接收的特定数据或者消息，也可以是手机的信号变化或电池的电量过低等信息。

Intent 是由组件名称、Action、Data、Category、Extra 及 Flag 六部分组成的。接下来将分别对其进行详细介绍。

组件名称实际上就是一个 ComponentName 对象，用于标识唯一的应用程序组件，即指明了期望的 Intent 组件，这种对象的名称是由目标组件的类名与目标组件的包名组合而成的。在 Intent 传递过程中，组件名称是一个可选项，当指定它时，便是显式的 Intent 消息，当不指定它时，Android 系统则会根据其他信息及 IntentFilter 的过滤条件选择相应的组件。

Action 实际上就是一个描述了 Intent 所触发动作名称的字符串，在 Intent 类中，已经定义好很多字符串常量来表示不同的 Action，当然，开发人员也可以自定义 Action，其定义规则同样非常简单。系统定义的 Action 常量有很多，下面只列出其中一些较常用的以供参考。

ACTION_CALL：拨出 Data 里封装的电话号码。

ACTION_EDIT：打开 Data 里指定数据所对应的编辑程序。

ACTION_VIEW：打开能够显示 Data 中封装的数据的应用程序。

ACTION_MAIN：声明程序的入口，该 Action 并不会接收任何数据，同时结束后也不会返回任何数据。

ACTION_BOOT_COMPLETED：BroadcastReceiverAction 的常量，表明系统启动完毕。

ACTION_TIME_CHANGED：BroadcastReceiverAction 的常量，表明系统时间通过设置而改变。

Data 主要是对 Intent 消息中数据的封装，主要描述 Intent 的动作所操作的数据的 URI 及类型。不同类型的 Action 会有不同的 Data 封装，例如打电话的 Intent 会封装 tel:格式的电话 URI，而 ACTION_VIEW 的 Intent 中的 Data 则会封装 http:格式的 URI。正确的 Data 封装对 Intent 匹配请求同样非常重要。

Category 是对目标组件类别信息的描述。同样作为一个字符串对象，一个 Intent 中可以包含多个 Category。与 Category 相关的方法有三个，addCategory()添加一个 Category，removeCategory()删除一个 Category，而 getCategory()得到一个 Category。Android 系统同样定义了一组静态字符常量来表示 Intent 的不同类型，下面列出一些常见的 Category 常量。

CATEGORY_GADGET：表明目标 Activity 是可以嵌入到其他 Activity 中的。

CATEGORY_HOME：表明目标 Activity 是 HOMEActivity。

CATEGORY_TAB：表明目标 Activity 是 TabActivity 的一个标签下的 Activity。

CATEGORY_LAUNCHER：表明目标 Activity 是应用程序中最先被执行的 Activity。

CATEGORY_PREFERNCE：表明目标 Activity 是一个偏好设置的 Activity。

Extra 中封装了一些额外的附加信息，这些信息是以键值对的形式存在的。Intent 可以通过 putExtras()与 getExtras()方法来存储和获取 Extra。在 Android 系统的 Intent 类中，同样对一些常用的 Extra 键值进行定义。

EXTRA_BCC：装有邮件密送地址的字符串数组。

EXTRA_EMAIL：装有邮件发送地址的字符串数组。

EXTRA_UID：使用 ACTION_UID_REMOVED 动作时，描述删除用户的 id。

EXTRA_TEXT：当使用 ACTION_SEND 动作时，描述要发送文本的信息。

Flag 是指一些有关系统如何启动组件的标志位，Android 同样对其进行了封装。

6.1.1 启动 Activity

在 Android 系统中，应用程序一般都有多个 Activity，Intent 可以实现不同 Activity 之间的切换和数据传递。Intent 启动 Activity 方式可以分为显式启动和隐式启动。显式启动必须在 Intent 中指明启动的 Activity 所在的类，而隐式启动则由 Android 系统，根据 Intent 的动作和数据来决定启动哪一个 Activity。也是说在隐式启动时，Intent 中只包含需要执行的动作和所包含的数据，而无须指明具体启动哪一个 Activity，选择权由 Android 系统和最终用户来决定。

使用 Intent 来显式启动 Activity，首先需要创造一个 Intent，并为它指定当前的应用程序上下文以及要启动的 Activity，把创建好的这个 Intent 作为参数传递给 startActivity()方法。具体格式如下：

```
Intent intent = new Intent(IntentDemo.this, ActivityToStart.class);
startActivity(intent);
```

下面用 IntentDemo 示例说明如何使用 Intent 启动新的 Activity。IntentDemo 示例包含两个 Activity 类，分别是 IntentDemo 和 ActivityToStart。程序启动是默认启动 IntentDemo 这个 Activity，如图 6-1 所示。在用户单击“启动 Activity”按钮后，程序启动 ActivityToStart 这个 Activity，如图 6-2 所示。

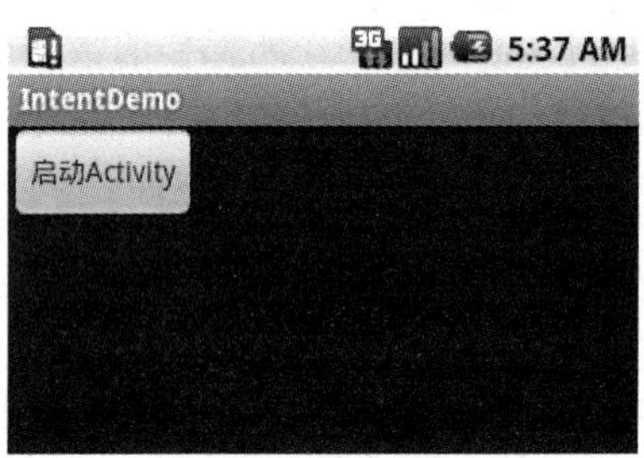

图 6-1 默认启动 IntentDemo

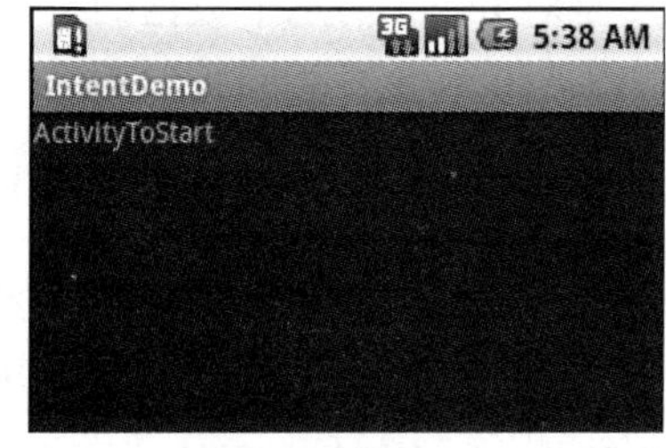

图 6-2 ActivityToStart

在 IntentDemo 示例中使用了两个 Activity，因此需要在 AndroidManifest.xml 文件中注册两个 Activity。注册 Activity 应使用<activity>标签，嵌套在<application>标签内部。

AndroidManifest.xml 文件代码如下：

```
<?xml version="1.0"encoding="utf-8"?>
<manifest xmlns:android="http://schemas.android.com/apk/res/android"
  package="xsc.xsc.IntentDemo"
  android:versionCode="1"
  android:versionName="1.0">
 <application android:icon="@drawable/icon"
android:label="@string/app_name">
   <activity android:name=".IntentDemo"
```

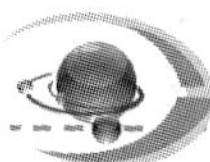

```
        android:label="@string/app_name">
        <intent-filter>
          <action android:name="android.intent.action.MAIN" />
        <category android:name="android.intent.category.LAUNCHER" />
        </intent-filter>
    </activity>
      <activity android:name=".ActivityToStart"
    android:label="@string/app_name"/>
     </application>
       <uses-sdk android:minSdkVersion="7" />
     </manifest>
```

在 IntentDemo.Java 文件中，包含了显示使用 Intent 启动 Activity 的核心代码如下：

```
Button button = (Button)findViewById(R.id.btn);
button.setOnClickListener(new OnClickListener(){
    public void onClick(View view){
         Intent intent = new Intent(IntentDemo.this,
ActivityToStart.class);
         startActivity(intent);
    } });
```

在点击事件的处理函数中，Intent 构造函数的第 1 个参数是应用程序上下文，程序中的应用程序上下文就是 IntentDemo；第 2 个参数是接收 Intent 的目标组件，使用的是显示启动方式，直接指明了需要启动的 Activity。

隐式启动的好处在于不需要指明要启动哪一个 Activity，而由 Android 系统来决定，这样有利于使用第三方组件。

隐式启动 Activity 时，Android 系统在应用程序运行时解析 Intent，并根据一定的规则对 Intent 和 Activity 进行配置，使 Intent 上的动作、数据与 Activity 完全吻合。匹配的 Activity 可以是应用程序本身的，也可以是 Android 系统内置的，还可以是第三方应用程序提供的。因此，这种方式更加强调了 Android 应用程序中组件的可复用性。

例如：用户希望启动一个浏览器，查看指定的网页内容，却不能确定具体应该启动哪一个 Activity，此时可以使用 Intent 的隐式启动方式，由 Android 系统在程序运行时决定具体启动哪一个应用程序的 Activity 来接收这个 Intent。程序开发人员可以将浏览动作和 Web 地址作为参数传递给 Intent，Android 系统则通过匹配动作和数据格式，找到最适合于此动作的组件。在默认情况下，Android 系统会调用内置的 Web 浏览器。如访问百度网页的小例子代码。

```
Intent intent = new Intent(Intent.ACTION_VIEW,
Uri.parse("http://www.baidu.com"));
startActivity(intent);
```

Intent 的动作是 Intent.ACTION_VIEW，根据 URI 的数据类型来匹配动作。数据部分的 URI 是 Web 地址，使用 Uri.parse(urlString)方法，可以简单地把一个字符串解释成 URI 对象。Intent 的语法如下:

```
Intent intent = new Intent(Intent.ACTION_VIEW, Uri.parse(urlString));
```

Intent 构造函数的第 1 个参数是 Intent 需要执行的动作，Android 系统支持的常见动作字符串常量可以参考表 6-1。第 2 个参数是 URI，表示需要传递的数据。

表 6-1　Intent 常用动作表

动　作	说　明
ACTION_ANSWER	打开接听电话的 Activity，默认为 Android 内置的拨号盘界面
ACTION_CALL	打开拨号盘界面并拨打电话，使用 URI 中的数字部分为电话号码
ACTION_DELETE	打开一个 Activity，对所提供的数据进行删除操作
ACTION_DIAL	打开内置拨号盘界面，显示 URI 中提供的电话号码
ACTION_EDIT	打开一个 Activity，对提供的数据进行编辑操作
ACTION_INSERT	打开一个 Activity，在提供的数据的当前位置插入新项
ACTION_PICK	启动一个子 Activity，从提供的数据列表中选取一项
ACTION_SEARCH	启动一个 Activity，执行搜索动作
ACTION_SENDTO	启动一个 Activity，向数据提供的联系人发送信息
ACTION_SEND	启动一个可以发送数据的 Activity
ACTION_VIEW	最常用的动作，对以 URI 方式传送的数据，根据 URI 协议部分以最佳方式启动相应的 Activity 进行处理。对于 http:address 将打开浏览器查看；对于 tel:address 将打开拨号呼叫指定的电话号码
ACTION_WEB_SERCH	打开一个 Activity，对提供的数据进行 Web 搜索

下面通过一个 WebViewIntentDemo 示例说明了如何隐式启动 Activity，用户界面如图 6-3 所示。

当用户在文本框中输入要访问网址后，通过单击“浏览器 URL”按钮，程序根据用户输入的网址生成一个 Intent，并以隐式启动调用 Android 内置 Web 浏览器，并打开指定的 Web 页面。本例输入的网址是百度地址，地址是 http://www.baidu.com，打开页面后的效果如图 6-4 所示。

图 6-3　用户界面图

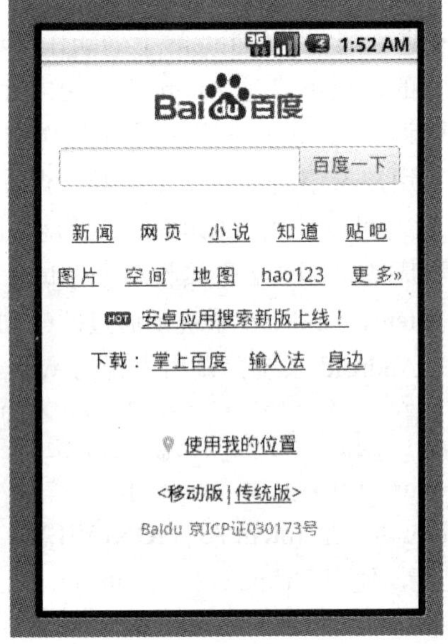

图 6-4　效果图

6.1.2　获取 Activity 返回值

在上节的 IntentDemo 示例中，通过 startActivity(Intent)方式启动 Activity，启动后的两个Activity 之间相互独立，没有任何的关联。在很多情况下，后启动的 Activity 是为了让用户对特定信息进行选择，在关闭这个 Activity 后，用户的选择信息需要返回给未关闭的那个 Activity。按照 Activity 启动的先后顺序，先启动的称为父 Activity，后启动的称为子 Activity。如果需要将子 Activity 的部分信息返回给父 Activity，则可以使用 Sub-Activity 的方式去启动子 Activity。

获取子 Activity 的返回值，一般可以分为以下三个步骤：

（1）以 Sub-Activity 的方式启动子 Activity。

（2）设置子 Activity 的返回值。

（3）在父 Activity 中获取返回值。

下面详细介绍每一个步骤的过程和代码实现。

1. 以 Sub-Activity 的方式启动子 Activity。

以 Sub-Activity 方式启动子 Activity，需要调用 startActivityForResult(Intent,requestCode)，参数 Intent 用于启动哪个 Activity，参数 requestCode 是唯一的标识子 Activity 的请求码。因为所有子 Activity 返回时，父 Activity 都调用相同的处理函数，因此父 Activity 使用 requestCode 来确定是哪一个子 Activity 的返回值。

显式启动子 Activity 的代码如下：

```
int SUBACTIVITY1 = 1;
Intent intent = new Intent(this, SubActivity1.class);
startActivityForResult(intent, SUBACTIVITY1);
```

隐式启动子 Activity 的代码如下：

```
int SUBACTIVITY2 = 2;
Uri uri = Uri.parse("content://contacts/people");
Intent intent = new Intent(Intent.ACTION_PICK, uri);
startActivityForResult(intent, SUBACTIVITY2);
```

2. 设置子 Activity 的返回值。

在子 Activity 调用 finish()函数关闭前，调用 setResult()函数将所需的数据返回给父 Activity。setResult()函数有两个参数，一个是结果码，一个是返回值。结果码表明了子 Activity 的返回状态，通常为 Activity.RESULT_OK 或者 Activity.RESULT_CANCELED，也可以是自定义的结果码，结果码均为整数类型。返回值封装在 Intent 中，也就是说子 Activity 通过 Intent 将需要返回的数据传递给父 Activity。数据主要以 URI 形式提供的返回给父 Activity，此外还可以附加一些额外信息，这些额外信息用 Extra 的集合表示。

以下代码说明如何在子 Activity 中设置返回值：

```
Uri data = Uri.parse("tel: " + tel_number);
Intent result = new Intent(null, data);
```

```
result.putExtra("address", "");
setResult(RESULT_OK, result);
finish();
```

3. **在父 Activity 中获取返回值。**

当子 Activity 关闭时，启动它的父 Activity 的 onActivityResult()函数将被调用。如果需要在父 Activity 中处理子 Activity 的返回值，则重载此函数即可。此函数的语法如下：

```
public void onActivityResult(int requestCode, int resultCode, Intent data);
```

其中第 1 个参数 requestCode，用来表示是哪一个子 Activity 的返回值，第 2 个参数 resultCode 用于表示子 Activity 的返回状态，第 3 个参数 data 是子 Activity 的返回数据，返回数据类型是 Intent。根据返回数据的用途不同，URI 数据的协议则不同，也可以使用 Extra 方法返回一些原始类型的数据。

以下代码说明如何在父 Activity 中处理子 Activity 的返回值：

```
private static final int SUBACTIVITY1 = 1;
private static final int SUBACTIVITY2 = 2;
@Override
public void onActivityResult(int requestCode, int resultCode, Intent data){
    Super.onActivityResult(requestCode, resultCode, data);
    switch(requestCode){
        case SUBACTIVITY1:
            if (resultCode == Activity.RESULT_OK){
                Uri uriData = data.getData();
            }else if (resultCode == Activity.RESULT_CANCEL){ }
        break;
case SUBACTIVITY2:
        if (resultCode == Activity.RESULT_OK){
            Uri uriData = data.getData(); }
    break;  }
  }
```

接下来我们通过 AcrivityCommunication 示例说明如何以 Sub-Activity 方式启动子 Activity，以及如何使用 Intent 进行组件间通信，该示例的用户界面如图 6-5 所示。

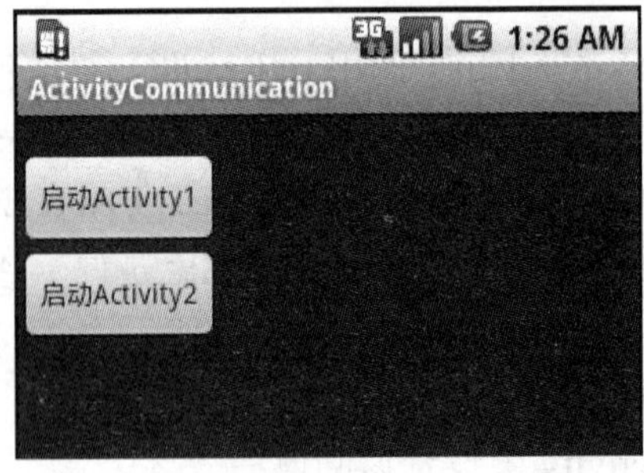

图 6-5　ActivityCommunication 用户界面图

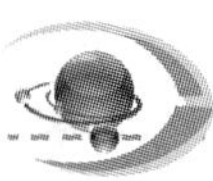

当用户单击“启动 Activiry1”和“启动 Activity2”按钮时，程序将分别启动子 SubActivity1 和 SubActivity2，如图 6-6 所示。SubActivity1 提供了一个输入框，以及“接收”和“撤销”两个按钮。如果在输入框中输入信息后单击“接受”按钮，程序会把输入框中的信息传递给其父 Activity，并在父 Activity 的界面上显示。而如果用户单击“撤销”按钮，则程序不会向父 Activity 传递任何信息。SubActivity2 主要是为了说明如何在父 Activity 中处理多个子 Activity，因此仅提供了用于关闭 AubActivity2 的“关闭”按钮，如图 6-7 所示。

图 6-6　SubActivity1 效果图

图 6-7　SubActivity2 效果图

ActivityCommunication 示例的文件结构如图 6-8 所示，父 Activity 的代码在 Activity Commmunication.java 文件中，界面布局在 main.xml 中：两个子 Sctivity 的代码分别在 SubActivity1.java 和 SubActivity2.java 文件中，界面布局分别在 subactivity1.xml 和 subactivity2.xml 中。

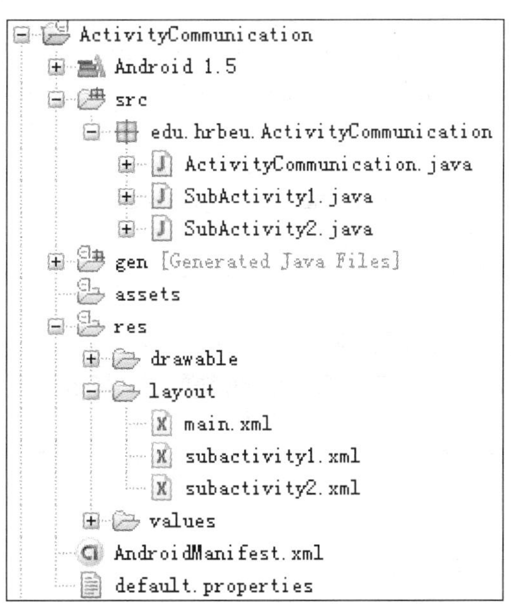

图 6-8　ActivityCommunication 示例项目结构图

ActivtityCommunicantion.java 文件的核心代码如下：

```
public class ActivityCommunication extends Activity {
//分别定义了两个子 Activity 的请求码
```

```
    private static final int SUBACTIVITY1 = 1;
    private static final int SUBACTIVITY2 = 2;
    TextView textView;
    @Override
    public void onCreate(Bundle savedInstanceState) {
            super.onCreate(savedInstanceState);
            setContentView(R.layout.main);
textView = (TextView)findViewById(R.id.textShow);
            final Button btn1 = (Button)findViewById(R.id.btn1);
            final Button btn2 = (Button)findViewById(R.id.btn2);
        btn1.setOnClickListener(new OnClickListener(){
        public void onClick(View view){
//以 Sub-Activity 的方式分别启动子 Activity
Intent intent = new Intent(ActivityCommunication.this,
SubActivity1.class);
        startActivityForResult(intent, SUBACTIVITY1);
            } });
    btn2.setOnClickListener(new OnClickListener(){
        public void onClick(View view){
    Intent intent = new Intent(ActivityCommunication.this,
 SubActivity2.class);
        startActivityForResult(intent, SUBACTIVITY2);
    } }); }
   @Override//子 Activity 关闭后的返回值处理函数
    protected void onActivityResult(int requestCode,int resultCode,Intent
    data){
    super.onActivityResult(requestCode, resultCode, data);
//对结果码进行判断，如果等于 RESULT_OK,获取子 Activity 的返回值中的数据
textView.setText("jytytjytyyjyyyjty");
  switch(requestCode){
    case SUBACTIVITY1:
if (resultCode == RESULT_OK){
// data 是返回值，子 Activity 需要返回的数据就是保存在 data 中
    Uri uriData = data.getData();
            textView.setText(uriData.toString());
                }
                break;
        case SUBACTIVITY2:
                 break;
        } }   }
```

SubActivity1.java 的核心代码如下：

```
public class SubActivity1 extends Activity {
    @Override
    public void onCreate(Bundle savedInstanceState) {
```

```
        super.onCreate(savedInstanceState);
        setContentView(R.layout.subactivity1);
        final EditText editText = (EditText)findViewById(R.id.edit);
        Button btnOK = (Button)findViewById(R.id.btn_ok);
        Button btnCancel = (Button)findViewById(R.id.btn_cancel);
        btnOK.setOnClickListener(new OnClickListener(){
            public void onClick(View view){
                String uriString = editText.getText().toString();
          //将EditText控件的内容作为数据保存在URI中
                Uri data = Uri.parse(uriString);
                Intent result = new Intent(null, data);
//将Intent作为返回值，RESUIT_OK作为结果码，通过调用setResult()函数，将//返回值
和结果码传递给父Activity
                setResult(RESULT_OK, result);
//调用finish()函数关闭当前的子Activity
                finish();   }});
    btnCancel.setOnClickListener(new OnClickListener(){
        public void onClick(View view){
            setResult(RESULT_CANCELED, null);
            finish();} });    }
    }
```

SubActivity2.java 的核心代码如下：

```
public class SubActivity2 extends Activity {
    @Override
    public void onCreate(Bundle savedInstanceState) {
        super.onCreate(savedInstanceState);
        setContentView(R.layout.subactivity2);
 Button btnReturn = (Button)findViewById(R.id.btn_return);
        btnReturn.setOnClickListener(new OnClickListener(){
            public void onClick(View view){
// setResult()函数仅设置结果码，第2个参数为null，表示数据不需要传递给父Activity
                setResult(RESULT_CANCELED, null);
                finish();
            }   }); } }
```

6.2 Intent 过滤器

在隐式启动 Activity 时，并没有在 Intent 中指明 Activity 所在的类，因此，Android 一定存在某种匹配机制，使 Android 系统能够根据在 Intent 中的数据信息，找到需要启动的 Activity。这种匹配机制的实现就靠 Android 系统中的 Intent 过滤器（IntentFilter）实现的。

Intent 过滤器是一种根据 Intent 中的动作（Action）、类别（Categorie）和数据（Data）等内容，对适合接收该 Intent 的组件进行匹配和筛选的机制。Intent 过滤器还可以匹配数据类型、路径和协议，还包括可以用来确定多个匹配项顺序的优先级（Priority）。应用程序的 Activity 组件、Service 组件和 BroadcastReceiver 都可以注册 Intent 过滤器。这样，这些组件在特定的数据格式上便可以产生相应的动作。

为了使组件能够注册 Intent 过滤器，通常在 AandroidManifest.xml 文件的各个组件的节点下定义<intent-filter>节点，然后在<intent-filter>节点声明该组件所支持的动作、执行的环境和数据格式等信息。当然，也可以在程序代码中动态地为组件设置 Intent 过滤器。<intent-filter>节点支持<action>标签、<category>标签和<data>标签，分别用来定义 Intent 过滤器的“动作”、“类别”和“数据”。IntentFilter 过滤 Intent 时，一般就是通过 Action、Data 及 Category 三方面进行检测的。一个 Intent 只能设置一种 Action，但是一个 IntentFilter 却可以设置多个 Action 过滤，当 IntentFilter 设置了多个 Action，只需一个满足即可完成 Action 验证。当 IntentFilter 中没有说明任何一个 Action 时，那么任何的 Action 都不会与之匹配，而如果 Intent 中没有包含任何 Action，那么只要 IntentFilter 中含有 Action 时，便会匹配成功。数据的检测只要包含两部分，即数据的 URI 及数据类型，而数据 URI 又被分成三部分进行匹配（scheme、authority、path），只有这些全部匹配时，Data 的验证才会成功。IntentFilter 同样可以设置多个 Category，当 Intent 中的 Category 与 IntentFilter 中的一个 Category 完全匹配时，便会通过 Category 的检查，而其他的 Category 并不受影响。但是当 IntentFilter 没有设置 Category 时，只能与没有设置 Category 的 Intent 相匹配。

<intent-filter>节点支持的标签和属性说明参考表如表 6-2。

表 6-2　<intent-filter>节点属性参考表

标　签	属　性	说　明
<action>	android:name	指定组件所能响应的动作，用字符串表示，通常使用 Java 类名和包的完全限定名构成
<category>	android:category	指定以何种方式去服务 Intent 请求的动作
<data>	android:host	指定一个有效的主机名
	android:mimetype	指定组件能处理的数据类型
	android:path	有效的 URI 路径名
	android:port	主机的有效端口号
	android:scheme	所需要的特定的协议

<category>标签用来指定 Intent 过滤器的服务方式，每个 Intent 过滤器可以定义多个<category>标签，程序开发人员可以使用自定义的类别，或使用 Android 系统提供的类别。Android 系统提供的类别可以参考表 6-3。

表 6-3　Android 系统提供的类别表

值	说　明
ALTERNATIVE	Intent 数据默认动作的一个可替换的执行方法
SELECTED_ALTERNATIVE	和 ALTERNATIVE 类似，但替换的执行方法不是指定的，而是被解析出来的

续表

值	说　明
BROWSABLE	声明 Activity 可以由浏览器启动
DEFAULT	为 Intent 过滤器中定义的数据提供默认动作
HOME	设备启动后显示的第一个 Activity
LAUNCHER	在应用程序启动时首先被显示

AndroidManifest.xml 文件中的每个组件的<intent-filter>都被解析成一个 Intent 过滤器对象。当应用程序安装到 Android 系统时，所有的组件和 Intent 过滤器都会注册到 Android 系统中。这样，Android 系统便知道了如何将任意一个 Intent 请求通过 Intent 过滤器映射到相应的组件上。

这种 Intent 到 Intent 过滤器的映射过程为“Intent 解析”。Intent 解析可以在所有的组件中，找到一个可以与请求的 Intent 达成最佳匹配的 Intent 过滤器。Android 系统中 Intent 解析的匹配规则如下：

（1）Android 系统把所有应用程序包中的 Intent 过滤器集合在一起，形成一个完整的 Intent 过滤器列表。

（2）在 Intent 与 Intent 过滤器进行匹配时，Android 系统将列表中所有 Intent 过滤器的“动作”和“类别”与 Intent 进行匹配，任何不匹配的 Intent 过滤器将过滤掉。没有指定“动作”的 Intent 过滤器可以匹配任何的 Intent，但没有指定“类别”的 Intent 过滤器只能匹配没有“类别”的 Intent。

（3）把 Intent 数据 URI 的每个子部与 Intent 过滤器的<data>标签中的属性进行匹配，如果<data>标签指定了协议、主机名、路径名或 MIME 类型，那么这些属性都要与 Intent 的 URI 数据部分进行匹配，任何不匹配的 Intent 过滤器均被过滤掉。

（4）如果 Intent 过滤器的匹配结果多于一个，则可以根据在<intent-filter>标签中定义的优先级标签来对 Intent 过滤器进行排序，优先级最高的 Intent 过滤器将被选择。

接下来我们通过 IntentResolutionDemo 示例来说明如何在 AndroidManifest.xml 文件中注册 Intent 过滤器，以及如何设置<intent-filter>节点属性来捕获指定的 Intent。AndroidManifest.xml 完整代码如下：

```
<?xml version="1.0" encoding="utf-8"?>
<manifest xmlns:android="http://schemas.android.com/apk/res/android"
    package="edu.hrbeu.IntentResolutionDemo"
    android:versionCode="1"
    android:versionName="1.0">
<application android:icon="@drawable/icon"
 android:label="@string/app_name">
      <activity android:name=".IntentResolutionDemo"
           android:label="@string/app_name">
<intent-filter>
    <action android:name="android.intent.action.MAIN" />
        <category android:name="android.intent.category.LAUNCHER" />
    </intent-filter>
```

```
            </activity>
            <activity android:name=".ActivityToStart"
                    android:label="@string/app_name">
              <intent-filter>
            // Activity 的 Intent 过滤器的动作是 android.intent.action.VIEW
                <action android:name="android.intent.action.VIEW" />
            //Activity 的 Intent 过滤器的类别，DEFAULT 表示数据的默认动作
                <category android:name="android.intent.category.DEFAULT" />
            // Activity 的 Intent 过滤器的数据的协议部分和数据的主机名称部分
                <data android:scheme="schemodemo" android:host="edu.hrbeu" />
              </intent-filter>
             </activity>
             </application>
              <uses-sdk android:minSdkVersion="7" />
             </manifest>
```

在 IntentResolutionDemo.java 文件中，定义了一个 Intent 用来启动一个新的 Activity，这个 Intent 与 Activity 设置的 Intent 过滤器是完全匹配的。IntentResolutionDemo.java 文件中 Intent 实例化和启动 Activity 的代码如下：

```
        Intent intent = new Intent(Intent.ACTION_VIEW, Uri.parse("schemodemo://edu.
hrbeu/path"));
        startActivity(intent);
```

代码所定义的 Intent，动作为 Intent.ACTION_VIEW，与 Intent 过滤器的动作 android.Intent.action.VIEW 匹配；URI 是“schemodemo://edu.hrbeu/path”，其中的协议部分为“schemodemo”，主机名部分为“edu.hrbeu”，也与 Intent 过滤器列表进行匹配时，会与 AndroidManifest.xml 文件中 ActivityToStart 定义的 Intent 过滤器完全匹配。

6.3 BroadcastReceive 组件应用

Intent 的另一种用途发送广播消息。应用程序和 Android 系统都可以使用 Intent 发送广播消息，广播消息的内容可以是与应用程序密切相关的数据信息，也可以是 Android 的系统信息，例如网络连接变化，电池电量变化、接收到短信和系统设置变化等。如果应用程序注册了 BroadcastReceiver，则可以接收到指定的广播消息。BroadcastReceiver 类位于 android.content 包下，是对广播消息进行过滤并响应的控件。

使用 Intent 广播消息非常简单，只需创建一个 Intent，并调用 sendBroadcast()函数就可把 Intent 携带的消息广播出去。但需要注意的是，在构造 Intent 时必须用一个全局唯一的字符串标识其执行的动作，通常使用应用程序包的名称。如果要在 Intent 传递额外数据，可以用 Intent 的 putExtra()方法。下面的代码构造用于广播消息的 Intent，并添加了额外的数据，然后调用 senBroadcast()发生了广播消息：

```
        String UNIQUE_STRING = "edu.hrbeu.BroadcastReceiverDemo";
        Intent intent = new Intent(UNIQUE_STRING);
        intent.putExtra("key1", "value1");
```

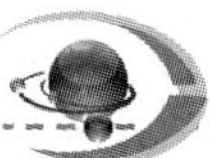

```
    intent.putExtra("key2", "value2");
    sendBroadcast(intent);
```

BroadcastReceiver 用于监听广播消息，可以在 AndroidManifest.xml 文件或在代码注册一个 BroadcastReceiver，并在其中使用 Intent 过滤器指定要处理的广播消息。创建 BroadcastReceiver 需要继承 BroadcastReceiver 类，并重载 onReceive()方法。示例代码如下：

```
    public class MyBroadcastReceiver extends BroadcastReceiver {
        @Override
        public void onReceive(Context context, Intent intent) {
            //TODO: React to the Intent received.
        }    }
```

注册 BroadcastReceiver 的应用程序不需要一直运行，当 Android 系统接收到与之匹配的广播消息时，会自动启动此 BroadcastReceiver，因此 BroadcastReceiver 非常适合做一些资源管理工作。在 BroadcastReceiver 接收到与之匹配的广播消息后，onReceive()方法会被调用，但 onReceive()方法必须要在 5 秒钟执行完毕，否则 Android 系统会认为该组件失去响应，并提示用户强行关闭该组件。

BroadcastReceiver 的运行机理比较简单，应用程序注册了 BroadcastReceiver 之后，当系统或其他应用程序发送广播时，所有已经注册的 BroadcastReceiver 就会检查注册时的 IntentFilter 是否与发送的 Intent 相匹配，若匹配则会调用 BroadcastReceiver 的 onReceive()方法进行处理，所以在开发与 BroadcastReceiver 相关的程序时，主要的工作是对 onReceive()方法的实现。BroadcastReceiver 发送广播有三种发送方式，各种发送方式的不同点如下所示。

（1）用 sendBroadcast 和 sendStickyBroadcast 发送方式发送的广播，所有满足条件的 BroadcastReceiver 都会执行其 onReceive()方法来处理响应，但若有多个满足条件的 BroadcastReceiver 时，其执行 onReceive()方法的顺序是没有保证的。

（2）通过 sendOrderedBroadcast 发送出去的 Intent，会根据 BroadcastReceive 注册时 IntentFilter 设置的优先级顺序来执行 onReceive()方法，而相同优先级的 BroadcastReceiver 会执行 onReceive()方法的顺序同样是没有保证的。

（3）sendStickyBroadcast 的主要不同是，Intent 在发送后一直存在，并且在以后调用 registerReceive 注册相匹配的 Receive 时会把这个 Intent 直接返回给新注册的 Receive。

然后就是接收方式，接收消息时，需要通过 IntentFilter 对象来过滤，然后交给相应的 BroadcastReceive 对象来处理。接下来简单介绍广播服务的接收程序的开发过程。

（1）开发 BroadcastReceive 类的子类，重写其中的 onReceive 方法。

（2）为应用程序添加适当的权限。

（3）注册 BroadcastReceive 对象，可以在 Java 代码中注册，也可以在 AndroidManifest.xml 中声明。

（4）等待接收广播。

接下来我们通过 BroadcastReceiverDemo 示例来具体说明了如何在应用程序中注册 BroadcastReceiver，并接收指定类型的广播消息。BroadcastReceiverDemo 示例的界面如图 6-9 所示，在单击“发送广播消息”按钮后，EditText 控件中内容将以广播消息的形式发送出去，示例内部的 BroadcastReceiver 将接收这个广播消息，并显示在用户界面的下方。

图 6-9　BroadcastReceiverDemo 示例界面图

BroadcastReceiverDemo.java 文件中包含发送广播消息的代码，其关键代码如下：

```
    button.setOnClickListener(new OnClickListener(){
        public void onClick(View view){
    //创建 Intent 时，将 edu.hrbeu. BroadcastReceiverDemo 作为识别广播消息的字符串标识
            Intent intent = new Intent("edu.hrbeu.BroadcastReceiverDemo");
    //添加额外信息
            intent.putExtra("message", entryText.getText().toString());
            sendBroadcast(intent);// 调用 sendBroadcast()函数发送广播消息
        }});
```

为了能够使用程序中的 BroadcastReceiver 接收指定的广播消息，首先要在 AndroidManifest.xml 文件中添加 Intent 过滤器，声明 BroadcastReceiver 可以接收的广播消息。AndroidManifest.xml 文件的完整代码如下：

```
    <?xml version="1.0" encoding="utf-8"?>
    <manifest xmlns:android="http://schemas.android.com/apk/res/android"
        package="edu.hrbeu.BroadcastReceiverDemo"
        android:versionCode="1"
        android:versionName="1.0">
        <application android:icon="@drawable/icon"
        android:label="@string/app_name">
            <activity android:name=".BroadcastReceiverDemo"
                    android:label="@string/app_name">
                <intent-filter>
                <action android:name="android.intent.action.MAIN" />
                    <category android:name="android.intent.category.LAUNCHER" />
                </intent-filter>
        </activity>
        <receiver android:name=".MyBroadcastReceiver">
```

```
            <intent-filter>
        //声明Intent过滤器的动作为"edu. hrbeu.  BroadcastReceiverDemo",这与Broadcast
ReceiverDemo.java 文件中 Intent 的动作相一致，声明这个 BroadcastReceiver 可以接收动作为
"edu.hrbeu. BroadcastReceiverDemo" 的广播消息。
            <action android:name="edu.hrbeu.BroadcastReceiverDemo" />
           </intent-filter>
         </receiver>
        </application>
        <uses-sdk android:minSdkVersion="7" />
      </manifest>
```

MyBroadcastReceiver.java 文件创建一个自定义的 BroadcastReceiver，其核心代码如下：

```
        //创建BroadcastReceiver的子类
        public class MyBroadcastReceiver extends BroadcastReceiver {
           @Override
        //重载onReveive()函数,当接收到AndroidManifest.xml文件定义的广播消息后,程序将自
        动调用onReveive()函数
           public void onReceive(Context context, Intent intent) {
        //通过调用getStringExtra()函数，从Intent中获取标识为message的字符串数据
              String msg = intent.getStringExtra("message");
        //提示信息
              Toast.makeText(context, msg, Toast.LENGTH_SHORT).show();
           }
        }
```

第 7 章

Service 后台服务

Service 是 Android 系统的后台服务组件，适用于开发无界面、长时间运行的应用功能。通过本章节的学习让我们了解后台服务的基本原理，掌握进程内服务与跨进程服务使用方法，有助于深入了解 Android 系统的进程间通信机制。

7.1 Service 组件应用

因为手机屏幕尺寸的限制，通常情况下在同一时刻仅有一个应用程序处于激活状态，并能够显示在手机屏幕上。因此，应用程序需要一种机制，在没有用户界面的情况下，使其能够长时间在后台运行，实现应用程序的特定功能，并能够处理事件或更新数据。Android 系统提供的 Service（服务）组件，不直接与用户进行交互，并能够长期在后台运行。有很多情况需要使用 Service，经常提到的例子就是 MP3 播放器，用户需要在关闭播放器界面后，仍然能够保持音乐继续播放，这就需要将音乐回放的功能在 Service 组件实现。

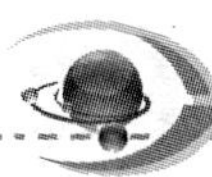

Service 非常适用于无须用户干预，且需要长期运行的后台功能。Service 没有用户界面，有利于降低系统资源的消耗，而且 Service 比 Activity 具有更高的优先级，因此在系统资源紧张时，Service 不会轻易被 Android 系统终止。即使 Service 被系统终止，在系统资源恢复后，Service 也将自动恢复运行状态，因此可以认为 Service 是在系统中永久运行的组件。Service 除了可以实现后台服务功能，还可以用于进程间通信（Inter Process Communication，IPC），解决两个不同 Android 应用程序之间的调用和通信问题。

Service 的生命周期比较简单，仅包括全生命周期和活动生命周期，还有三个事件回调函数，分别是 onCreate()、onStart()和 onDestroy()，如图 7-1 所示。

图 7-1　Service 生命周期图

Service 生命周期从 onCreate()开始到 onDestroy()结束，在 onCreate()中完成 Service 初始化工作，在 onDestroy()中释放所有占用的资源。活动生命周期从 onStart()开始，但没有与之对应的“停止”函数，因此可以近似认为活动周期也是以 onDestroy()标志结束。

Service 的使用方式一般有两种，一种是启动方式，另一种是绑定方式。在启动方式中，通过调用 Context.startService()启动 Service，通过调用 Context.stopService()或 Service.stopSefl()停止 Service。因此，Service 一定是由其他的组件启动的，但停止过程可以通过其他组件或自身完成。在启动方式中，启动 Service 的组件不能获取到 Service 对象，因此无法调用 Service 中实现的方法，也不能获取 Service 中任何状态和数据信息。因此，如果仅以启动方式使用 Service，这个 Service 需要具备自管理的能力，且不需要通过函数调用向外部组件提供数据或功能。

在绑定方式中，Service 的使用是通过服务链接（Connection）实现的，服务链接能获取 Service 对象，因此绑定 Service 组件可以调用 Service 中实现的函数，或直接获取 Service 中状态和数据信息。使用 Service 的组件通过 Context.bindService()建立服务链接，通过 Context.unbindService()停止服务链接。如果在绑定过程中 Service 没有启动，Context.bindService()会自动启动 Service。而且同一个 Service 可以绑定多个服务链接，这样可以同时为多个不同的组件提供服务。

当然，这两种使用方法并不是完全独立的，可以在某些情况下混合使用启动方式和绑定方式。还是以 MP3 播放器为例，在后台工作的 Service 通过 Context.startService()启动某个特定音乐播放，但在播放过程中如果用户需要暂停音乐播放，则需要通过 Context.bindService()获取服务链接和 Service 对象，进而通过调用 Service 对象中的函数，暂停音乐播放过程，并保存相关信息。在这种情况下，如果调用 Context.stopService()不能停止 Service，需要在所有服务链接关闭后，Service 才能真正停止。

7.2 进程内服务

7.2.1 服务管理

服务管理主要指服务的启动和停止，在介绍如何启动和停止服务前，首先说明如何在代码中实现 Service。Service 是一段在后台运行、没有用户界面代码。其最小代码集如下：

```
import android.app.Service;
import android.content.Intent;
import android.os.IBinder;
//声明 RandomService 继承了 android.app.Service 类
public class RandomService extends Service{
   @Override
// onBind()函数是在 Service 被绑定后调用的函数，能够返回 Service 的对象
public IBinder onBind(Intent intent) {
return null;
}  }
```

这个 Service 最小代码集并不能完成任何实际的功能，为了使 Service 具有实际意义，一般需要重载 onCreate()、onStart()和 onDestroy()。Android 系统在创建 Service 时，会自动调用 onCreate()，用户一般在 onCreate()方法中完成必要的初始化工作，例如创建线程、建立数据库链接等。在 Service 没有必要再存在时，系统会自动调用 onDestroy()，用户在 onDestroy()释放所有占用的资源。通过 Context.startService(Intent)启动 Service 时，onStart()则会被系统调用，Intent 会传递给 Service 一些重要的参数。当然，不是所有的 Service 都需要重载这三个函数，完全可以根据实际情况选择需要重载的那个函数。

```
public class RandomService extends Service{
       @Override
   public void onCreate() {
        super.onCreate();  }
       @Override
   public void onStart(Intent intent, int startId) {
        super.onStart(intent, startId); }
       @Override
   public void onDestroy() {
        super.onDestroy(); }
}
```

重载 onCreate()、onStart()和 onDestroy()三个函数时，务必要在代码中调用父函数。完成 Service 类后，需要在 AndroidManifest.xml 文件注册这个 Service。这个注册过程非常重要，如果用户不注册，则这个 Service 根本无法启动。AndroidManifest.xml 文件中注册 Service 的代码如下：

```
<service android:name=".RandomService"/>
```

使用<service>标签声明服务，其中的 android：name 表示的是 Service 的类名称，一定要与

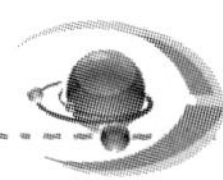

用户建立的 Service 类名称一致。

在完成 Service 代码和注册后，下一步来说明如何启动和停止 Service。有两种方法启动 Service，显式启动和隐式启动。显式启动需要在 Intent 中指明 Service 所在的类，并调用 startService(Intent)函数启动 Service，示例代码如下：

```
final Intent serviceIntent = new Intent(this, RandomService.class);
startService(serviceIntent);
```

在上面的代码中，在 Intent 中指明了启动的 Service 在 RandomService.class 中。隐式启动则需要在注册 Service 时，声明 Intent-filter 的 action 属性。

```
<service android:name=".RandomService">
    <intent-filter>
        <action android:name="edu.hrbeu.RandomService" />
    </intent-filter>
</service>
```

在隐式启动 Service 时，需要设置 Intent 的 action 属性，这样则可以在不声明 Service 所在类的情况下启动服务。隐式启动的代码如下：

```
final Intent serviceIntent = new Intent();
serviceIntent.setAction("edu.hrbeu.RandomService");
startService(serviceIntent);
```

如果 Service 和调用服务的组件在同一个应用程序中，可以使用显式启动或隐式启动，显式启动更加易于使用，且代码简单。但如果服务和调用服务的组件在不同的应用程序中，则只能使用隐式启动。

无论是显式启动还是隐式启动，停止 Service 的方法都是相同的，将启动 Service 的 Intent 传递给 stopService(Intent)函数即可，示例代码如下：

```
stopService(serviceIntent);
```

在调用 startService(Intent)函数首次启动 Service 后，系统会先后调用 onCreate()和 onStart()，如果再次调用 startService(Intent)函数，则系统仅调用 onStart()。在调用 stopService(Intent)函数停止 Service 时，系统会调用 onDestroy()。无论调用过多少次 startService(Intent)，在调用 stopService(Intent)函数时，系统仅调用 onDestroy()一次。

接下来我们通过一个示例来学习进程内服务，SimpleRandomServiceDemo 案例是在应用程序中建立 Service，并以显式启动服务的示例。在工程中创建了 RandomService 服务，该服务启动后会产生一个随机数，并使用 Toast 显示在屏幕上，如图 7-2 所示。

图 7-2 SimpleRandomServiceDemo 用户界面

在界面上单击“启动 Service”按钮调用 startService(Intent)函数，启动 RandomService 服务。“停止 Service”按钮调用 stopService(Intent)函数，停止 RandomService 服务。为了在每个函数中都使用 Toast，在界面上产生提示信息。

RandomService.java 文件代码如下：

```
public class RandomService extends Service{
    @Override
  public void onCreate() {
    super.onCreate();
    Toast.makeText(this, "(1) 调用 onCreate()",Toast.LENGTH_LONG).show();
     }
  @Override
public void onStart(Intent intent, int startId) {
    super.onStart(intent, startId);
  Toast.makeText(this, "(2) 调用 onStart()", Toast.LENGTH_SHORT).show();
    double randomDouble = Math.random();
    String msg = "随机数: "+ String.valueOf(randomDouble);
    Toast.makeText(this,msg, Toast.LENGTH_SHORT).show();
     }
  @Override
    public void onDestroy() {
      super.onDestroy();
    Toast.makeText(this, "(3) 调用 onDestroy()", Toast.LENGTH_SHORT).show();
    }
  @Override
 public IBinder onBind(Intent intent) {return null;  }
}
```

在 AndroidManifest.xml 文件注册 Service 的代码如下：

```
<service android:name=".RandomService"/>
```

调用 AndroidManifest.xml 文件，在<application>标签下，包含一个<activaty>标签和一个<service>标签，在<service>标签中，声明了 RandomService 所在的类。SimpleRandomServiceDemo.java 文件是应用程序中有启动和停止 Service 功能的代码。

SimpleRandomServiceDemo.java 文件的代码如下：

```
public class SimpleRandomServiceDemo extends Activity {
  @Override
public void onCreate(Bundle savedInstanceState) {
super.onCreate(savedInstanceState);
 setContentView(R.layout.main);
  Button startButton = (Button)findViewById(R.id.start);
Button stopButton = (Button)findViewById(R.id.stop);
final Intent serviceIntent = new Intent(this, RandomService.class);
startButton.setOnClickListener(new Button.OnClickListener(){
public void onClick(View view){
```

```
startService(serviceIntent); } });
stopButton.setOnClickListener(new Button.OnClickListener(){
public void onClick(View view){
stopService(serviceIntent);}   }); }
  }
```

7.2.2 使用线程

在 Android 系统中，Activity、Service 和 BroadcastReceiver 都是工作在主线程上，因此任何耗时的处理都会降低用户界面的响应速度，甚至导致用户界面失去响应。当用户界面失去响应超过 5 秒钟，Android 系统会允许用户强行关闭应用程序。因此，较好的解决方法是将耗时的处理过程转移到子线程上，这样可以避免负责界面更新的主线程无法处理界面事件，从而避免用户界面长时间失去响应。耗时的处理过程除了指运算量巨大的复杂运算外，还包括大量的文件操作、网络操作和数据库操作等。

线程是独立的程序单元，多个线程可以并行工作。在多处理器系统中，每个中央处理器（CPU）单独运行一个线程，因此线程是并行工作的。但在单处理器系统中，处理器会给每个线程一个小段时间，在这个时间内，线程是被执行的，然后处理器执行下一个线程，这样就产生了线程并行运行的假象。无论线程是否真的并行工作，在宏观上可以认为子线程是独立于主线程，且能与主线程并行工作的程序单元。

在 Java 语言中，建立和使用线程比较简单，首先需要实现 Java 的 Runnable 接口，并重载 run()方法。在 run()中放置代码的主体部分。

```
private Runnable backgroudWork = new Runnable(){
    @Override
    public void run() {
  //过程代码
    }   };
```

然后创建 Thread 对象，并将上面实现的 Runnable 对象作为参数传递给 Thread 对象。在 Thread 的构造函数中，第 1 个参数用来表示线程组，第 2 个参数是需要执行的 Runnable 对象，第 3 个参数是线程的名称。

```
private Thread workThread;
workThread = new Thread(null,backgroudWork,"WorkThread");
```

最后，调用 start()方法启动线程。

```
workThread.start();
```

当线程在 run()方法返回后，线程就自动终止了。当然，也可以调用 stop()方法在外部终止线程，但这种方法并不推荐使用，因为这种方法并不安全，有可能会产生异常。最好的方法是通知线程自行终止，一般调用 interrupt()方法通告线程准备终止，线程会释放它正在使用的资源，在完成所有的清理工作后自行关闭。

```
workThread.interrupt();
```

其实 interrupt()方法并不能直接终止线程，仅是改变了线程内部的一个布尔字段，run()方法能够检测到这个布尔字段，从而知道何时应该释放资源和终止线程。在 run()方法的代码，一般

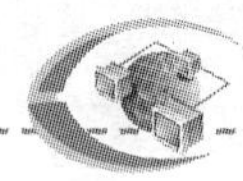

通过 Thread.interrupted()方法查询线程是否被中断。在很多情况下，子线程需要无限运行，除非外部调用 Thread.interrupted()方法判断线程是否应被中断。下面代码中，以 1 秒为间隔循环检测线程是否应被中断。

```
public void run() {
while(!Thread.interrupted()){
//过程代码
Thread.sleep(1000); //使线程休眠 1000 毫秒
}   }
```

当线程在休眠过程中被中断，则会产生 InterruptedException。在中断的线程上调用 sleep()方法，同样会产生 InterruptedException。因此除了使用 Thread.interrupted()方法判断线程是否应被中断，还可以通过捕获 InterruptedException 判断线程是否应被中断，并且在捕获到 InterruptedException 后，安全终止线程。参考代码如下：

```
public void run() {
     try {
       while(true){
          //过程代码
             Thread.sleep(1000);}
} catch (InterruptedException e) { e.printStackTrace();  }   }
```

目前，我们已经可以设计自己的线程，但还存在一个不可回避的问题，即在图形用户界面中使用线程，如何使用线程中的数据更新用户界面。Android 系统提供了多种方法解决这个问题，下面仅介绍如何使用 Handler 更新用户界面。

Handler 允许将 Runnable 对象发送到线程的消息队列中，每个 Handler 对象绑定到一个单独的线程和消息队列上。当用户建立一个新的 Handler 对象，通过 post()方法将 Runnable 对象从后台线程发送到 GUI 线程的消息队列中，在 Runnable 对象通过消息队列后，这个 Runnable 对象将被运行。

```
private static Handler handler = new Handler(); //建立一个私有的静态的 Handler
对象
public static void UpdateGUI(double refreshDouble){
 //公有的界面更新函数,后台线程通过调用该函数，将后台产生的数据 refreshDouble
//传递到 UpdateGUI()函数内部，然后并直接调用 post()方法
handler.post(RefreshLable);    }
private static Runnable RefreshLable = new Runnable(){
  @Override
  public void run() {
  //过程代码
}   };
```

接下来我们还是通过一个实例来具体学习，ThreadRandomServiceDemo 是使用线程待续产生随机数的示例。当用户单击“启动 Service”按钮后，将启动后台线程，单击“停止 Service”按钮后，将关闭后台线程。后台线程每 1 秒钟产生一个 0～1 之间的随机数，并通过 Handler 将该随机数显示在用户界面上。ThreadRandomServiceDemo 的用户界面如图 7-3 所示。

图 7-3　ThreadRandomServiceDemo 用户界面图

在 ThreadRandomServiceDemo 示例中，RandomService.java 文件是描述 Service 的文件，用来创建线程、产生的随机数和调用界面更新函数。ThreadRandomServiceDemo.java 文件是界面的 Activity 文件，封装 Handler 的界面更新函数就在这个文件中。下面给出 RandomService.java 和 ThreadRandomServiceDemo.java 文件完整代码。

RandomService.java 文件完整代码如下：

```
public class RandomService extends Service{
private Thread workThread;
@Override
public void onCreate() {
  super.onCreate();
 Toast.makeText(this, "(1) 调用 onCreate()", Toast.LENGTH_LONG).show();
  workThread = new Thread(null,backgroudWork,"WorkThread");  }
@Override
public void onStart(Intent intent, int startId) {
   super.onStart(intent, startId);
  Toast.makeText(this, "(2) 调用 onStart()", Toast.LENGTH_SHORT).show();
    if (!workThread.isAlive()){
      workThread.start();
      }  }
@Override
public void onDestroy() {
   super.onDestroy();
 Toast.makeText(this, "(3) 调用 onDestroy()", Toast.LENGTH_SHORT).show();
workThread.interrupt();   }
@Override
public IBinder onBind(Intent intent) {
return null;  }
private Runnable backgroudWork = new Runnable(){
@Override
public void run() {
  try {
  while(!Thread.interrupted()){
```

```
 double randomDouble = Math.random();
ThreadRandomServiceDemo.UpdateGUI(randomDouble);
  Thread.sleep(1000);    }
 } catch (InterruptedException e) {   e.printStackTrace();    }
   }  };
  }
```

ThreadRandomServiceDemo.java 文件完整代码如下：

```
public class ThreadRandomServiceDemo extends Activity {
  private static Handler handler = new Handler();
  private static TextView labelView = null;
  private static double randomDouble ;
  public static void UpdateGUI(double refreshDouble){
randomDouble = refreshDouble;
handler.post(RefreshLable);   }
    private static Runnable RefreshLable = new Runnable(){
    @Override
public void run() {
labelView.setText(String.valueOf(randomDouble));
}   };
@Override
public void onCreate(Bundle savedInstanceState) {
super.onCreate(savedInstanceState);
setContentView(R.layout.main);
labelView = (TextView)findViewById(R.id.label);
Button startButton = (Button)findViewById(R.id.start);
Button stopButton = (Button)findViewById(R.id.stop);
final Intent serviceIntent = new Intent(this, RandomService.class);
    startButton.setOnClickListener(new Button.OnClickListener(){
    public void onClick(View view){
startService(serviceIntent); }  });
stopButton.setOnClickListener(new Button.OnClickListener(){
  public void onClick(View view){
stopService(serviceIntent);     } });
 }
}
```

7.2.3 服务绑定

以绑定方式使用 Service，能够获取到 Service 对象，不仅能够正常启动 Service，而且能够调用正在运行中的 Service 实现的公有方法和属性。为了使 Service 支持绑定，需要在 Service 类中重载 onBind()方法，并在 onBind()方法中返回 Service 对象，示例代码如下：

```
public class MathService extends Service{
```

```
private final IBinder mBinder = new LocalBinder();
 public class LocalBinder extends Binder{
MathService getService() {
return MathService.this;    } }
  @Override
public IBinder onBind(Intent intent) {
return mBinder;
}  }
```

当 Service 被绑定时，系统会调用 onBind()函数，通过 onBind()函数的返回值，将 Service 对象返回给调用者。onBind()函数的返回值必须是符合 IBind()函数接口，因此在代码中声明一个接口变量 mBinder，mBinder 符合 onBind()函数返回值的要求，因此将 mBinder 传递给调用者。IBind()函数是用于进程内部和进程间过程中调用的轻量级接口，定义了与远程对象交互的抽象协议，使用时通过继承 Binder 的方法实现。LocalBinder 是继承 Binder 的一个内部类，当调用者获取到 mBinder 后，通过调用 getService()即可获取到 Service()的对象。

调用都通过 bindService()函数绑定服务，调用格式如下：

```
final Intent serviceIntent = new Intent(this, MathService.class);
bindService(serviceIntent, mConnection, Context.BIND_AUTO_CREATE);
```

bindService()函数中的第 1 个参数将 Intent 传递给 bindService()函数，声明需要启动的 Servic。第 3 个参数 Context.BIND_AUTO_CREATE 表明只要绑定存在，就自动建立 Service；同时也告知 Android 系统，这个 Service 的重要程度与调用者相同，除非考虑终止调用者，否则不要关闭这个 Service。第 2 个参数是 ServiceConnection，当绑定成功后，系统将调用 ServiceConnection 的 onServiceConnected()方法；而当绑定意外断开后，系统将调用 ServiceConnection 中的 onServiceDisconnected 方法。因此，以绑定的方式使用 Service，调用者需要声明一个 ServiceConnection，并重载内部的 onServiceConnected()方法和 onServiceConnected()方法。

```
private ServiceConnection mConnection = new ServiceConnection() {
@Override
public void onServiceConnected(ComponentName name, IBinder service) {
//绑定成功后通过 getService()获取 Service 对象，这样便可以调用 Service 中的方法和属性。
mathService = ((MathService.LocalBinder)service).getService();}
@Override
public void onServiceDisconnected(ComponentName name) {
// Service 对象设置为 null，表示绑定意外失效，Service 实例不再可用。
mathService = null; }  };
```

取消绑定仅需要使用 unbindService()方法，并将 ServiceConnection 传递给 unbindService()方法。但需要注意的是，unbindService()方法成功后，系统并不会调用 onServiceConnected()，因为 onServiceConnected()仅在意外断开绑定时才被调用。

```
unbindService(mConnection);
```

在绑定方式中，当调用者通过 bindService()函数绑定 Service 时，onCreate()函数和 onBind()函数将先后被调用。当调用者通过 unbindService()函数取消绑定 Service 时，onBnbind()函数将被调用。如果 onBnbind()函数返回 true，则表示在新调用者绑定服务时，onRebind()函数将被调用。绑定方式的函数调用顺序如图 7-4 所示。

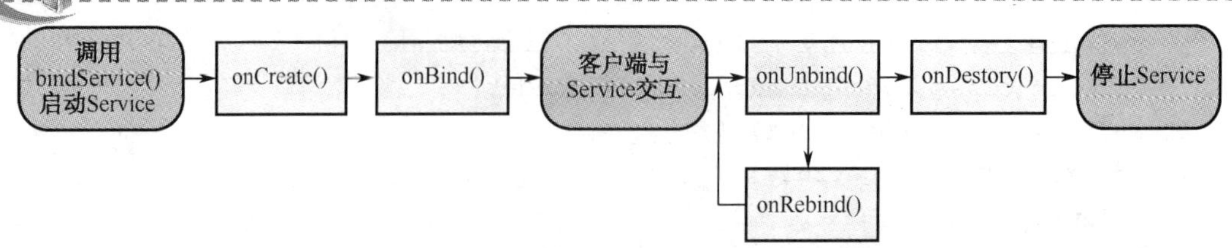

图 7-4　绑定方式的函数调用顺序图

下面的 SimpleMathServiceDemo()是使用绑定方式使用 Service 的示例。在示例中创建了 MathService 服务，用来完成简单的数学运算，这里的数学运算仅包括加法运算，虽然没有实际意义，但足以说明如何使用绑定方式调用 Service 实例中的公有方法。服务绑定后，用户可以单击"加法运算"按钮，将两个随机产生的数值传递给 MathService 服务，并从 MathService 对象中获取到加法运算的结果，然后显示在屏幕的上方。单击"取消绑定"按钮可以解除与 MathService 的绑定关系，在取消绑定后，无法通过"加法运算"按钮获取加法运算结果。SimpleMathServiceDemo 的用户界面如图 7-5 所示。

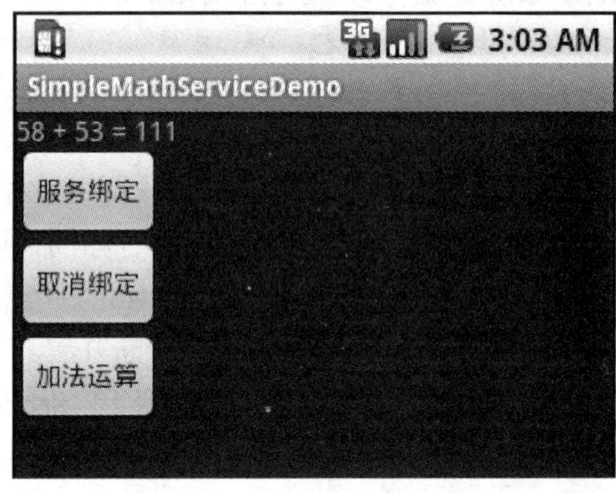

图 7-5　SimpleMathServiceDemo 用户界面

在 SimpleMathServiceDemo 示例中，MathService.java 文件是描述 Service 的文件。SimpleMathServiceDemo.java 文件是界面的 Activity 文件，绑定和取消绑定服务的代码在这个文件中。下面给出 MathService.java 和 SimpleMathServiceDemo.java 文件的完整代码。

MathService.java 文件的完整代码如下：

```
public class MathService extends Service{
private final IBinder mBinder = new LocalBinder();
 public class LocalBinder extends Binder{
MathService getService() {
return MathService.this;
}}
@Override
public IBinder onBind(Intent intent) {
 Toast.makeText(this, "本地绑定：MathService", Toast.LENGTH_SHORT).show();
return mBinder;  }
@Override
```

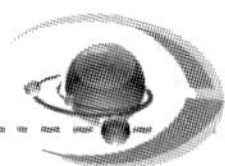

```
public boolean onUnbind(Intent intent){
 Toast.makeText(this, "取消本地绑定：MathService", Toast.LENGTH_SHORT).
 show();
return false; }
  public long Add(long a, long b){
 return a+b;  }
  }
```

SimpleMathServiceDemo.java 文件的完整代码如下：

```
public class SimpleMathServiceDemo extends Activity {
private MathService mathService;
private boolean isBound = false;
TextView labelView;
 @Override
 public void onCreate(Bundle savedInstanceState) {
  super.onCreate(savedInstanceState);
  setContentView(R.layout.main);
  labelView = (TextView)findViewById(R.id.label);
 Button bindButton = (Button)findViewById(R.id.bind);
 Button unbindButton = (Button)findViewById(R.id.unbind);
Button computButton = (Button)findViewById(R.id.compute);
 bindButton.setOnClickListener(new View.OnClickListener(){
@Override
public void onClick(View v) {
if(!isBound){
  final Intent serviceIntent = new  Intent(SimpleMathServiceDemo.this,
  MathService.class);
bindService(serviceIntent,mConnection,Context.BIND_AUTO_CREATE);
 isBound = true;  }     }   });
 unbindButton.setOnClickListener(new View.OnClickListener(){
@Override
public void onClick(View v) {
if(isBound){
isBound = false;
    unbindService(mConnection);
mathService = null;  }  }     });
computButton.setOnClickListener(new View.OnClickListener(){
   @Override
   public void onClick(View v) {
 if (mathService == null){
   labelView.setText("未绑定服务");
    return;  }
long a = Math.round(Math.random()*100);
long b = Math.round(Math.random()*100);
      long result = mathService.Add(a, b);
```

```
        String msg = String.valueOf(a)+" + "+String.valueOf(b)+ " = "+String.
valueOf(result);
        labelView.setText(msg);    }     });   }
        private ServiceConnection mConnection = new ServiceConnection() {
        @Override
        public void onServiceConnected(ComponentName name, IBinder service) {
          mathService = ((MathService.LocalBinder)service).getService();
        }
        @Override
         public void onServiceDisconnected(ComponentName name) {
            mathService = null;}   };
          }
```

7.3 Handler 消息传递机制

在 Android 平台中，新启动的线程是无法访问 Activity 里的 Widget 的，当然也不能将运行状态外送出来，这就需要有 Handler 机制进行消息的传递了，Handler 类位于 android.os 包下，主要的功能是完成 Activity 的 Widget 与应用程序中线程之间的交互。接下来对该类中常用的方法进行介绍，如表 7-1 所示。

表 7-1　Handler 类的常用方法

方法签名	描　述
public void handleMessage (Message msg)	子类对象通过该方法接收信息
public final boolean sendEmptyMessage (int what)	发送一个只含有 what 值的消息
public final boolean sendMessage (Message msg)	发送消息到 Handler，通过 handleMessage 方法接收
public final boolean hasMessages (int what)	监测消息队列中是否还有 what 值的消息
public final boolean post (Runnable r)	将一个线程添加到消息队列
public final boolean postDelayed (Runnable r, long delayMillis)	每隔一定时间将一个线程添加到消息队列
public final void removeCallbacks (Runnable r)	将一个线程从消息队列中移除
public final Message obtainMessage ()	得到一个 Message 类型的消息对象

开发带有 Handler 类的程序步骤如下。

在 Activity 或 Activity 的 Widget 中开发 Handler 类的对象，并重写 handleMessage 方法。在新启动的线程中调用 sendEmptyMessage 或者 sendMessage 方法向 Handler 发送消息。Handler 类的对象用 handleMessage 方法接收消息，然后根据消息的不同执行不同的操作。

接下来我们通过一个 Handler 控制进度条的案例来具体学习 Handler 类的使用方法。该案例的具体开发步骤如下。

步骤一：项目创建，资源准备。在 Eclipse 中新建一个名称为 TestBarHandler 的 Android 项目。打开 res\layout 目录下的布局描述文件 main.xml 文件。其代码添加如下：

```
        <ProgressBar android:id="@+id/bar"
```

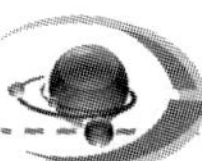

```
        style="?android:attr/progressBarStyleHorizontal"
        android:layout_width="200dp"
        android:layout_height="wrap_content"
        android:visibility="gone"/>
    <Button android:id="@+id/startButton"
        android:layout_width="fill_parent"
        android:layout_height="wrap_content"
        android:text="开始"/>
    <Button android:id="@+id/removeButton"
        android:layout_width="fill_parent"
        android:layout_height="wrap_content"
    android:text="重置"/>
```

步骤二：编写主逻辑文件 TestBarHandler.java。其代码如下：

```
    public class TestBarHandler extends Activity {
        //声明控件变量
        ProgressBar bar = null;
        Button startButton = null;
        Button removeButton = null;
        int i = 0 ;
        @Override
        public void onCreate(Bundle savedInstanceState) {
            super.onCreate(savedInstanceState);
            setContentView(R.layout.main);
            //根据控件的 id 得到代表控件的对象,并为按钮设置监听器
            bar = (ProgressBar)findViewById(R.id.bar);
            startButton = (Button)findViewById(R.id.startButton);
            startButton.setOnClickListener(new startButtonListener());
            removeButton = (Button)findViewById(R.id.removeButton);
            removeButton.setOnClickListener(new removeButtonListener());}
        //当单击 startButton 按钮时，就会执行 ButtonListener 的 onClick 方法
        class startButtonListener implements OnClickListener{
            @Override
            public void onClick(View v) {
                bar.setVisibility(View.VISIBLE);
                updateBarHandler.post(updateThread);}}
        class removeButtonListener implements OnClickListener{
            @Override
            public void onClick(View v) {
                // TODO Auto-generated method stub
                bar.setVisibility(View.GONE);
                i=0;
                updateBarHandler.removeCallbacks(updateThread);}}
          //使用匿名内部类来复写 Handler 当中的 handleMessage 方法
          Handler updateBarHandler = new Handler(){
```

```
        @Override
        public void handleMessage(Message msg) {
            bar.setProgress(msg.arg1);
            System.out.println("msg.arg1---="+msg.arg1);
            //updateBarHandler.post(updateThread);
            updateBarHandler.postDelayed(updateThread, 2000);}};
    //线程类，该类使用匿名内部类的方式进行声明
    Runnable updateThread = new Runnable(){
        @Override
        public void run() {
            if( i >= 100){
                //如果当 i 的值为 100 时，就将线程对象从 handler 当中移除
                updateBarHandler.removeCallbacks(updateThread);}
            else{
            i = i + 5 ;
            //得到一个消息对象，Message 类是有 Android 操作系统提供
            Message msg = updateBarHandler.obtainMessage();
            //将 msg 对象的 arg1 参数的值设置为 i,用 arg1 和 arg2 这两个成员变量传递
            消息，优点是系统性能消耗较少
            msg.arg1 = i ;
            //将 msg 对象加入到消息队列当中
            updateBarHandler.sendMessage(msg);}} };}
```

步骤三：运行程序，查看结果，如图 7-6 所示。

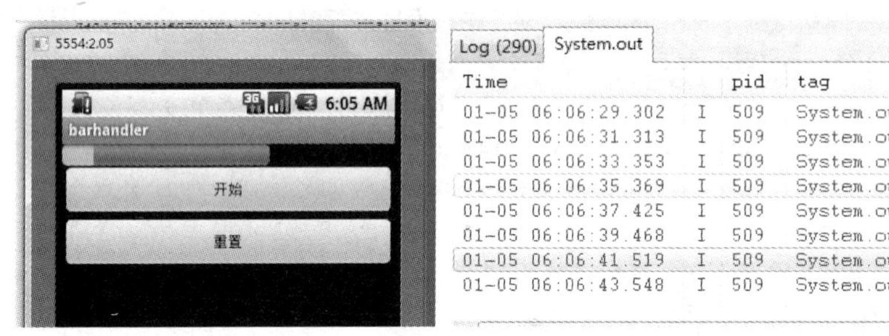

图 7-6　Handler 控制进度条案例运行效果

下面我们再介绍一个 Handler 与 Bundle 类配合使用传递消息的案例，其案例开发代码如下：

```
public class HandlerTest extends Activity {
    @Override
    protected void onCreate(Bundle savedInstanceState) {
        // TODO Auto-generated method stub
        super.onCreate(savedInstanceState);
        setContentView(R.layout.main);
        //打印了当前线程的 id
        System.out.println("Activity-->" + Thread.currentThread().getId());
```

```
        //生成一个 HandlerThread 对象，实现了使用 Looper 来处理消息队列的功能，这个
类由 Android 应用程序框架提供
        HandlerThread handlerThread = new HandlerThread("handler_thread");
        //在使用 HandlerThread 的 getLooper()方法之前，必须先调用该类的 start();
        handlerThread.start();
        MyHandler myHandler = new MyHandler(handlerThread.getLooper());
        Message msg = myHandler.obtainMessage();
        //将 msg 发送到目标对象，所谓的目标对象，就是生成该 msg 对象的 handler 对象
        Bundle b = new Bundle();
        b.putInt("age", 20);
        b.putString("name", "Jhon");
        msg.setData(b);
        msg.sendToTarget();}
       class MyHandler extends Handler{
         public MyHandler(){}
         public MyHandler(Looper looper){super(looper);}
          @Override
         public void handleMessage(Message msg) {
            Bundle b = msg.getData();
            int age = b.getInt("age");
            String name = b.getString("name");
            System.out.println("age is " + age + ", name is" + name);
            System.out.println("Handler--->" + Thread.currentThread().
            getId());}}}
```

其案例的运行结果如图 7-7 所示。

01-05 06:21:49.978	I	536	System.out	Activity-->1
01-05 06:21:49.998	I	536	System.out	age is 20, name isJhon
01-05 06:21:50.008	I	536	System.out	Handler--->8

图 7-7　运行结果

7.4　单机版音乐盒实例

我们在本章前面的章节中详细介绍了 Android 消息机制和后台服务，在本节中，我们通过介绍一个完整的音乐播放盒实例，来理解和掌握 Service 组件、BoradcastReceiver 组件、Intent 和菜单对话框等的综合应用。该音乐盒可以后台播放音乐，即当前台的用户界面退出后，后台的音乐还会继续播放，并且在播放过程中可以通过前台的按钮控制声音的播放、暂停和关闭。接下来我们开始介绍该案例的详细开发步骤。

步骤一：项目的创建与资源的准备。在 Eclipse 开发工具中建立一个名为 Example_7_4 的 Android 项目，准备如图 7-8 所示项目应用的图片资源和音乐资源，并且存放到相应的目录下。音乐资源放在 raw 文件夹中，如果 res 目录下没有 raw 文件夹，读者可以自行创建一个。

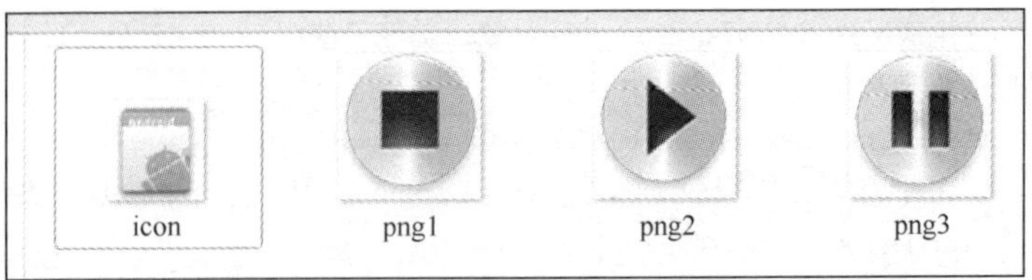

图 7-8　项目需用的图片

步骤二：布局界面的设计。界面布局结果如图 7-9 和图 7-10 所示，涉及到线性布局的嵌套，这里的代码比较简单，读者可以自行设计，在这里就不详细介绍了。

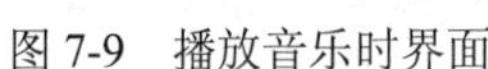

图 7-9　播放音乐时界面　　　　图 7-10　暂停音乐时界面

步骤三：后台服务的开发。该类便为音乐盒的后台服务类，负责声音的播放等操作。在 src 的相关目录下创建一个名称为 MyService.java 的文件，该类的主要代码如下：

```
public class MyService extends Service{
    MediaPlayer mp;//声音播放类对象的创建
    ServiceReceiver serviceReceiver;
    int status = 1;//当前的状态,1 没有声音播放 ,2 正在播放声音,3 暂停
    @Override
    public IBinder onBind(Intent intent) {//重写的 onBind 方法, 因为本案例并不
    适用绑定启动 Service, 所以该方法的返回值为空
        return null;}
    //该方法在 Service 创建时被调用
    public void onCreate() {//重写的 onCreate 方法
        // TODO Auto-generated method stub
        status = 1;
        serviceReceiver = new ServiceReceiver();//创建 BroadcastReceiver
        IntentFilter filter = new IntentFilter();//创建过滤器
        filter.addAction("xsc.xsc.control");//添加 Action
```

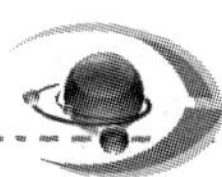

```
            registerReceiver(serviceReceiver, filter);//注册BroadcastReceiver
            super.onCreate();}
        @Override
        public void onDestroy() {//重写的onDestroy方法,在Service摧毁时被调用
            // TODO Auto-generated method stub
            unregisterReceiver(serviceReceiver);//取消注册，即取消监听
            super.onDestroy();}
        public class ServiceReceiver extends BroadcastReceiver{//自定义
        BroadcastReceiver
            @Override
    public void onReceive(Context context, Intent intent) {//重写的响应方法
        int action = intent.getIntExtra("ACTION", -1);//得带需要的数据
            switch(action){
            case 1://播放或暂停声音
            if(status == 1){//当前没有声音播放
            mp = MediaPlayer.create(context, R.raw.nx);//初始化声音资源
            status = 2;//设置状态值
            Intent sendIntent = new Intent("xsc.xsc.update");//创建Intent
            sendIntent.putExtra("update", 2);//存入数据
            sendBroadcast(sendIntent);//发送广播
            mp.start();}//播放声音
                    else if(status == 2){//正在播放声音
                        mp.pause(); //停止播放
                        status = 3;//改变状态
                        Intent sendIntent = new Intent("xsc. xsc.update");
                        sendIntent.putExtra("update", 3);//存放数据
                        sendBroadcast(sendIntent); }//发送广播
                    else if(status == 3){//暂停中
                        mp.start();//播放声音
                        status = 2;//改变状态
                        Intent sendIntent = new Intent("xsc. xsc.update");
                        sendIntent.putExtra("update", 2);//存放数据
                        sendBroadcast(sendIntent); }//发送广播
                    break;
                case 2://停止声音
                    if(status == 2 || status == 3){//播放中或暂停中
                        mp.stop();//停止播放
                        status = 1;//改变状态
                        Intent sendIntent = new Intent("xsc. xsc.update");
                        sendIntent.putExtra("update", 1);//存放数据
                        sendBroadcast(sendIntent);//发送广播
                    } } } }
    }
```

步骤四：Activity 视图界面开发。该类的功能主要是把我们第二步设计的界面显示在屏幕

上，并创建IntentFilter过滤器，实现在不同播放情况下，界面图片的相互转换。其主要代码如下：

```
public class Example_7_4 extends Activity implements OnClickListener{
    ImageButton start;//播放、暂停按钮
    ImageButton stop;//停止按钮
    ActivityReceiver activityReceiver;
    int status = 1;//当前的状态,1 没有声音播放,2 正在播放声音,3 暂停
    /** Called when the activity is first created. */
    @Override
    public void onCreate(Bundle savedInstanceState) {//重写的 onCreate 方法
        super.onCreate(savedInstanceState);
        setContentView(R.layout.main);//设置当前的用户界面
        start = (ImageButton) this.findViewById(R.id.start);//得到 start 按
        钮的引用
        stop = (ImageButton) this.findViewById(R.id.stop);//得到 stop 按钮的
        引用
        start.setOnClickListener(this);//为按钮添加监听
        stop.setOnClickListener(this);//为按钮添加监听
        activityReceiver = new ActivityReceiver();//创建 BroadcastReceiver
        IntentFilter filter = new IntentFilter();//创建 IntentFilter 过滤器
        filter.addAction("xsc. xsc.update");//添加 Action
        registerReceiver(activityReceiver, filter);//注册监听
        Intent intent = new Intent(this, MyService.class);//创建 Intent
        startService(intent);//启动后台 Service
    }
    public class ActivityReceiver extends BroadcastReceiver{//自定义的
    BroadcastReceiver
        @Override
        public void onReceive(Context context, Intent intent) {//重写的
        onReceive 方法
            // TODO Auto-generated method stub
    int update = intent.getIntExtra("update", -1);//得到 intent 中的数据
            switch(update){//分支判断
            case 1://没有声音播放
                status = 1; //设置当前状态
                break;
            case 2://正在播放声音
                start.setImageResource(R.drawable.png3);//更换图片
                status = 2; //设置当前状态
                break;
            case 3://暂停中
                start.setImageResource(R.drawable.png2);//更换图片
                status = 3; //设置当前状态
                break;
            } } }
```

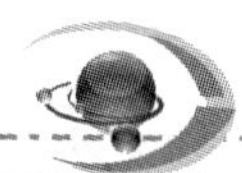

```
@Override
public void onClick(View v) {//接口中的方法
    // TODO Auto-generated method stub
    Intent intent = new Intent("xsc. xsc.control");//创建 Intent
    switch(v.getId()){//分支判断
    case R.id.start://按下播放、暂停按钮
        intent.putExtra("ACTION", 1);//存放数据
        sendBroadcast(intent);//发送广播
        break;
    case R.id.stop://按下停止按钮
        intent.putExtra("ACTION", 2);//存放数据
        sendBroadcast(intent);//发送广播
        break; } }
@Override
protected void onDestroy() {//释放时被调用
    // TODO Auto-generated method stub
    super.onDestroy();
   Intent intent = new Intent(this, MyService.class);//创建 Intent
   stopService(intent); }//停止后台的 Service
//创建停止后台服务的菜单和对话框
public boolean onCreateOptionsMenu(Menu menu){//弹出菜单
    menu.add(0,Menu.FIRST,0,"退出")
        .setIcon(android.R.drawable.ic_menu_delete);//设置图标
    return true; }
@Override
public boolean onOptionsItemSelected(MenuItem item){//选择的菜单项
    switch(item.getItemId()){//分支判断
    case Menu.FIRST:
        showDialog(1);//显示对话框
        break;
    }
    //将来可在此进行扩展
    return false;}
@Override
protected Dialog onCreateDialog(int id){//创建对话框
    switch(id){//判断
    case 1:
        return new AlertDialog.Builder(this)
            .setTitle("您确定退出？")
            .setPositiveButton("确定",new android.content.
            DialogInterface. OnClickListener(){
                @Override
                public void onClick(DialogInterface dialog, int which)
                {  System.exit(0); } })//直接退出
```

```
                .setNegativeButton("取消", null)//取消按钮
                .create();
        default:
            return null;
        } }
    }
```

步骤五：在程序声明文件 AndroidManifest.xml 文件中添加后台服务配置条件如下：

```
    <service
            android:name=".MyService"
            android:process=":remote">
        </service>
```

至此，该案例开发结束，接下来我们可以运行该案例，可以通过单击“播放”按钮图片来播放声音，也可以通过单击“停止”按钮来停止声音的播放，或按手机键盘中的菜单键，用户根据提示确认是否退出当前界面。需要注意的是，当我们从 Activity 退出时，在后台播放的音乐服务并不会退出，而是会继续播放，直到再次进入该程序并单击“停止”按钮才停止声音播放。

第 8 章

Android 数据存储与访问

Android 平台提供多种数据存储方法，包括易于使用的 SharedPreferences，经典的文件存储和轻量级的 SQLite 数据库。通过本章的学习，读者可以了解 Android 平台各种组件数据存储方法的特点和使用方法，掌握跨进度的数据共享方法。

8.1 简单存储

在应用程序的使用过程中，用户经常会根据自己的习惯，更改应用程序的设置，或设定个性化的内容。为了能够保存用户的设置和个性化内容，应用程序一般在文件系统中保存了一个配置文件，并在每次应用程序启动时，读取这个配置文件中的内容。在文件系统中使用配置文件，需要注意配置文件的格式，一般使用 INI 文件或 XML 文件，当然也可以自定义文件格式。INI 文件格式简单，容易读懂，但需要写代码实现文件读取和写入。XML 有成熟的类支持，在代码方面更容易实现，但可读性要比 INI 文件差一些。无论是使用 INI 文件，还是使用 XML 文件保存用户的配置和个性化内容，都是通过程序的开发人员进行烦琐的编码实现的。

Android 为开发人员提供了更为简单的数据存储方法 SharedPreferences。这是一种轻量级的数据保存方式，通过 SharedPreferences 开发人员可以将 NVP（Name/Value/Pair，名称/值对）保存在 Android 的文件系统中，而且 SharedPreferences 完全屏蔽对文件系统的操作过程，开发人员仅是通过调用 SharedPreferences 对 NVP 进行保存和读取。

SharedPreferences 不仅能够保存数据，还能够实现不同应用程序间的数据共享。SharedPreferences 支持三种访问模式：私有（MODE_PRIVATE），全局读（MODE_WORLD_READABLE）和全局写（MODE_WORLD_WRLTEABLE）。如果将 SharedPreferences 定义为私有模式，仅有创建程序有权限对其进行读取或写入；如果将 SharedPreferences 定义为全局读模式，不仅创建程序有权限对其进行读取或写入，其他应用程序也有读取操作的权限，但没有写入操作的权限；如果将 SharedPreferences 定义为全局写模式，则创建程序和其他程序都可以对其进行写入操作，但没有读取的权限。

在使用 SharedPreferences 前，先定义 SharedPreferences 的访问模式。下面的代码将访问模式定义为私有模式。

```
    public static int MODE=MODE_PRIVATE;
```

有的时候需要将 SharedPreferences 的访问模式设定为既可以全局读，也可以全局写，这样就需要将两种模式写成下面的方式。

```
    public static int MODE = Context.MODE_WORLD_READABLE +
    Context.MODE_WORLD_WRITEABLE;
```

除了定义 SharedPreferences 的访问模式，还要定义 SharedPreferences 的名称，这个名称与在 Android 文件系统中保存的文件同名。因此，只要具有相同的 SharedPreferences 名称的 NVP 内容，都会保存在同一个文件中。

```
    public static final String PREFERENCE_NAME = "SaveSetting";
```

为了可以使用 SharedPreferences，需要将访问模式和 SharedPreferences 名称作为参数，传递到 getSharedPreferences()函数，并获取到 SharedPreferences 对象。

```
    SharedPreferences sharedPreferences = getSharedPreferences(PREFERENCE_NAME,
MODE);
```

在获取到 SharedPreferences 对象后，则可以通过 SharedPreferences.Editor 类对 SharedPreferences 进行修改，最后调用 commit()函数保存修改内容。SharedPreferences 广泛支持各种基本数据类型，包括整型，布尔型，浮点型和长型等。

```
    SharedPreferences.Editor editor = sharedPreferences.edit();
    editor.putString("Name", "Tom");
    editor.putInt("Age", 20);
    editor.putFloat("Height", );
    editor.commit();
```

如果需要从已经保存的 SharedPreferences 中读取数据，同样是调用 getSharedPreferences()函数，并在函数的第一个参数中指明需要访问的 SharedPreferences 名称，最后通过 get<Type>()函数获取保存在 SharedPreferences 中的 NVP。Get<Type>()函数的第一个参数是 NVP 的名称，第二个参数是在无法获取到数值的时候使用的默认值。

```
    SharedPreferences sharedPreferences = getSharedPreferences(PREFERENCE_NAME,
MODE);
```

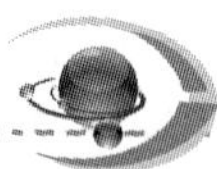

```
    String name = sharedPreferences.getString("Name","Default Name");
    int age = sharedPreferences.getInt("Age", 20);
    float height = sharedPreferences.getFloat("Height",);
```

在前面我们介绍了 SharedPreferences 的使用方法，下面将通过 SimplePreferenceDemo 示例具体介绍说明 SharedPreferences 的文件保存位置和保存格式。SimplePreferenceDemo 示例的用户界面如图 8-1 所示，用户在界面上输入的信息，将通过 SharedPreferences 在 Activity 关闭时进行保存。当应用程序重新开启时，保存在 SharedPreferencse 的信息将被读取出来，并重新呈现在用户界面上。

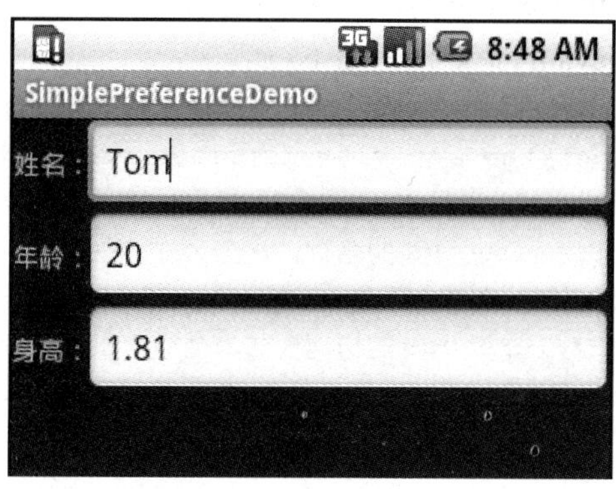

图 8-1 SimplePreferenceDemo 用户界面图

SimplePreferenceDemo 示例运行后，通过 FileExplorer 查看\data\data 下的数据，Android 为每个应用程序建立了与包同名的目录，用来保存应用程序产生的数据，这些数据包括文件，SharedPreferences 文件和数据库等。SharedPreferences 文件就保存在\data\data\<packagename>\shared_prefs 目录下。

在本例中，shared_prefs 目录下生成了一个名为 SaveSetting.xml 的文件，该文件的保存路径如图 8-2 所示。这个文件就是保存 SharedPreferences 的文件，文件大小为 170 字节，在 Linux 下的权限为“-rw-rw-rw”。 在 Linux 系统中，文件权限分别描述了创建者，同组用户和其他用户对文件的操作权限。X 表示可执行，y 表示可读，w 表示可写，d 表示目录，-表示普通文件。因此，“-rw-rw-rw”表示 SaveSetting.xml 可以被创建者、同组用户和其他用户进行读取和写入操作，但不可以执行。产生这样的文件权限与程序人员设定的 SharedPreferences 的访问模式有关，“-rw-rw-rw”的权限是“全局读+全局写”的结果。如果将 SharedPreferences 的访问模式设置为私有，则文件的权限将成为“-rw-rw---”，表示仅有创建者和同组用户具有读写文件的权限。

⊟ edu.hrbeu.SimplePreferenceDemo		2009-07-10	02:18	drwxr-xr-x
⊞ lib		2009-07-10	02:18	drwxr-xr-x
⊟ shared_prefs		2009-07-10	03:01	drwxrwx--x
SaveSetting.xml	170	2009-07-15	08:45	-rw-rw-rw-
⊞ edu.hrbeu.SimpleRandomServiceDemo		2009-06-30	12:17	drwxr-xr-x
⊞ edu.hrbeu.SpinnerDemo		2009-06-21	07:01	drwxr-xr-x

图 8-2 SaveSetting.xml 文件保存路径图

SaveSetting.xml 文件是以 XML 格式保存的信息，内容如下：

```
<?xml version='1.0' encoding='utf-8' standalone='yes' ?>
<map>
    <float name="Height" value="1.81" />
    <string name="Name">Tom</string>
    <int name="Age" value="20" />
</map>
```

SimplePreferenceDemo 示例在 onStart()函数中调用 loadSharedPreference()函数，读取保存在 SharedPreferences 中的姓名、年龄和身高信息，并显示在用户界面上。当 Activity 关闭时，在 onStop()函数调用 saveSharedPreferences()，保存界面上的信息。下面给出示例的完整代码。

SimplePreferenceDemo.java 的完整代码如下：

```
public class SimplePreferenceDemo extends Activity {
    /** Called when the activity is first created. */
    private EditText nameText,ageText,heightText;
    //定义 SharedPreferences 的名称，这个名称与在 Android 文件系统中保存的文件同名
    public static final String PREFERENCE_NAME="SaveSetting";
    //设置 SharedPreferences 的访问模式为全局读和全局写
    public static int MODE=Context.MODE_WORLD_READABLE+Context.MODE_WORLD_
    WRITEABLE;
    @Override
    public void onCreate(Bundle savedInstanceState) {
        super.onCreate(savedInstanceState);
        setContentView(R.layout.main);
        nameText=(EditText)findViewById(R.id.name);
        ageText=(EditText)findViewById(R.id.age);
        heightText=(EditText)findViewById(R.id.height); }
    //回调方法,当运行该 Activity 的时候，则运行此方法
    @Override
    protected void onStart() {
        // TODO Auto-generated method stub
        super.onStart();
        //调用自定义的方法
        loadSharedPreferences();    }
     //回调方法,当关闭该 Activity 的时候，则运行此方法
    @Override
    protected void onStop() {
        super.onStop();
        //调用自定义的方法
        saveSharedPreferences(); }
    //读取保存在 SharedPreferences 中的姓名、年龄和身高信息，并显示在用户界面上
    private void loadSharedPreferences() {
//为了可以使用 SharedPreferences,需要将访问模式和 SharedPreferences 名称作为参数,
传递到getSharedPreferences()函数,并获取到SharedPreferences 对象SharedPreferences
sharedPreferences=getSharedPreferences(PREFERENCE_NAME, MODE);
```

```
            String name=sharedPreferences.getString("Name", "Tom");
            int age=sharedPreferences.getInt("Age", 20);
            float height=sharedPreferences.getFloat("Height", 1.80f);
            nameText.setText(name);
            ageText.setText(age+"");
            heightText.setText(height+""); }
        //保存界面上的信息到 SharedPreferences 中
        private void saveSharedPreferences() {
        //为了可以使用 SharedPreferences，需要将访问模式和 SharedPreferences 名称作为参数，传递到 getSharedPreferences()函数，并获取到 SharedPreferences 对象
            SharedPreferences
        sharedPreferences=getSharedPreferences(PREFERENCE_NAME, MODE);
    /* 在获取到 SharedPreferences 对象后，则可以通过 SharedPreferences.Editor 类
     * 对 SharedPreferences 进行修改，最后调用 commit()函数保存修改内容*/
        SharedPreferences.Editor editor=sharedPreferences.edit();
        editor.putString("Name", nameText.getText().toString());
        editor.putInt("Age",
    Integer.parseInt(ageText.getText().toString()));
    editor.putFloat("Height",
Float.parseFloat(heightText.getText().toString()));
            editor.commit();
        }
    }
```

示例 SharedPreferencesDemo 将说明如何读取其他应用程序保存的 SharedPreferences 数据。SharedPreferencesDemo 示例的用户界面如图 8-3 所示，示例将读取 SimplePreferenceDemo 示例保存的信息，并在程序启动时显示在用户界面上。

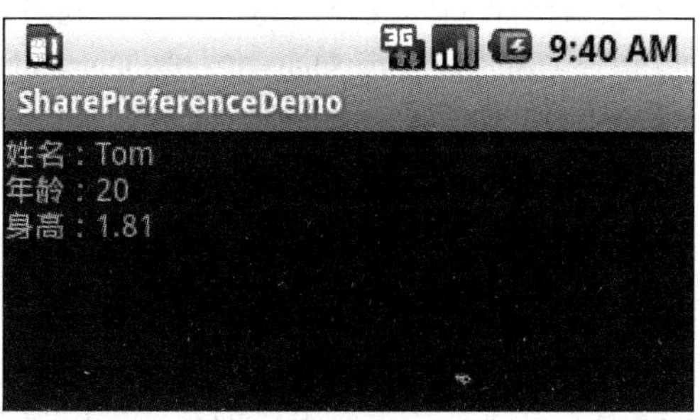

图 8-3 SharedPreferenceDemo 用户界面图

下面给出 SharedPreferenceDemo 示例的核心代码：

```
    public class SharePreferenceDemo extends Activity {
    // xsc.xsc 为 SimplePreferenceDemo 示例的包名
        public static final String PREFERENCE_PACKAGE="xsc.xsc";
        public static final String PREFERENCE_NAME="SaveSetting";
    public static int MODE=Context.MODE_WORLD_READABLE+
    Context.MODE_WORLD_WRITEABLE;
```

```
        TextView textView;
    @Override
        public void onCreate(Bundle savedInstanceState) {
            super.onCreate(savedInstanceState);
            setContentView(R.layout.main);
            Context c=null;
            try {
          /*调用 createPackageContext()获取到了 SimplePreferenceDemo 示例的 Context,
第 1 个参数是 SimplePreferenceDemo 的包名称，第 2 个参数 Context.CONTEXT_IGNORE_ SECURITY
表示忽略所有可能产生的安全问题
              这段代码可能引发异常，因此必须防止在 try/catch 中*/
          c=this.createPackageContext(PREFERENCE_PACKAGE,
Context.CONTEXT_IGNORE_SECURITY);
              } catch (Exception e) {  e.printStackTrace(); }
              /*通过 Context 得到了 SimplePreferenceDemo 示例的 SharedPreferences 对
象，同样在 getSharedPreferences()函数中，需要将正确的 SharedPreferences 名称传递给函数*/
              SharedPreferences sharedPreferences=
    c.getSharedPreferences(PREFERENCE_NAME, MODE);
              String name=sharedPreferences.getString("Name", "Tom");
              int age=sharedPreferences.getInt("Age", 20);
              float height=sharedPreferences.getFloat("Height", 1.81f);
              textView=(TextView)findViewById(R.id.textview);
              textView.setText(name+"\n"+age+"\n"+height);
                  }
    }
```

由此可见，访问其他应用程序的 SharedPreferences 必须满足 3 个条件：（1）共享者须将 SharedPreferences 的访问模式设置为全局读或全局写；（2）访问者需要知道共享者的包名称和 SharedPreferences 的名称，以通过 Context 获得 SharedPreferences 对象；（3）访问者需要确切知道每个数据的名称和数据类型，用以正确读取数据。

8.2 文件存储

虽然 SharedPreferences 能够为程序开发人员简化数据存储和访问的过程，但使用文件系统直接保存数据仍然是数据存储中不可缺少的重要组成部分。Android 使用的是基于 Linux 的文件系统，程序开发人员可以建立和访问程序自身的私有文件，也可以访问保存在资源目录中的原始文件和 XML 文件，还可以在 SD 卡等外部存储设备中保存文件。

8.2.1 内部存储

Android 系统允许应用程序创建仅能够自身访问的私有文件，文件保存在设备的内部存储器上，在 Linux 系统下的\data\data\<package name>\files 目录中。Android 系统不仅支持标准 Java

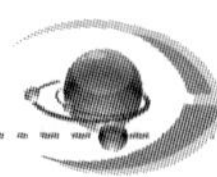

的 IO 类和方法，还提供了能够简化读写流式文件过程的方法。下面主要介绍两个方法：openFileOutput()方法和 openFileInput()方法。

openFileOutput()方法是为写入数据做准备而打开的应用程序私有文件，如果指定的文件不存在，则创建一个新的文件。openFileOutput()方法的语法格式如下：

```
public FileOutputStream openFileOutput(String name, int mode)
```

openFileOutput()方法中的第 1 个参数是文件名称，这个参数不可以包含描述路径的斜杠。第 2 个参数是操作模式，Android 系统支持 4 种文件操作模式，文件操作模式的说明参照表 8-1。方法的返回值是 FileOutputStream 类型。

表 8-1　4 种文件操作模式表

模　式	说　明
MODE_PRIVATE	私有模式，缺陷模式，文件仅能够被文件创建程序访问，或具有相同的 UID 的程序访问
MODE_APPEND	追加模式，如果文件已经存在，则在文件的结尾处添加新数据
MODE_WORLD_READABLE	全局读模式，允许任何程序读取私有文件
MODE_WORLD_WRITEABLE	全局写模式，允许任何程序写入私有文件

使用 openFileOutput()函数建立新文件的示例代码如下：

```
String FILE_NAME = "fileDemo.txt";
FileOutputStream fos=openFileOutput(FILE_NAME,Context.MODE_PRIVATE)
String text = "Some data";
fos.write(text.getBytes());
fos.flush();
fos.close();
```

代码首先定义建立文件的名称 fileDemo.txt，然后使用 openFileOutput()函数以私有模式建立文件，并调用 write()函数将数据写入文件，调用 flush()函数将所有剩余的数据写入文件，最后调用 close()函数关闭 FileOutputStream ()。为了提高文件系统的性能，一般调用 write()函数时，如果写入的数据量较小，系统会把数据保存在数据缓冲区中，等数据量累积到一定程度时再一次性写入文件中。因此，在调用 close()函数关闭文件前，务必要调用 flusu()函数，将缓冲区内所有的数据写入文件。

openFileInput()方法为读取数据做准备而打开的应用程序私有文件。openFileOutput()方法的语法格式如下：

```
public FileInputStream openFileInput (String name)
```

第 1 个参数也是文件名称，同样不允许包含描述路径的斜杠。使用 openFileInput()方法打开已有文件的示例代码如下：

```
String FILE_NAME = "fileDemo.txt";
FileInputStream fis = openFileInput(FILE_NAME);
byte[] readBytes = new byte[fis.available()];
while(fis.read(readBytes) != -1){  }
```

上面的两部分代码在实际使用过程中会遇到错误提示，因为文件操作可能会遇到各种问题而最终导致操作失败，因此代码应该使用 try/catch 捕获可能产生的异常。

接下来我们通过 InternalFileDemo 示例来具体学习在内部存储器上进行文件写入和读取的操作。InternalFileDemo 示例用户界面如图 8-4 所示，用户将需要写入的数据添加在 EditText 中，通过“写入文件”按钮将数据写入到\data\data\edu.hrbeu.InternalFileDemo.text 文件中。如果用户选择“追加模式”，数据将会添加到 fileDemo.text 文件的结尾处。通过“读取文件”按钮程序会自动读取 fileDemo.text 文件的内容，并显示在界面下方的白色区域中。

图 8-4　InternalFileDemo 示例界面图

下面给出 InternalFileDemo 示例的核心代码：

```
public class InternalFileDemo extends Activity implements OnClickListener{
    TextView lableView,diaplayview;
    Button writeButton,readButton ;
    EditText enteyEditText ;
    CheckBox checkBox;
    @Override
    public void onCreate(Bundle savedInstanceState) {
        super.onCreate(savedInstanceState);
        setContentView(R.layout.main);
        lableView=(TextView)findViewById(R.id.lableview);
        diaplayview=(TextView)findViewById(R.id.diaplayview);
        enteyEditText=(EditText)findViewById(R.id.entryText);
        checkBox=(CheckBox)findViewById(R.id.check);
        writeButton=(Button)findViewById(R.id.Button01);
        writeButton.setOnClickListener(this);
        readButton=(Button)findViewById(R.id.Button02);
        readButton.setOnClickListener(this); }
    @Override
    public void onClick(View v) {
        //写入文件的数据流对象
        FileOutputStream fos=null;
        if(v==writeButton){
```

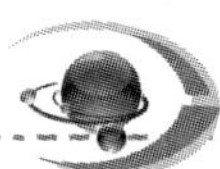

```
        if(checkBox.isChecked()){try {
        fos=openFileOutput("fileDemo.txt", Context.MODE_APPEND);
        } catch (FileNotFoundException e) {
            // TODO Auto-generated catch block
        e.printStackTrace();}}
        else{try {
        fos=openFileOutput("fileDemo.txt", Context.MODE_PRIVATE);
        } catch (FileNotFoundException e) {
            // TODO Auto-generated catch block
        e.printStackTrace();}}
        String text=enteyEditText.getText().toString();
        try {
        fos.write(text.getBytes());
        lableView.setText("文件写入成功，写入文件的长度为:
        "+text.length());
            diaplayview.setText("");
            fos.flush();
            fos.close();
          } catch (IOException e) {
            // TODO Auto-generated catch block
            e.printStackTrace();  }  }
    if(v==readButton){
        diaplayview.setText("");
        FileInputStream fis=null;
        try {  fis=openFileInput("fileDemo.txt");
            // available()就是返回的实际可读字节数，也就是总大小
            System.out.println(fis.available());
            if(fis.available()==0)return;
            byte[] fileread=new byte[fis.available()];
            while(fis.read(fileread)!=-1){}
            String textString=new String(fileread);
            diaplayview.setText(textString);
            lableView.setText("文件读取成功，文件的长度为:
            "+textString.length());
         } catch (Exception e) {
            // TODO Auto-generated catch block
            e.printStackTrace();
        }   }   } }
```

程序运行后，在\data\data\edu.hrbeu.InternalFileDemo\files 目录下，找到了新建立的 fileDemo.txt 文件，如图 8-5 所示。fileDemo.txt 从文件权限上进行分析，“-rw-rw---”表明文件允许文件创建者和同组用户读写，其他用户无权使用。文件的大小为 9 个字节，保存的数据为 Some data。

edu.hrbeu.InternalFileDemo		2009-07-16 02:28	drwxr-xr-x
files		2009-07-16 02:31	drwxrwx--x
fileDemo.txt	9	2009-07-16 03:24	-rw-rw----
lib		2009-07-16 02:28	drwxr-xr-x

图 8-5　fileDemo.txt 文件保存路径状态图

8.2.2　外部存储

Android 的外部存储设备指的是 SD 卡，是一种广泛使用于数码设备上的记忆卡，如图 8-6 所示。虽然并不是所有的 Android 手机都有 SD 卡，但 Android 系统提供了对 SD 卡的便捷的访问方法。

SD 卡适用于保存大尺寸的文件或者是一些无须设置访问权限的文件。如果用户希望保存录制的视频文件和音频文件，因为设备的内部存储空间有限，所以 SD 卡则是非常适合的选择。另一方面，如果需要设置文件的访问权限，则不能够使用 SD 卡，因为 SD 卡支持使用的 FAT（File Allocation Table）的文件系统，不支持访问模式和权限控制，但内部存储器使用的是 Linux 文件系统，可以通过文件访问权限来控制和保证文件的私密性。

Android 模拟器支持 SD 卡，但模拟器中没有默认的 SD 卡，开发人员必须在模拟器中手工添加 SD 卡的映像文件。使用<Android SDK>\tools 目录下的 mksdcard 工具创建 SD 卡的映像文件，命令如下：

```
mksdcard -l 256M SDCARD E:\android\sdcard_file
```

第 1 个参数-l 表示后面的字符串是 SD 卡的标签，这个新建立的 SD 卡的标签是 SDCARD。第 2 个参数 256M 表示 SD 卡的容量是 256MB。最后一个参数表示 SD 卡的映像文件的保存位置，上面的文件将命令保存在 E:\android 目录下的 sdcard_file 文件中。在 CMD 中执行该命令后，则可在所指定的目录中找到生产的 SD 卡映像文件。

如果希望 Android 模拟器启动时能够自动加载指定的 SD 卡，还需要在模拟器的“运行设置”中添加 SD 卡加载命令。SD 卡加载命令中只要指明映像文件位置即可，命令如图 8-7 所示。

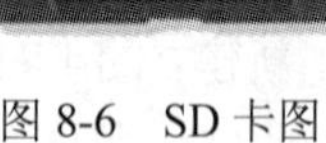

图 8-6　SD 卡图

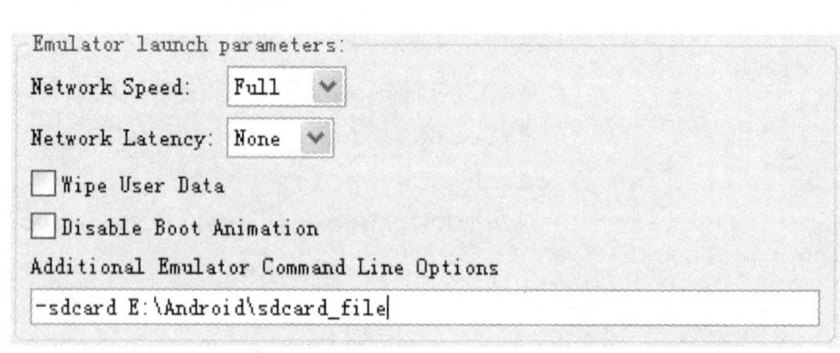

图 8-7　SD 卡加载命令图

为了测试 SD 卡映像文件是否加载正确，在模拟器启动后，使用 FileExplorer 向 SD 卡中随意上传一个文件，如果文件上传成功，则表明 SD 卡映像文件已经成功加载。

在 Android 系统编程访问 SD 卡非常简单。首先需要检测系统\sdcard 目录是否可用，如果

不可用，则说明设备中的 SD 卡已经被移除，在 Android 模拟器则表明 SD 卡映像没有被正确加载。如果可用，则直接通过使用标准的 java.io.File 类进行访问。

SDcardFileDemo 示例说明了如何将数据保存在 SD 卡中。用户首先通过“生成随机数列”按钮生产 10 个随机小数，然后通过“写入 SD 卡”按钮将生产的数据保存在 SD 卡的目录下。SDcardFileDemo 示例的用户界面如图 8-8 所示。

SDcardFileDemo 示例运行后，在每次单击“写入 SD 卡”按钮后，都会在 SD 卡中生产一个新文件，文件名各不相同，如图 8-9 所示。

图 8-8　SDcardFileDemo 用户界面图

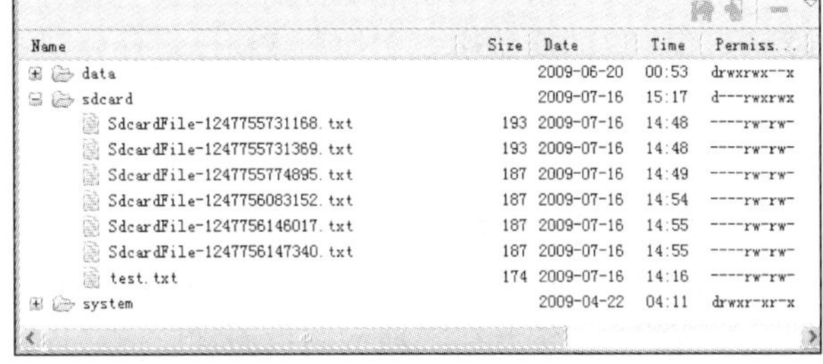

图 8-9　SD 卡中生产的文件图

SDcardFileDemo 示例与 InternalFileDemo 示例的核心代码比较相似，不同之处在于代码中添加了\sdcard 目录存在性检查，并使用“绝对目录+文件名”的形式表示新建立的文件，并写入文件前对文件的存在性和可写入性进行检查，为了保证在 SD 卡中多次写入时文件名不会重复，在文件名中使用了唯一且不重复的标识，这个标识通过调用 System.currentTimeMillis()函数获得，表示从 1970 年到 00：00：00 到当前所经过的毫秒数。SDcardFileDemo 示例的核心代码如下：

```
private static String randomNumbersString ="";
OnClickListener writeButtonListener = new OnClickListener() {
    @Override
    public void onClick(View v) {
    String fileName = "SdcardFile-"+System.currentTimeMillis()+".txt";
        File dir = new File("/sdcard/");
        if (dir.exists() && dir.canWrite()) {
          File newFile = new File(dir.getAbsolutePath() + "/" + fileName);
          FileOutputStream fos  = null;
         try {
           newFile.createNewFile();
           if (newFile.exists() && newFile.canWrite()) {
           fos  = new FileOutputStream(newFile);
           fos.write(randomNumbersString.getBytes());
       TextView labelView = (TextView)findViewById(R.id.label);
           labelView.setText(fileName + "文件写入SD卡"); }
         } catch (IOException e) {  e.printStackTrace();
```

```
            } finally {   if (fos != null) {
           try{    fos.flush();
                   fos.close();  }
           catch (IOException e) { }   }
            }  }  }
    };
```

8.2.3　资源文件

程序开发人员除了可以在内部存储设备和外部存储设备上使用文件以外，还可以将程序开发阶段已经准备好的原始格式文件和 XML 文件分别存放在\res\raw 和\res\xml 目录下，供应用程序在运行时进行访问。

原始格式文件可以是任何格式的文件，例如视频格式文件，音频格式文件，图像文件和数据文件等，在应用程序编译和打包时，\res\raw 目录下所有文件都会保留原有格式不变。而\res\xml 目录下的 XML 文件，一般用来保存格式化的数据，在应用程序编译和打包时会将 XML 文件转换为高效的二进制格式，应用程序运行时会以特殊的方式进行访问。

ResourceFileDemo 示例演示了如何在程序运行时访问资源文件。当用户单击“读取原始文件”按钮时，程序将读取\res\raw\raw_file.txt 文件，并将内容显示在界面上，如图 8-10 所示。当用户单击“读取 XML 文件”按钮时，程序将读取\res\xml\people.xml 文件，并将内容显示在界面上，如图 8-11 所示。

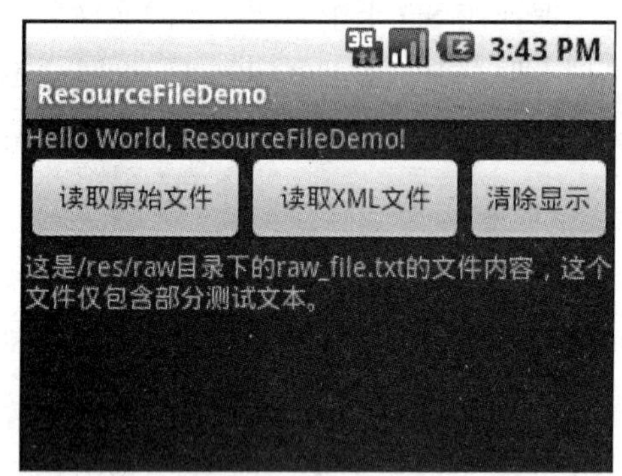

图 8-10　读取原始文件图

图 8-11　读取 XML 文件图

读取原始格式文件，首先需要调用 getResource()函数获得资源对象，然后通过调用资源对象的 openRawResource()函数，以二进制流的形式打开指定的原始格式文件。在读取文件结束后，调用 close()函数关闭文件流。

ResourceFileDemo 示例中关于读取原始格式文件的核心代码如下：

```
Resources resources = this.getResources();
InputStream inputStream = null;
try {
    inputStream = resources.openRawResource(R.raw.raw_file);
    byte[] reader = new byte[inputStream.available()];
```

```
        while (inputStream.read(reader) != -1) { }
        displayView.setText(new String(reader,"utf-8"));
    } catch (IOException e) {
        Log.e("ResourceFileDemo", e.getMessage(), e);
    } finally {
        if (inputStream != null) {
        try {
            inputStream.close();    }
        catch (IOException e) { }
        }
```

代码中的 new String(reader,"utf-8")，表示以 UTF-8 的编码方式，从字节组中实例化一个字符串。程序开发人员需要确定\res\raw\raw_file.txt 文件使用的 UTF-8 的编码方式，否则程序运行时会产生乱码。确认的方法是，在 raw_file.txt 文件上右击，选择 Properties，打开 raw_file.txt 文件的属性设置框，然后在 Resource 栏下的 Text file encoding 中，选择 Other 为 UTF-8，如图 8-12 所示。

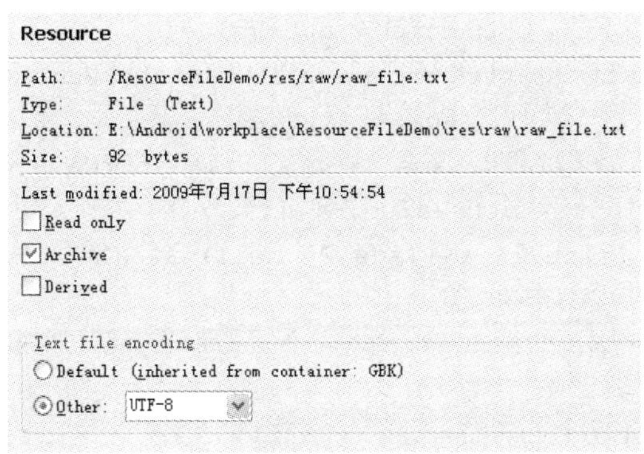

图 8-12　更改 raw_file.txt 文件编码方式图

\res\xml 目录下的 XML 文件，与其他资源文件有所不同，程序开发人员不能够以流的方式进行读取，其主要原因在于 Android 系统为了提高读取效率，减少占用的存储空间，将 XML 格式转换为高效的二进制格式。为了说明如何在程序运行时读取\res\xml 目录下的 XML 文件，首先在\res\xml 目录下创建一个名为 people.xml 的文件。XML 文件定义了多个<person>元素，每个<person>元素都包含三个属性，即 name，age 和 height，分别表示姓名，年龄和身高。\res\xml\people.xml 文件代码如下：

```
<people>
    <person name="李某某" age="21" height="1.81" />
    <person name="王某某" age="25" height="1.76" />
    <person name="张某某" age="20" height="1.69" />
</people>
```

读取 XML 格式文件，首先通过调用资源对象的 getXML()函数，获取到 XML 解析器 XmlPullParser。XmlPullParser 是 Android 平台标准的 XML 解析器，这项技术来自于一个开源的 XML 解析 API 项目 XMLPULL。

ResourcesFileDemo 示例关于读取 XML 文件的核心代码如下：

```
XmlPullParser parser = resources.getXml(R.xml.people);
String msg = "";
try {
    while (parser.next() != XmlPullParser.END_DOCUMENT) {
        String people = parser.getName();
        String name = null;
String age = null;
        String height = null;
        if ((people != null) && people.equals("person")) {
            int count = parser.getAttributeCount();
            for (int i = 0; i < count; i++) {
                String attrName = parser.getAttributeName(i);
                String attrValue = parser.getAttributeValue(i);
//通过分析属性名称获取到正确的属性值
        if ((attrName != null) && attrName.equals("name")){
                    name = attrValue;
        } else if ((attrName != null) && attrName.equals("age")){
                    age = attrValue;
        } else if ((attrName != null) && attrName.equals("height")){
            height = attrValue; } }
    if ((name != null) && (age != null) && (height != null)) {
//将属性值整理成需要显示的信息
      msg += "姓名："+name+"，年龄："+age+"，身高："+height+"\n";
        } } }
} catch (Exception e) {
Log.e("ResourceFileDemo", e.getMessage(), e); }
displayView.setText(msg);
```

代码中通过资源对象的 getXml()函数获取到 XML 解析器。parser.next()方法可以获取到高等级的解析事件，并通过对比确定事件类型，XML 事件类型参考表 8-2。

表 8-2　XmlPullParser 的 XML 事件类型表

事件类型	说　明	事件类型	说　明
START_TAG	读取到标签开始标志	END_TAG	读取到标签结束标志
TEXT	读取文本内容	END_DOCUMENT	文档末尾

使用 getName()函数获得元素的名称，使用 getAttributeCount()函数获取元素的属性数量，通过 getAttributeName()函数得到属性名称。

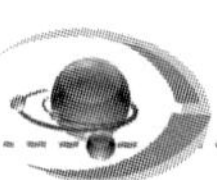

8.3 SQLite 数据库存储

8.3.1 SQLite 数据库

SQLite 是一个开源的嵌入式关系数据库，2000 年由 D.Richard Hipp 发布。自几十年前出现的商业应用程序以来，数据库就成为了应用程序的主要组成部分，同时数据库管理系统也变得非常庞大和复杂，并占用了相当多的系统资源。随着嵌入式应用程序的大量出现，一种新型的轻量级数据库 SQLite 也随之产生。SQLite 数据库比传统的数据库更加适用于嵌入式系统，因为它占用空间非常少，运行高效可靠，可移植性好，并且提供了零配置（zero-configuration）运行模式。

SQLite 数据库的优势在于其嵌入到使用它的应用程序中。这样不仅提高了运行效率，而且屏蔽了数据库使用和管理的复杂性，程序仅需要进行最基本的数据操作，其他操作可以交给进程内部的数据库引擎完成。同时因为客户端和服务器在同一进程空间运行，不需要进行网络配置和管理，因此减少了网络调用所造成的额外开销，简化的数据库的管理过程，使应用程序更加易于部署和使用。程序开发人员仅需要把 SQLite 数据库正确编译到应用程序即可。

SQLite 数据库采用了模块化设计，模块将复杂的查询过程分解为细小的工作进行处理。SQLite 数据库有 8 个独立的模块构成，这些独立的模块又构成了三个主要的子系统。SQLite 数据库体系结构如图 8-13 所示。

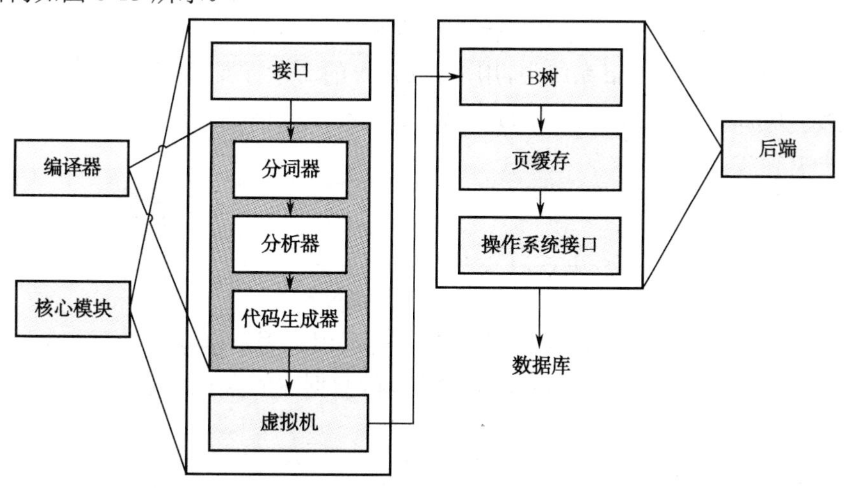

图 8-13 SQLite 数据库体系结构图

接口由 SQLite C API 组成，因此无论是应用程序、脚本，还是库文件，最终都是通过接口与 SQLite 交互。

在编译器中，分词器和分析器对 SQL 语句进行语法检测，然后把 SQL 语句转化为底层能更方便处理的分层的数据结构，这种分层的数据结构称为“语法树”。然后把语法树传给代码生成器进行处理，生成一种针对 SQLite 的汇编代码，最后由虚拟机执行。

SQLite 数据库体系结构中最核心的部分是虚拟机，也称为虚拟数据库引擎。与 Java 虚拟机相似，虚拟数据库引擎用来解释执行字节代码。虚拟数据库引擎的字节代码由 128 个操作码构成，这些操作码主要用以对数据库进行操作，每一条指令都可以完成特定的数据库操作，或以特定的方式处理栈的内容。

后端由 B-树，页缓存和操作系统接口构成。B-树和页缓存共同对数据进行管理。B-树的主要功能就是索引，它维护着各个页面之间的复杂的关系，便于快速找到所需数据。页缓存的主要作用就是通过操作系统接口在 B-树和磁盘之间传递页面。

SQLite 数据库具有很强的移植性，可以运行在 Windows、Linux、BSD、Mac OS X 和一些商用 UNIX 系统，比如 Sun 公司的 Solaris，IBM 公司的 AIX。同样也可以工作在许多嵌入式操作系统下，例如 QNX、VxWorks、Palm OS、Symbin 和 Windows CE。SQLite 的核心大约有 3 万行标准 C 代码，模块化的设计使这些代码更加易于理解。

8.3.2 手动建库

在 Android 系统中，每个应用程序的 SQLite 数据库被保存在各自的\data\data\<package name>\databases 目录下。默认情况下，所以数据库都是私有的，仅允许创建数据库的应用程序访问，如果需要共享数据库则可以使用 ContentProvider。虽然应用程序完全可以在代码中动态建立 SQLite 数据库，但通过命令行手工建立和管理数据库仍然是非常重要的内容，对于调用具有数据库的应用程序非常有用。

手动建立数据库指的是使用 sqlite3 工具，通过手工输入命令行完成数据库的建立过程。sqlite3 是 SQLite 数据库自带的一个基于命令行的 SQL 命令执行工具，并可以显示命令执行结果。sqlite3 工具被集成在 Android 系统中，用户在 Linux 的命令行界面中输入 sqlite3 可启动 sqlite3 工具，并得到工具的版本信息，如下面的代码所示。启动 Linux 的命令行界面的方法是在 CMD 中输入 adb shell 命令。

```
# sqlite3
SQLite version
Enter ".help" for instructions
sqlite>
```

在启动 sqlite3 工具后，提示符从“#”变为“sqlite>”，表示命令行界面进入与 SQLite 数据库的交互模式，此时可以输入命令建立，删除或修改数据库的内容。正确退出 sqlite3 工具的方法是使用.exit 命令。

```
sqlite> .exit
#
```

原则上，每个应用程序的数据库都保存在各自的\data\data\<package name>\databases 目录下，但如果使用手工方式建立数据库，则必须手工建立数据库目录，目前版本无须修改数据库目录的权限。

```
# mkdir databases
# ls -l
drwxrwxrwx root     root            2009-07-18 15:43 databases
drwxr-xr-x system   system           2009-07-18 15:31 lib
#
```

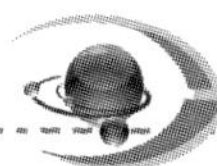

在 SQLite 数据库中，每个数据库保存在一个独立的文件中，使用 sqlite3 工具后加文件名的方式打开数据库文件，如果指定的文件不存在，sqlite3 工具则自动创建新文件。下面的代码将创建名为 people 的数据库，在文件系统中将产生一个名为 people.db 的数据库文件。

```
# sqlite3 people.db
SQLite version
Enter ".help" for instructions
sqlite>
```

下面的代码在新创建数据库中，构造了一个名为 peopleinfo 的表，使用 Create table 命令，关系模式为 peopleinfo(_id,name,age,height)。表包含 4 个属性，_id 是整型的主键；name 表示姓名，字符型；not null 表示这个属性一定要填写，不可以为空值；age 表示年龄，整数型；height 表示身高，浮点型。

```
sqlite> create table peopleinfo
...> (_id integer primary key autoincrement,
...> name text not null,
...> age integer,
...> height float);
sqlite>
```

为了确认数据表是否创建成功，可以使用.table 命令，显示当前数据库中的所有表。从下面的代码中可以观察到，当前数据库只有一个名为 peopleinfo 的表。

```
sqlite> .tables
poepleinfo
sqlite>
```

当然也可以使用.schema 命令查看建立表时所使用的 SQL 命令。如果当前数据库中包含多个表，则可以使用[.schema 表名]的形式，显示指定表的建立命令。

```
sqlite>.schema
CREATE TABLE peopleinfo (_id integer primary key autoincrement,
name text not null, age integer, height float);
sqlite>
```

下一步是向 peopleinfo 表中添加数据库，使用 insert into...values 命令。在下面的代码运行成功后，数据库 peopleinfo 表将有三条数据，内容如表 8-3 所示。因为_id 是自添加的主键，因此输入 null 后，SQLite 数据库会自动填写该项的内容。

```
sqlite> insert into peopleinfo values(null,'Tom',21,1.81);
sqlite> insert into peopleinfo values(null,'Jim',22,1.78);
sqlite> insert into peopleinfo values(null,'Lily',19,1.68);
```

表 8-3 peopleinfo 表内容

_id	name	age	height
1	Tom	21	1.81
2	Jim	22	1.78
3	Lily	19	1.68

在数据添加完毕后，使用 select 命令，显示指定数据表中的所有数据信息，命令格式为[select 属性 form 表名]。下面的代码用来显示 peopleinfo 表的所有数据。

```
select * from peopleinfo;
1|Tom|21|1.81
2|Jim|22|1.78
3|Lily|19|1.68
sqlite>
```

上面的查询结果看起来不是非常直观，可以使用 mode 命令将结果输出格式更改为“表格”方式。mode 命令除了支持常见的 column 格式，还支持 csv 格式、html 格式、insert 格式、line 格式、list 格式、tabs 格式和 tcl 格式。

```
sqlite> .mode column
sqlite> select * from peopleinfo;
1          Tom        21          1.81
2          Jim        22          1.78
3          Lily       19          1.68
sqlite>
```

更新数据可以使用 update 命令，命令格式为[update 表名 set 属性= “新值” where 条件]。更新数据后同样使用 select 命令显示数据，则可以确定数据是否正确更新。下面的代码将姓名为 Lily 数据中的高度值更新为 1.88。

```
sqlite> update peopleinfo set height=1.88 where name="Lily";
sqlite> select * from peopleinfo;
select * from peopleinfo;
1          Tom        21          1.81
2          Jim        22          1.78
3          Lily       19          1.88
sqlite>
```

删除数据可以使用 delete 命令，命令格式为[delete form 表名 where 条件]。下面的代码将 _id 为 3 的数据从表 peopleinfo 中删除。

```
sqlite> delete from peopleinfo where _id=3;
sqlite> select * from peopleinfo;
select * from peopleinfo;
1          Tom        21          1.81
2          Jim        22          1.78
sqlite>
```

sqlite3 工具还支持大量的命令，可以使用.help 命令查询 sqlite3 的命令列表，也可以参考表 8-4。

表 8-4　sqlite3 命令列表

编　号	命　令	说　明
1	.bail ON|OFF	遇到错误时停止，默认为 OFF
2	.databases	显示数据库名称和文件位置

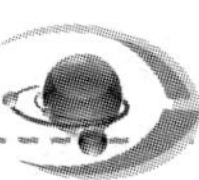

续表

编　号	命　令	说　明
3	.dump?TABLE?...	将数据库以 SQL 文本形式导出
4	.echo ON|OFF	开启和关闭回显
5	.exit	退出
6	.explain ON|OFF	开启或关闭适当输出模式，如果开启模式将更改为 column，并自动设置宽度
7	.header(s)ON|OFF	开启或关闭标题显示
8	.help	显示帮助信息
9	.import FILE TABLE	将数据从文件导入表中
10	.indices TABLE	显示表中所有的列名
11	.load FILE?ENTRY?	导入扩展库
12	.mode MODE?TABLE?	设置输入格式
13	.nullvalue STRING	打印是使用 STRING 代替 NULL
14	.output FILENAMG	将输入保存到文件
15	.output stdout	将输入显示在屏幕上
16	.prompt MAIN CONTINUE	替换标准提示符
17	.quit	退出
18	.read FILENAME	在文件中执行 SQL 语句
19	.schema ? TABLE	显示表的创建语句
20	.separator STRING	更改输入和输出的分隔符
21	.show	显示当前设置变量值
22	.tables ? PATTERN?	显示符合匹配模式的表名
23	.timeout MS	尝试打开被锁定的表 MS 毫秒
24	.timer ON|OFF	开启或关闭 CPU 计时器
25	.width NUM NUM…	设置 column 模式的宽度

8.3.3　代码建库

在代码中动态建立数据库是比较常用的方法。例如在程序运行过程中，当需要进行数据库操作时，应用程序会首先尝试打开数据库，此时如果数据库不存在，程序则会自动建立数据库，然后再打开数据库。

在编程实现时，一般将所有对数据库的操作都封装在一个类中，因此只要调用这个类，就可以完成对数据库的添加、更新、删除和查询等操作。下面内容是 DBAdapter 类的部分代码，封装了对数据库的建立，打开和关闭等操作：

```
public class DBAdapter {
    private static final String DB_NAME = "people.db";
```

```
        private static final String DB_TABLE = "peopleinfo";
        private static final int DB_VERSION = 1;
        public static final String KEY_ID = "_id";
        public static final String KEY_NAME = "name";
        public static final String KEY_AGE = "age";
        public static final String KEY_HEIGHT = "height";
        private SQLiteDatabase db;
        private final Context context;
        private DBOpenHelper dbOpenHelper;
    private static class DBOpenHelper extends SQLiteOpenHelper {}
        public DBAdapter(Context _context) {
               context = _context; open(); }
    public void open() throws SQLiteException {
    dbOpenHelper = new DBOpenHelper(context,DB_NAME,null,DB_VERSION);
     try {db = dbOpenHelper.getWritableDatabase();
          }catch (SQLiteException ex) {
        db = dbOpenHelper.getReadableDatabase();
               }}
    public void close() {
           if (db != null){
        db.close();db = null; }}
    }
```

从代码中可以看出，在DBAdapter类中首先声明了数据库的基本信息，包括数据库文件的名称，数据库表格名称和数据库版本，以及数据库表中的属性名称。从这些基本信息上不难发现，这个数据库与8.2.3节手动建立的数据库是完全相同的。

代码中声明了SQLiteDatabase对象db。SQLiteDatabase类封装了非常多的方法，用以建立，删除数据库，执行SQL命令，对数据库进行管理等工作。声明了一个非常重要的帮助类SQLiteOpenHelper，这个帮助类可以辅助建立，更新和打开数据库。虽然在代码中也自定义了open()函数用来打开数据库，但open()函数中没有任何对数据库进行操作的代码，而是调用了SQLiteOpenHelper类的getWritableDatabase()函数和getReadableDatabase()函数。这两个函数会根据数据库是否存在，版本号和是否可写等情况，决定在返回数据库对象前，是否需要建立数据库。

在代码中的close()函数中，调用了SQLiteDatabase对象的close()方法关闭数据库。这是上面的代码中，唯一直接调用了SQLiteDatabase对象方法的地方。SQLiteDatabase中也封装了打开数据库的函数openDatabases()和创建数据库函数openOrCreate Databases()，因为代码中使用了帮助类SQLiteOpenHelper，从而避免直接调用SQLiteDatabase中的打开和创建数据库的方法，简化了数据库打开过程中烦琐的逻辑判断过程。

代码中实现了内部静态类DBOpenHelper，继承了帮助类SQLiteOpenHelper，该类中需要重载onCreate()函数和onUpgrade()函数，具体代码如下：

```
    private static class DBOpenHelper extends SQLiteOpenHelper {
    public DBOpenHelper(Context context, String name,
```

```
 CursorFactory factory, int version){
super(context, name, factory, version);}
 private static final String DB_CREATE = "create table " +
DB_TABLE + " (" + KEY_ID + " integer primary key autoincrement,
" + KEY_NAME+ " text not null, " + KEY_AGE+ " integer," + KEY_HEIGHT
 + " float);";
  @Override
 public void onCreate(SQLiteDatabase _db) {
      _db.execSQL(DB_CREATE); }
@Override
public void onUpgrade(SQLiteDatabase _db, int _oldVersion,
int _newVersion) {
_db.execSQL("DROP TABLE IF EXISTS " + DB_TABLE);
    onCreate(_db);
    }
}
```

代码开始是构造方法和创建表的SQL命令。然后分别重载了onCreate()函数和onUpgrade()函数，这是继承SQLiteOpenHelper类必须重载的两个函数。onCreate()函数在数据库的第一次建立时被调用，一般用来创建数据库中的表，并做适当的初始化工作。在代码中，通过调用SQLiteDatabase对象的execSQL()方法，执行建立表的SQL命令。onUpgrade()函数在数据库需要升级时被调用，一般用来删除旧的数据库表，并将数据转移到新版本的数据库表中。在代码中，为了简单起见，并没有做任何数据转移，而仅仅删除原有的表后建立新的数据库表。

程序开发人员不应直接调用onCreate()函数和onUpgrade()函数，而应该由SQLiteOpenHelper类来决定何时调用这两个函数。SQLiteOpenHelper类的getWritableDatabase()函数和getReadableDatabase()函数是可以直接调用的函数。getWritableDatabase()函数用来建立或打开可读写的数据库对象，一旦函数调用成功，数据库对象将被缓存，任何需要使用数据库对象时，都可以调用这个方法获取到数据库对象，但一定要在不使用时调用close()函数关闭数据库。如果保存数据库磁盘空间已满，调用getWritableDatabase()函数则无法获得可读写的数据库对象，这时可以调用getReadableDatabase()函数，获得一个只读的数据库对象。

当然，如果程序开发人员不希望使用SQLiteOpenHelper类，同样可以直接创建数据库。首先调用openOrCreateDatabases()函数创建数据库对象，然后执行SQL命令建立数据库中的表和直接的关系，示例代码如下：

```
private static final String DB_CREATE = "create table " +
DB_TABLE + " (" + KEY_ID + " integer primary key autoincrement, "
 +KEY_NAME+ " text not null, " + KEY_AGE+ " integer," + KEY_HEIGHT
 + " float);";
public void create() {
    db.openOrCreateDatabases(DB_NAME, context.MODE_PRIVATE, null)
    db.execSQL(DB_CREATE);
}
```

8.3.4 数据操作

数据操作指的是对数据的添加，删除，查找和更新的操作，虽然程序开发人员完全可以通过执行 SQL 命令完成数据库操作，但还是推荐使用 Android 提供的专用类和方法，这些类和方法更加简洁，易用。

为了使 DBAdapter 支持对数据的添加、删除、查找和更新等功能，在 DBAdapter 类中增加下面的这些函数。其中，insert(People people)用来添加一条数据，queryAllData()用来获取全部数据，queryOneData(long id)根据 id 获取一条数据，deleteAllData()用来删除全部数据，deleteOneData(long id) 根据 id 删除一条数据，updateOneData(long id,People people) 根据 id 更新一条数据。

```
public class DBAdapter {
    public long insert(People people) {}
    public long deleteAllData() { }
    public long deleteOneData(long id) { }
    public People[] queryAllData() {}
    public People[] queryOneData(long id) { }
    public long updateOneData(long id , People people){ }
    private People[] ConvertToPeople(Cursor cursor){}
}
```

ConvertToPeople(Cursor cursor)是私有函数，作用是将查询结果转换为用来存储数据自定义的 People 类对象。People 类包含 4 个公共属性，分别为 ID、Name、Age 和 Height，对应数据库中的 4 个属性值。重载 toString()函数，主要是便于界面显示需要。People 类的代码如下：

```
public class People {
        public int ID = -1;
        public String Name;
        public int Age;
        public float Height;
        @Override
        public String toString(){
            String result = "";
            result += "ID: " + this.ID + ", ";
            result += "姓名: " + this.Name + ", ";
            result += "年龄: " + this.Age + ",  ";
            result += "身高: " + this.Height + ", ";
            return result;
        }
}
```

SQLiteDatabase 类的公共函数 insert()、delete()、update()和 query()，封装了执行的添加、删除、更新和查询功能的 SQL 命令。下面分别介绍如何使用 SQLiteDatabase 类的公共函数，完成数据的添加、删除、更新和查询等操作。

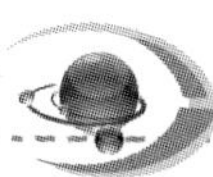

1. 添加功能。

为了添加一条新数据，首先构造一个 ContentValues 对象，然后调用 ContentValues 对象的 put()方法，将每个属性的值写入到 ContentValues 对象中，最后使用 SQLiteDatabase 对象的 insert()函数，将 ContentValues 对象中的数据写入指定的数据库表中。insert()函数的返回值是新数据插入的位置，即 ID 值。ContentValues 类是一个数据承载容器，主要用来向数据库表中添加一条数据。示例代码如下。

```
public long insert(People people) {
    ContentValues newValues = new ContentValues();
    newValues.put(KEY_NAME, people.Name);
    newValues.put(KEY_AGE, people.Age);
    newValues.put(KEY_HEIGHT, people.Height);
    return db.insert(DB_TABLE, null, newValues);
}
```

代码中向 ContentValues 对象 newValues 中添加一个名称/值对，put()函数的第 1 个参数是名称，第 2 个参数是值。在 insert()函数中，第 1 个参数是数据表的名称，第 2 个参数是在 NULL 时的替换数据，第 3 个参数是需要向数据库表中添加的数据。

2. 删除功能。

删除数据比较简单，只需调用当前数据库对象 delete()函数，并指明表名称和删除条件即可。示例代码如下。

```
public long deleteAllData() {
    return db.delete(DB_TABLE, null, null);}
public long deleteOneData(long id) {
    return db.delete(DB_TABLE,  KEY_ID + "=" + id, null);}
```

delete()函数的第 1 个参数是数据库的表名称，第 2 个参数是删除条件。删除条件为 null，表示删除表中的所有数据。而代码中指明了需要删除数据的 id 值，因此 deleteOneData()函数仅是删除一条数据，此时 delete()函数返回值表示被删除的数据的数量。

3. 更新功能。

更新数据同样要使用 ContentValues 对象，首先构造 ContentValues 对象，然后调用 put()函数将属性值写入到 ContentValues 对象中，最后使用 SQLiteDatabase 对象的 update()函数，并指定数据的更新条件。示例代码如下。

```
public long updateOneData(long id , People people){
    ContentValues updateValues = new ContentValues();
    updateValues.put(KEY_NAME, people.Name);
    updateValues.put(KEY_AGE, people.Age);
    updateValues.put(KEY_HEIGHT, people.Height);
    return db.update(DB_TABLE, updateValues,  KEY_ID + "=" + id, null);
}
```

update()函数的第 1 个参数是数据库表名称，第 2 个参数是更新数据，第 3 个参数是更新条件。update ()函数返回值表示数据库表中被更新的数据的数量。

4. 查询功能。

介绍查询功能前，先介绍 Cursor 类。在 Android 系统中，数据库查询结果的返回值并不是数据集合的完整复本，而是返回数据集的指针，这个指针就是 Cursor 类。Cursor 类支持在数据查询的数据集合中多种移动方式，并能够获取数据集合的属性名称和序号，具体的方法和说明可以参考表 8-5。

表 8-5　Cursor 类的方法和说明表

函　　数	说　　明
moveToFirst	将指针移动到第一条数据上
moveToNext	将指针移动到下一条数据上
moveToPrevious	将指针移动到上一条数据上
getCount	获取集合的数据数量
getColumnlnIndexOrThrow	返回指定属性名称的序号，如果属性不存在则产生异常
getColumnName	返回指定序号的属性名称
getColumnNames	返回属性名称的字符串数组
getColumnIndex	根据属性名称返回序号
moveToPosition	将指针移动到指定的数据上
getPosition	返回当前指针的位置

从 Cursor 中提取数据可以参考 ConvertToPeople()函数的实现方法。在提取 Cursor 数据中的数据前，推荐测试 Cursor 中的数据数量，避免在数据获取中产生异常，从 Cursor 中提取数据使用安全类型的 get<Type>()函数，函数的输入值为属性的序号，为了获取属性的序号，可以使用 getColumnIndex()函数获取指定属性的序号。

```
private People[] ConvertToPeople(Cursor cursor){
    int resultCounts = cursor.getCount();
    if (resultCounts == 0 || !cursor.moveToFirst()){
            return null; }
    People[] peoples = new People[resultCounts];
    for (int i = 0 ; i<resultCounts; i++){
        peoples[i] = new People();
        peoples[i].ID = cursor.getInt(0);
     peoples[i].Name=cursor.getString(cursor.getColumnIndex
(KEY_NAME));
       peoples[i].Age= cursor.getInt(cursor.getColumnIndex(KEY_AGE));
peoples[i].Height=cursor.getFloat(cursor.getColumnIndex
(KEY_HEIGHT));
        cursor.moveToNext();  }
```

```
        return peoples;
    }
```

要进行数据查询就需要调用 SQLiteDatabase 类的 query()函数，这个函数的参数较多，可以参照参数说明表 8-6，query()函数的语法如下：

```
    Cursor android.database.sqlite.SQLiteDatabase.query(String table, String[]
columns, String selection, String[] selectionArgs, String groupBy, String having,
String orderBy)
```

表 8-6 query()函数的参数说明表

位 置	类型+名称	说 明
1	String table	表名称
2	String[] columns	返回的属性列名称
3	String selection	查询文件
4	String[] selectionArgs	如果在查询条件中使用问号，则需要定义替换符的具体内容
5	String groupBy	分组方式
6	String having	定义组的过滤器
7	String orderBy	排序方式

下面给出根据 id 查询数据的代码:

```
    public People[] getOneData(long id) {
        Cursor results =  db.query(DB_TABLE, new String[] { KEY_ID, KEY_NAME,
     KEY_AGE, KEY_HEIGHT}, KEY_ID + "=" + id, null, null, null, null);
        return ConvertToPeople(results);
    }
```

下面是查询全部数据的代码:

```
    public People[] getAllData() {
        Cursor results = db.query(DB_TABLE, new String[] { KEY_ID, KEY_NAME,
     KEY_AGE, KEY_HEIGHT}, null, null, null, null, null);
        return ConvertToPeople(results);
    }
```

SQLiteDemo 是对数据库操作的一个示例，如图 8-14 所示。在这个示例中，用户可以在界面的上方输入数据信息，通过“添加数据”按钮将数据写入数据库。“全部显示”按钮相当于查询数据库中的所有数据，并将数据显示在界面下方。“清除显示”按钮清除界面下面显示的数据，而不对数据库中的所有数据进行任何操作。在界面中部，以 ID+功能命名的按钮，分别是根据 ID 删除数据，根据 ID 进行数据查询，根据 ID 更新数据，而这个 ID 值就取自本行的 EditText 控件。

图 8-14 SQLiteDemo 用户界面图

8.4 内容提供器——Content Providers

Android 操作系统采用内容提供器机制来公开自己私有的数据到数据内容容器，通过授权机制，可以供其他应用程序来读取数据。内容提供器是应用程序的一个可选的机制，它提供标准语法来读取与写入数据。

一般而言，Android 操作系统的应用程序所建立的数据只允许自己使用，应用程序彼此间无法借助公用存储器来共享数据，Android 操作系统提供一个机制称为内容提供器来公开数据，如图 8-15 所示，通过数据内容提供机制，应用程序 B 可以读出和写入应用程序 A 的数据库数据。

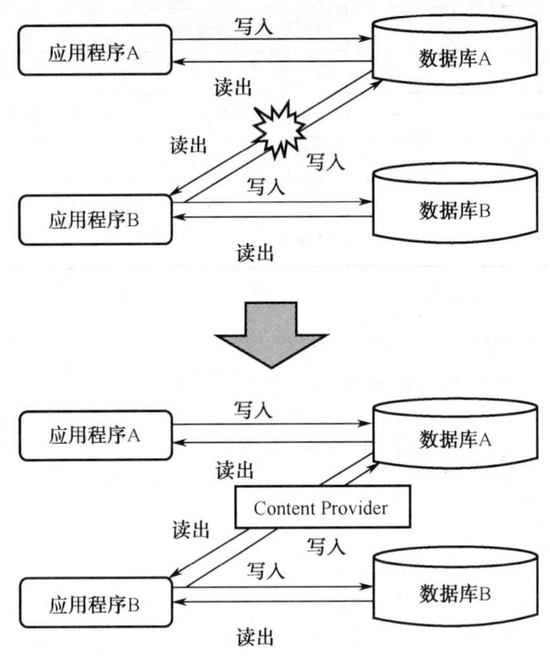

图 8-15　Content Providers 读出与写入图

实现内容提供器和实现 SQLite 数据有什么不同呢？应用程序实现 SQLite 数据库时，由应用程序直接和数据库接口，所以在应用程序中需要实现如图 8-16 所示的 SQLite 的接口 db.onCreate()、db.insert()、db.update()、db.delete()、db.query()和 db.close()。在实现内容提供器时，在应用程序和数据库之间要实现一个 Content Providers 程序，这个 Content Providers 程序会直接和数据库接口，此时应用程序需要实现和 Content Providers 程序接口的方法。新建立的类继承 ContentProvider 后，共有 6 个函数需要重载，分别是 delete()、getType()、insert()、onCreate()、query()和 update()。其中 delete()、insert()、query()和 update()分别用于对数据集的删除、添加、查询和更新操作，程序开发人员根据底层数据的存储方式不同，使用不同方式实现数据操作函数。而 onCreate()一般用来初始化底层数据集和建立数据连接等工作。getType()函数用来返回指定 URI 的 MIME 数据类型。

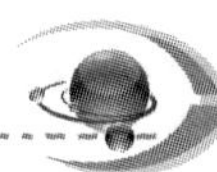

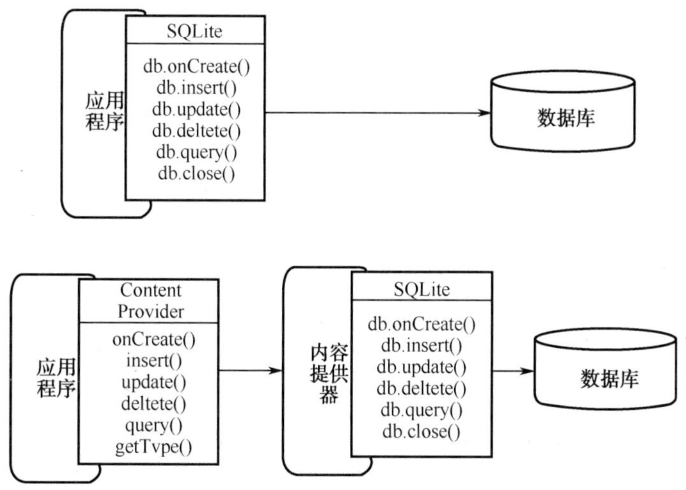

图 8-16 实现 Content Providers 和实现 SQLite 数据库图

新建立的类继承 ContentProvider 后，Eclipse 会提示程序开发人员需要重载部分代码，并自动生成需要重载的代码框架。下面的代码是 Eclipse 自动生成的代码框架：

```
public class PeopleProvider extends ContentProvider{
    @Override
    public int delete(Uri uri, String selection, String[] selectionArgs){
        // TODO Auto-generated method stub
        return 0;}
@Override
    public String getType(Uri uri) {
        // TODO Auto-generated method stub
        return null;}
    @Override
public Uri insert(Uri uri, ContentValues values) {
        // TODO Auto-generated method stub
        return null;}
    @Override
    public boolean onCreate() {
        // TODO Auto-generated method stub
        return false;}
@Override
    public Cursor query(Uri uri, String[] projection, String selection,
            String[] selectionArgs, String sortOrder) {
        // TODO Auto-generated method stub
        return null;}
    @Override
    public int update(Uri uri, ContentValues values, String selection,
            String[] selectionArgs) {
        // TODO Auto-generated method stub
```

```
        return 0;}
    }
```

建立内容提供器的实现，首先完成下列三项工作：

（1）建立一个系统来存储数据，大部分内容提供器程序采用 Android 文件存储方法或以 SQLite 数据库来管理。Android 操作系统提供 SQLiteOpenHelper 类协助建立数据库和 SQLite Database 类来管理数据库。

（2）继承 ContentProvider 类来读取数据，要读取数据的应用程序需要通过 ContentResolver 和 Cursor 类来实现。共有 6 个 ContentProvider 子类方法要声明：query()、insert()、update()、delete()、getType()和 onCreate()。

（3）在要读取数据的应用程序的清单文件中声明一个公开的 URI。

那应用程序又是如何找到内容提供器程序呢？每一个内容提供器程序会在 AndroidManifiest.xml 配置文件一个公开的 URI，这个 URI 是唯一的，相对应于某一个特定的数据库。下面代码是本节案例在 AndroidManifest.xml 配置文件内声明的 URI。

```
<provider android:name="TestProvider"
android:authorities="com.example.android.provider.testprovider" />
```

所有 URI 提供给应用程序的字符串前头是"content://"，content:是表示内容提供器程序所控制数据的位置。在应用程序中要定义 URI 所在的位置，然后要设置一个 URI 类变量 uri_test，来找到内容提供器程序的接口，这里的案例所连接上的接口是 TestProvider 内容提供器程序。参考代码如下：

```
getIntent().setData(Uri.parse("content://com.example.android.provider.t
estprovider"));
 Uri uri_test = getIntent().getData();
```

内容提供器所管理的数据，应用程序可以利用 ContentProvider 子类方法来添加、更新、删除和查找。

（1）添加一行新数据。应用程序通过内容提供器调用 ContentResolver.insert()方法来添加。

```
public final Uri insert (Uri url, ContentValues values)
```

url：URI 所在的位置。

values：添加一行数据的字段的初始值。设成空白字段是会添加一行空白行。

（2）更新一行数据。应用程序通过内容提供器调用 ContentResolver.update()方法来更新。

```
public int update(Uri uri, ContentValues values, String selection,
        String[] selectionArgs)
```

url：URI 所在的位置。

values：对应到一行数据的字段的修改值。

selection：设置查询的条件，与 SQL WHERE 子句相同，但没有 WHERE 关键字。设置成 null 时表示所有行都要返回。

selectionArgs：定义 SQL WHERE 子句的相关查询参数，在语句中有"？"参数。

（3）删除一行数据。应用程序通过内容提供器调用 ContentResolver.delete()方法来删除。

```
public int delete(Uri uri, String selection, String[] selectionArgs)
```

url：URI 所在的位置。

selection：设置查询的条件，和 SQL WHERE 子句相同，但没有 WHERE 关键字。设置成 null 时表示所有列都要删除。

selectionArgs：定义 SQL WHERE 子句的相关查询参数，在语句中有"？"参数。

（4）检索数据。应用程序通过内容提供器调用 ContentResolver.query()或 managedQuery()方法来检索。因检索提供了 ContentResolver.query()或 managedQuery()两种方法。

```
        public Cursor query(Uri uri, String[] projection, String selection,
String[] selectionArgs, String sortOrder)
```

url：URI 所在的位置。

projection：检索后的返回字段，设置为 null 时返回所有字段，不建议设成 null，一定要定义要求返回的字段。

selection：设置查询的条件，和 SQL WHERE 子句相同，但没有 WHERE 关键字。设置成 null 时表示所有行都要查询。

selectionArgs：定义 SQL WHERE 子句的相关查询参数，在语句中有"？"参数。

sortOrder：设置排序的条件，和 SQL ORDER BY 子句相同，但没有 ORDER BY 关键字。设置为 null 时表示不需要排序。

（5）取得 URI 的种类。应用程序通过内容提供器调用 ContentResolver.getType()方法来取得返回的 URI 的 MINE 种类。

```
      public String getType(Uri uri)
```

url：URI 所在的位置。

接下来，我们利用实际案例程序 ProviderTest.java 和内容提供器程序 TestProvider.java 来说明如何建立内容提供器的机制，以及如何通过内容提供器程序来添加和检索数据的方法。本案例程序主要是说明内部提供器的构建方法，建立一个 SQLite 数据库系统来存储和管理数据，同时利用 SQLiteOpenHelper 类协助建立数据库和 SQLiteDatabase 类来管理数据库，其项目结构如图 8-17 所示。本案例直接在应用程序内生成测试用的数据，然后检索存储在数据库内的所有行，并用列表字段显示出来。程序运行效果如图 8-18 所示。

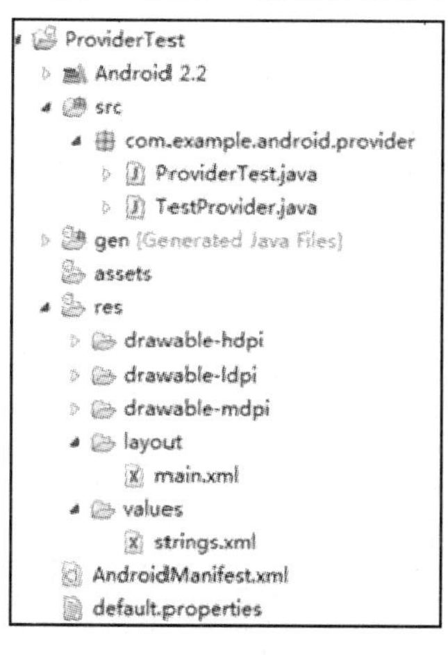

图 8-17　项目结构图

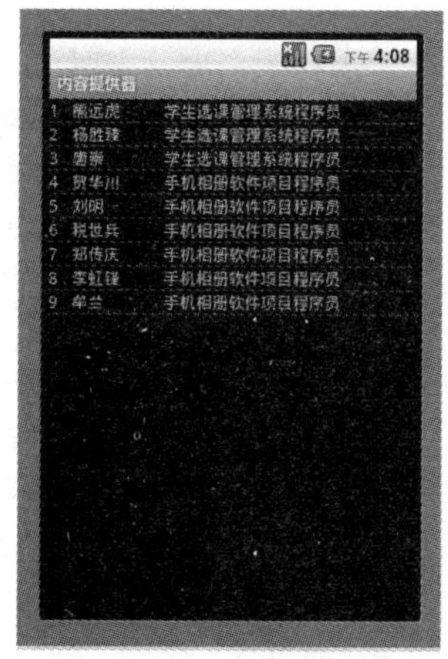

图 8-18　效果图

实际案例定义了一张表 test，共有三个字段。表的结构如表 8-7 所示。

表 8-7　表 test 的数据字典

数据名称	数据类型	数据属性
_id	INTEGER	主键，自动生成
name	TEXT	
description	TEXT	

案例 ProviderTest.java 的 AndroidManifest.xml 文件的参考代码如下：

```
<?xml version="1.0" encoding="utf-8"?>
<manifest xmlns:android="http://schemas.android.com/apk/res/android"
    android:versionCode="1"
    android:versionName="1.0" package="com.example.android.provider">
  <application android:icon="@drawable/icon" android:label="@string/
  app_name">
      <provider android:name="TestProvider"
 android:authorities="com.example.android.provider.testprovider" />
      <activity android:label="@string/app_name" android:name="ProviderTest">
   <intent-filter>
      <action android:name="android.intent.action.MAIN" />
      <category android:name="android.intent.category.LAUNCHER" />
    </intent-filter>
  </activity>
  </application>
  <uses-sdk/>
</manifest>
```

TestProvider.java 类的代码如下：

```
public class TestProvider extends ContentProvider {
   //SQLiteOpenHelper-建立数据库 test.db 和 Table:test
  private static class DatabaseHelper extends SQLiteOpenHelper{
      //建立 test.db 数据库
   DatabaseHelper(Context context) {
         super(context, "test.db", null, 1);
      }
   //建立 test 表
      @Override
      public void onCreate(SQLiteDatabase db) {
         db.execSQL("CREATE TABLE test (" + BaseColumns._ID
               + " INTEGER PRIMARY KEY," + "name TEXT,"
               + "description TEXT" + ");");
      }
      //更新新版本
      @Override
      public void onUpgrade(SQLiteDatabase db, int oldVersion, int
      newVersion) {
```

```
            // TODO Auto-generated method stub
            db.execSQL("DROP TABLE IF EXISTS test");
            onCreate(db);
        }
    }
    //定义 DatabaseHelper 类变量 databaseHelper
    DatabaseHelper databaseHelper;
    //实现 Content Providers 的 onCreate()
    @Override
    public boolean onCreate() {
        databaseHelper = new DatabaseHelper(getContext());
        return true;
    }
    //实现 Content Providers 的 insert()
    @Override
    public Uri insert(Uri uri, ContentValues values) {
        SQLiteDatabase db = databaseHelper.getWritableDatabase();
        db.insert("test", null, values);
        return null;
    }
    //实现 Content Providers 的 query()
    @Override
    public Cursor query(Uri uri, String[] projection, String selection,
            String[] selectionArgs, String sortOrder) {
        SQLiteDatabase db = databaseHelper.getReadableDatabase();
        SQLiteQueryBuilder qb = new SQLiteQueryBuilder();
        qb.setTables("test");
        Cursor c = qb.query(db, projection, selection, selectionArgs, null,
                null, null);
        return c;
    }
    //实现 Content Providers 的 delete()
    @Override
public int delete(Uri uri, String selection, String[] selectionArgs){
        // TODO Auto-generated method stub
        return 0;
    }
    //实现 Content Providers 的 getType()
    @Override
    public String getType(Uri uri) {
        // TODO Auto-generated method stub
        return null;
    }
    //实现 Content Providers 的 update()
```

```
    @Override
    public int update(Uri uri, ContentValues values, String selection,
            String[] selectionArgs) {
        // TODO Auto-generated method stub
        return 0;
    }
}
```

ProviderTest.java 类的参考代码如下：

```
public class ProviderTest extends Activity {
//ProvideTest 主程序
    /** Called when the activity is first created. */
    @Override
    public void onCreate(Bundle savedInstanceState) {
        super.onCreate(savedInstanceState);
        //取得 Content Provider 的 Uri
      getIntent().setData(Uri.parse("content://com.example.android
.provider.testprovider"));
        Uri uri_test = getIntent().getData();
        //建立二行测试用的数据
        ContentValues values = new ContentValues();
        values.put("name", "郑传庆");
        values.put("description", "手机相册软件项目程序员");
        getContentResolver().insert(uri_test, values);
        values.put("name", "李虹锋");
        values.put("description", "手机相册软件项目程序员");
        getContentResolver().insert(uri_test, values);
        //经 Content Provider 来检索
        Cursor cur = managedQuery(uri_test, null, null, null, null);
        cur.moveToFirst();
        //设定 ArrayList 显示下拉列表
        ArrayList<Map<String, Object>> data = new ArrayList<Map<String,
        Object>>();
        Map<String, Object> item;
        //将自数据库读出的数据整理到 ArrayList data 容器内
         for (int i = 0; i < cur.getCount(); i++) {
             item = new HashMap<String, Object>();
             item.put("column00", cur.getString(0));
             item.put("column01", cur.getString(1));
             item.put("column02", cur.getString(2));
             data.add(item);
             cur.moveToNext();
         }
         cur.close();
         //ArrayList data 容器内的数据放到 mListView01
```

```
            ListView mListView01 = new ListView(this);
            SimpleAdapter adapter = new SimpleAdapter(this, data,
            R.layout.main, new String[] {"column00", "column01", "column02" },
            new int[] {
            R.id.TextView01, R.id.TextView02, R.id.TextView03 });
            mListView01.setAdapter(adapter);
            setContentView(mListView01);
        }
    }
```

8.5 实训

在前面的章节中，我们学习了 SQLite Database 数据存储方法，也学习了内容提供器（Content Provider）对数据的共享。在本节中我们通过实训来完成如何建立内容提供器机制，如何通过内容提供器程序来添加、更新、删除和检索数据的方法。本实训应用程序主要是说明内容提供器的构造方法，建立一个 SQLite 数据库系统来存储和管理数据，同时利用 SQLiteOpenHelper 类协助建立数据库和 SQLiteDatabase 类来管理数据库。

本实训的应用程序的显示窗体包含有一个下拉列表（Spinner），4 个文本框 TextView，4 个编辑框 EditText 和 4 个按钮 Button。启动程序时，程序会处理下拉列表的内容，同时在编辑文本框内显示第一条数据的相关信息；单击下拉列表会弹出一个下拉查询表，单击选项后，会在编辑框中显示该数据的相关信息；如果想删除这条数据，直接单击“删除”按钮就可以完成；如果想更新该行数据，在输入字段文本框中修改后，单击“更新”按钮就可以完成；如果需要添加一行新数据，可以先单击“清除”按钮，清空所有编辑文本框中的信息，输入新的数据后，单击“添加”按钮就可以完成。该实训运行效果如图 8-19 和图 8-20 所示。

图 8-19　项目运行效果图

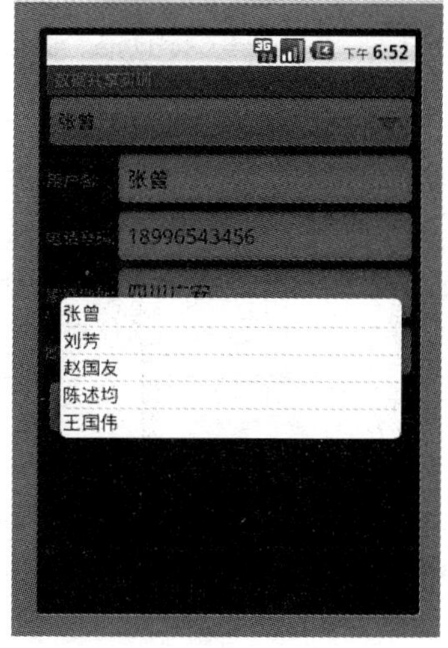

图 8-20　下拉查询表图

本实训定义了一张表 Users，表中有 5 个字段项目，其中“_id”是主键，数据项如表 8-8 所示。

表 8-8　Users 表数据项表

数据名称	数据类型	数据描述
_id	INTEGER	用户编号，主键，自动生成
user_name	TEXT	用户姓名
telephone	TEXT	用户电话号码
address	TEXT	用户家庭地址
mail_address	TEXT	用户邮箱地址

实训主体类 TestProvider.java 类的参考代码如下：

```
public class TestProvider extends ContentProvider {
    //SQLiteOpenHelper-建立数据库 PhoneContentDB 和 Table:Users
private static class DatabaseHelper extends SQLiteOpenHelper {
    private static final String DATABASE_NAME = "PhoneContentDB";
        private static final int DATABASE_VERSION = 1;
        //建立 PhoneContentDB 数据库
        private DatabaseHelper(Context ctx) {
            super(ctx, DATABASE_NAME, null, DATABASE_VERSION);
        }
        //建立 Users 表
        @Override
        public void onCreate(SQLiteDatabase db) {
String sql = "CREATE TABLE " + UserSchema.TABLE_NAME + " ("
+ UserSchema.ID + " INTEGER primary key autoincrement, "
            + UserSchema.USER_NAME + " text not null, "
            + UserSchema.TELEPHONE + " text not null, "
            + UserSchema.ADDRESS + " text not null, "
            + UserSchema.MAIL_ADDRESS + " text not null "+ ");";
            db.execSQL(sql);
        }
        //更新版本
        @Override
    public void onUpgrade(SQLiteDatabase db, int oldVersion, int newVersion) {
            db.execSQL("DROP TABLE IF EXISTS test");
          onCreate(db);
        }
      }
    //定义 DatabaseHelper 类变量 databaseHelper
    static DatabaseHelper databaseHelper;
    //实现 Content Providers 的 onCreate()
    @Override
    public boolean onCreate() {
    databaseHelper = new DatabaseHelper(getContext());
```

```
        return true;
    }
    public interface UserSchema {
        String TABLE_NAME = "Users";          //表名
        String ID = "_id";                 //ID
        String USER_NAME = "user_name";        //User Name
        String ADDRESS = "address";          //Address
        String TELEPHONE = "telephone";        //Phone Number
        String MAIL_ADDRESS = "mail_address"; //Mail Address
    }
    //实现Content Providers的insert()
    @Override
    public Uri insert(Uri uri, ContentValues values) {
    SQLiteDatabase db = databaseHelper.getWritableDatabase();
        db.insert(UserSchema.TABLE_NAME, null, values);
        db.close();
        return null;
    }
    //实现Content Providers的query()
    @Override
    public int update(Uri uri, ContentValues values, String selection,
    String[] selectionArgs) {
    SQLiteDatabase db = databaseHelper.getWritableDatabase();
        db.update(UserSchema.TABLE_NAME, values, selection ,null);
        db.close();
        return 0;
    }
    //实现Content Providers的delete()
    @Override
    public int delete(Uri uri, String selection, String[] selectionArgs) {
    SQLiteDatabase db = databaseHelper.getWritableDatabase();
        db.delete(UserSchema.TABLE_NAME, selection ,null);
        db.close();
        return 0;
    }
    //实现Content Providers的update()
    @Override
public Cursor query(Uri uri, String[] projection,String selection,  String[]
selectionArgs, String sortOrder) {
        SQLiteDatabase db = databaseHelper.getReadableDatabase();
        SQLiteQueryBuilder qb = new SQLiteQueryBuilder();
        qb.setTables(UserSchema.TABLE_NAME);
        Cursor c = qb.query(db, projection, selection, selectionArgs, null,
        null, null);
```

```
            return c;
        }
        //实现 Content Providers 的 getType()
        @Override
        public String getType(Uri uri) {
            // TODO Auto-generated method stub
            return null;
        }
    }
```

实训主体类 ProviderSQLite.java 类的参考代码如下：

```
public class ProviderSQLite extends Activity {
    OnClickListener listener_add = null;
    OnClickListener listener_update = null;
    OnClickListener listener_delete = null;
    OnClickListener listener_clear = null;
    Button button_add;
    Button button_update;
    Button button_delete;
    Button button_clear;
    public int id_this;
    public interface UserSchema {
            String TABLE_NAME = "Users";           //表名
            String ID = "_id";                    //ID
            String USER_NAME = "user_name";        //User Name
            String ADDRESS = "address";           //Address
            String TELEPHONE = "telephone";        //Phone Number
            String MAIL_ADDRESS = "mail_address"; //Mail Address
        }
    //SQLiteTest 主程序
    @Override
    public void onCreate(final Bundle savedInstanceState) {
        super.onCreate(savedInstanceState);
        //取得 Content Provider Uri
        getIntent().setData(Uri.parse("content://com.example.android.
        provider02.testprovider02"));
        final Uri uri_test = getIntent().getData();
        setContentView(R.layout.main);
        final EditText mEditText01 = (EditText)findViewById(R.id.EditText01);
        final EditText mEditText02 = (EditText)findViewById(R.id.EditText02);
        final EditText mEditText03 = (EditText)findViewById(R.id.EditText03);
        final EditText mEditText04 = (EditText)findViewById(R.id.EditText04);
        //建立数据库 PhoneBookDB 和表 Table:Users
        final String[] FROM = {
        UserSchema.ID,
```

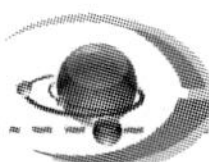

```
 UserSchema.USER_NAME,
 UserSchema.TELEPHONE,
 UserSchema.ADDRESS,
 UserSchema.MAIL_ADDRESS
 };
//取得所有数据的USER_NAME，放置在list[]上
Cursor c = managedQuery(uri_test, new String[] {UserSchema.USER_NAME},
null, null, null);
c.moveToFirst();
 CharSequence[] list = new CharSequence[c.getCount()];
 for (int i = 0; i < list.length; i++) {
        list[i] = c.getString(0);
        c.moveToNext();
 }
 c.close();
 //显示USER_NAME在Spinner下拉列表-spinner上
 Spinner spinner = (Spinner)findViewById(R.id.Spinner01);
 spinner.setAdapter(new ArrayAdapter<CharSequence>(this, android.R.
 layout.simple_spinner_item, list));
 //在Spinner下拉列表-spinner上选定查询数据，显示所有数据在画面上
 spinner.setOnItemSelectedListener(new OnItemSelectedListener(){
     public void onItemSelected(AdapterView<?> parent, View view,
     int position, long id) {
     String user_name = ((Spinner)parent).getSelectedItem().
     toString();
 Cursor c = managedQuery(uri_test, FROM , "user_name='" + user_name
 + "'", null, null);
     c.moveToFirst();
     id_this = Integer.parseInt(c.getString(0));
     String user_name_this = c.getString(1);
     String telephone_this = c.getString(2);
     String address_this = c.getString(3);
     String mail_address_this = c.getString(4);
     c.close();
     mEditText01.setText(user_name_this);
     mEditText02.setText(telephone_this);
     mEditText03.setText(address_this);
     mEditText04.setText(mail_address_this);
     }
     public void onNothingSelected(AdapterView<?> parent) {
     } });
//按下Add按钮时，新增一行数据
listener_add = new OnClickListener() {
 public void onClick(View v) {
```

```
                ContentValues values = new ContentValues();
            values.put(UserSchema.USER_NAME,
mEditText01.getText().toString());
            values.put(UserSchema.TELEPHONE,
mEditText02.getText().toString());
                values.put(UserSchema.ADDRESS,
mEditText03.getText().toString());
                values.put(UserSchema.MAIL_ADDRESS,
mEditText04.getText().toString());
                getContentResolver().insert(uri_test, values);
                onCreate(savedInstanceState);
            }
        };
        //按下 Update 按钮时，更新一行数据
        listener_update = new OnClickListener() {
            public void onClick(View v) {
                ContentValues values = new ContentValues();
                values.put(UserSchema.USER_NAME,
 mEditText01.getText().toString());
                values.put(UserSchema.TELEPHONE,
mEditText02.getText().toString());
                values.put(UserSchema.ADDRESS,
 mEditText03.getText().toString());
                values.put(UserSchema.MAIL_ADDRESS,
mEditText04.getText().toString());
                String where = UserSchema.ID + " = " + id_this;
     getContentResolver().update(uri_test, values, where, null);
            onCreate(savedInstanceState);
            }
        };
        //按下 Delete 按钮时，删除一行数据
        listener_delete = new OnClickListener() {
            public void onClick(View v) {
                String where = UserSchema.ID + " = " + id_this;
                getContentResolver().delete(uri_test, where, null);
                onCreate(savedInstanceState);
            }
        };
        //按下 Clear 按钮时，清空编辑框
        listener_clear = new OnClickListener() {
            public void onClick(View v) {
                mEditText01.setText("");
                mEditText02.setText("");
                mEditText03.setText("");
```

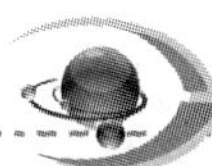

```
                mEditText04.setText("");
            }
        };
        //设定 BUTTON0i,i=1,2,3,4 的 OnClickListener
        button_add = (Button)findViewById(R.id.Button01);
        button_add.setOnClickListener(listener_add);
        button_update = (Button)findViewById(R.id.Button02);
        button_update.setOnClickListener(listener_update);
        button_delete = (Button)findViewById(R.id.Button03);
        button_delete.setOnClickListener(listener_delete);
        button_clear = (Button)findViewById(R.id.Button04);
        button_clear.setOnClickListener(listener_clear);
    }
}
```

第 9 章

Android 位置服务与地图应用

位置服务和地图应用是发展最为迅速，有着大量潜在需求的领域，通过本章的学习可以让读者简单地了解位置服务和地图应用的概念、方法和技巧。读者可以使用 Google 提供的地图服务，构建提供位置服务的应用程序。

9.1 位置服务

位置服务（Location Based Services，LBS），又称定位服务或基于位置的服务，融合了 GPS 定位、移动通信、导航等多种技术，提供了与空间位置相关的综合应用服务。位置服务最先是在日本得到商业化的应用。2001 年 7 月，DoCoMo 发布了第一款具有三角定位功能的手持设备，2001 年 12 月，KDDI 发布第一款具有 GPS 功能的手机。近些年来，基于位置的服务发展更加迅速，涉及到商务、医疗、工作和生活等各个方面，为用户提供定位、追踪和敏感区域警告等一系列服务。

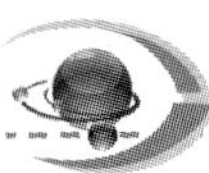

Android平台支持提供服务的API，在开发过程中主要用到LocationManager和LocationProviders对象。LocationManager可以用来获取当前的位置，追踪设备的移动路线，或设定敏感区域，在进入或者离开敏感区域时设备会发出特定警报。LocationProviders则是能够提供定位功能的组件集合，集合中的每种组件以不同的技术提供设备的当前位置，区别在于定位的精度、速度和成本等方面。

为了使开发的程序能够提供位置服务，首先需要获得LocatioManager对象。获取LocatioManager可以通过调用android.app.Activity.getSystemService()函数实现，代码如下：

```
String serviceString = Context.LOCATION_SERVICE;
LocationManager LocationManager =
(LocationManager)getSystemService(serviceString);
```

代码中的Context.LOCATION_SERVICE指明获取的服务是位置服务，getSystemService()函数，可以根据服务名称获取Android提供的系统级服务。Android支持的系统级服务如表9-1所示。

表9-1　Android支持的系统级服务表

Context类的静态常量	值	返回对象	说　明
LOCATION_SERVICE	location	LocationManager	控制位置等设备的更新
WINDOW_SERVICE	window	WindowManager	最顶层的窗口管理器
LAYOUT_INFLATER_SERVICE	layout_inflater	LayoutInflater	将XML资源实例化为View
POWER_SERVICE	power	PowerManager	电源管理
ALARM_SERVICE	alarm	AlarmManager	在指定时间接受Intent
NOTIFICATION_SERVICE	notification	NotificationManager	后台事件通知
KEYGUARD_SERVICE	keyguard	KeyguardManager	锁定或解锁键盘
SEARCH_SERVICE	search	SearchManager	访问系统的搜索服务
VIBRATOR_SERVICE	vibrator	Vibrator	访问支持振动的硬件
CONNECTIVITY_SERVICE	connection	ConnectivityManager	网络连接管理
WIFI_SERVICE	wifi	WifiManager	Wi-Fi连接管理
INPUT_METHOD_SERVICE	input_method	InputMethodManager	输入法管理

在获得LocationManager后，还需要指定LocationManager的定位方法，然后才能够调用LocationManager.getLastKnowLocation()方法获取当前位置。目前LocationManager支持的定位方法有两种，分别是使用GPS定位和使用网络定位。GPS定位可以提供更加精确的位置信息，但定位速度和质量受到卫星数量和环境情况的影响；网络定位提供的位置信息精度较差，但速度较GPS定位快。LocationManager支持定位方法参考表9-2。

表 9-2　LocationManager 支持定位方法表

LocationManager 类的静态常量	值	说　　明
GPS_PROVIDER	gps	使用 GPS 定位，利用卫星提供精确的位置信息，需要 android.permissions.ACCESS_FINE_LOCATION 用户权限
NETWORK_PROVIDER	network	使用网络定位，利用基站或 Wi-Fi 提供近似的位置信息，需要具有如下权限：android.permission.ACCESS_COARSE_LOCATION 或 android.permission.ACCESS_FINE_LOCATION.

在指定 LocationManager 的定位方法后，则可以调用 getLastKnowLocation()方法获取当前的位置信息。以使用 GPS 定位为例，获取位置信息的代码如下：

```
String provider = LocationManager.GPS_PROVIDER;
Location location = locationManager.getLastKnownLocation(provider);
```

代码中返回的 Location 对象中，包含了可以确定位置的信息，如经度、纬度和速度等，用户可以通过调用 Location 中的 getLatitude()和 getLonggitude()方法分别获取位置信息中的纬度和经度，示例代码如下：

```
double lat = location.getLatitude();
double lng = location.getLongitude();
```

在很多提供定位服务的应用程序中，不仅需要获取当前的位置信息，还需要监视位置的变化，在位置变化时调用特定的处理方法。LocationManager 提供了一种便捷、高效的位置监视方法 requestLocationUpdates()，可以根据位置的距离变化和时间间隔设定产生位置改变事件的条件，这样可以避免因微小的距离变化而产生大量的位置改变事件。LocationManaget 中设定监听位置变化的代码如下：

```
locationManager.requestLocationUpdates(provider, 2000, 10,
 locationListener);
```

方法中的第 1 个参数是定位的方法，GPS 定位或网络定位；第 2 个参数是产生位置改变事件的时间间隔，单位为微秒；第 3 个参数是距离条件，单位是米；第 4 个参数是回调函数，是在满足条件后的位置改变事件的处理函数。上面的代码将产生位置改变的距离改变为 10 米，时间间隔为 2 秒。实现 locationListener 代码如下：

```
LocationListener locationListener = new LocationListener(){
 public void onLocationChanged(Location location) { }
 public void onProviderDisabled(String provider) {}
 public void onProviderEnabled(String provider) { }
 public void onStatusChanged(String provider, int status,
 Bundle extras) { }
 };
```

代码中的 onLocationChanged()在设备的位置改变时被调用；onProviderDisabled()在用户禁用具有定位功能的硬件时被调用；onProviderEnabled()在用户启用具有定位功能的硬件时被调用；onStatusChanged()在提供定位功能的硬件的状态改变时被调用，如从不可获取位置信息状态到可以获取位置信息的状态，反之亦然。

最后，为了使 GPS 定位功能生效，还需要在 AndroidManifest.xml 文件中加入用户许可，代码如下：

```
<uses-permission
 android:name="android.permission.ACCESS_FINE_LOCATION"/>
```

CurrentLocationDemo 是一个提供位置服务的基本示例，提供了显示当前位置新的功能，并能够监视设备的位置变化。CurrentLocationDemo 的用户界面如图 9-1 所示。

位置服务一般都需要使用设备上的硬件，最理想的调试方式是将程序上传到物理设备上运行，但在没有物理设备的情况下，也可以使用 Android 模拟器提供的虚拟方式模拟设备的位置变化，调用具有位置服务的应用程序。首先打开 DDMS 中的模拟器控制，在 Location Controls 中输入设备当前的经度和纬度，然后单击 Send 按钮，就将虚拟的位置信息发送到 Android 模拟器中，如图 9-2 所示。

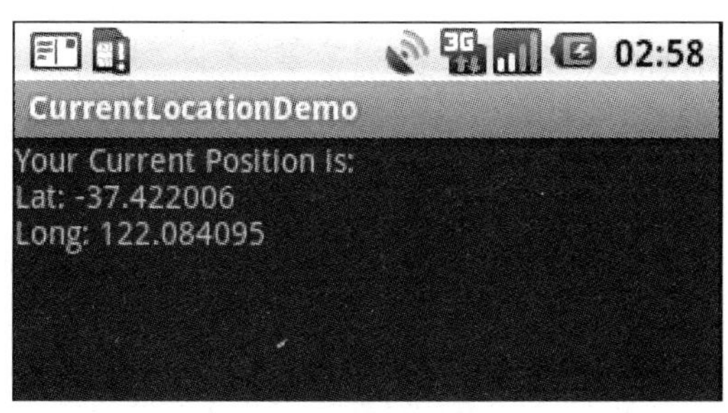

图 9-1　示例界面图

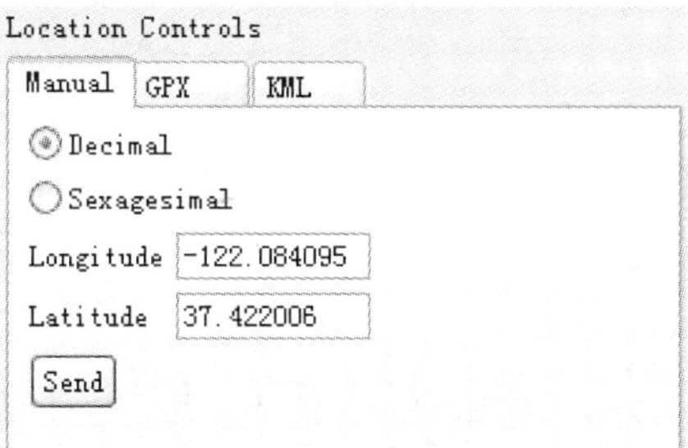

图 9-2　模拟器控制器图

在程序运行过程中，可以在模拟器控制器中改变经度和纬度坐标值，程序在检测到位置的变化后，会将最新的位置信息显示在界面上。

下面给出 CurrentLocationDemo 示例 LocationBasedServiceDemo.java 文件的完整代码：

```
public class LocationBasedServiceDemo extends Activity {
 @Override
 public void onCreate(Bundle savedInstanceState) {
     super.onCreate(savedInstanceState);
     setContentView(R.layout.main);
     String serviceString = Context.LOCATION_SERVICE;
  LocationManager locationManager =
(LocationManager)getSystemService(serviceString);
     String provider = LocationManager.GPS_PROVIDER;
     Location location =
     locationManager.getLastKnownLocation(provider);
     getLocationInfo(location);
locationManager.requestLocationUpdates(provider, 2000, 0,
     locationListener); }
private void getLocationInfo(Location location){
     String latLongInfo;
     TextView locationText =
          (TextView)findViewById(R.id.txtshow);
```

```
        if (location != null){
            double lat = location.getLatitude();
            double lng = location.getLongitude();
            latLongInfo = "Lat: " + lat + "\nLong: " + lng; }
        else{  latLongInfo = "No location found"; }
        locationText.setText("Your Current Position is:\n" +
        latLongInfo);   }
    private final LocationListener locationListener =
          new LocationListener(){
       @Override
       public void onLocationChanged(Location location) {
           getLocationInfo(location);  }
              @Override
        public void onProviderDisabled(String provider) {
            getLocationInfo(null); }
              @Override
        public void onProviderEnabled(String provider) {
            getLocationInfo(null); }
              @Override
        public void onStatusChanged(String provider, int status,
           Bundle extras) {}
        };
    }
```

9.2 Google 地图应用

9.2.1 申请地图密钥

为了在手机中更直观地显示地理信息，程序开发人员可以直接使用 Google 提供的地图服务，实现地理信息的可视化开发。只要使用 MapView(com.google.android.maps.MapView)就可以将 Google 地图嵌入到 Android 应用程序中。但在使用 MapView 进行开发前，必须向 Google 申请一组经过验证的“地图密钥（Map API KEY）”，才能正常使用 Google 的地图服务。“地图密钥”是访问 Google 地图数据的密钥，无论是模拟器还是在真实设备中都需要使用这个密钥。

注册“地图密钥”的第一步是申请一个 Google 账户，也就是 Gmail 电子邮箱，申请地址 https://www.google.com/accounts/Login。

下一步工作是找到保存 Debug 证书的 keystore 的保存位置，并获取证书的 MD5 散列值。keystore 是一个密码保护的文件，用来存储 Android 提供的用于调试的证书，获取 MD5 散列值的主要目的是为下一步申请“地图密钥”做准备。获取证书的保存地址如图 9-3 所示，首先打开 Eclipse，通过 Windows→Preferences 打开配置窗体，在 Android→Build 栏中的 Default debug keystore 中可以找到。

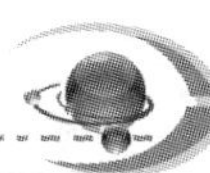

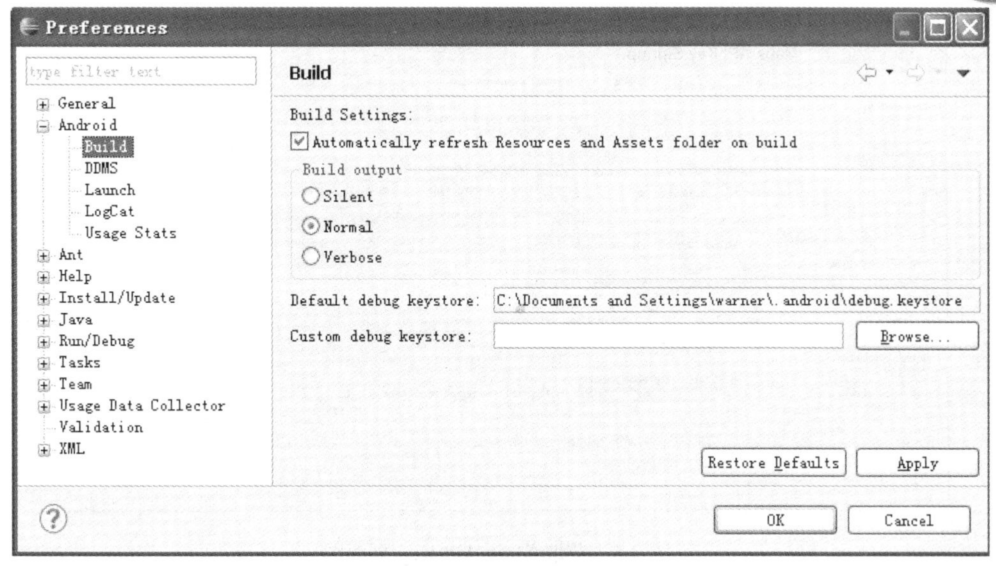

图 9-3　证书保存地址位置图

为了获取 Debug 证书的 MD5 散列值，需要打开命令行工具 CMD，然后切换到 keystore 的目录，输入如下命令：

```
keytool -list -keystore debug.keystore
```

如果提示无法找到 keytool，可以将<Java SDK>\bin 的路径添加到系统的 PATH 变量中。在提示输入 Keystore 密码时，输入默认密码 android，MD5 散列值将显示在最下方。如图 9-4 所示，每台电脑的 MD5 散列值都不一样，这个一定要引起读者注意。笔者的 MD5 散列值为 68:76:89:C8:A4:24:61:F9:EA:F3:F7:70:CC:FD:C8:15。

申请“地图密钥”的最后一步是打开申请页面，输入 MD5 散列值。申请页面的地址是：http://code.google.com/intl/zh-CN/android/google-apis/maps-api-signup.html，如图 9-5 所示。

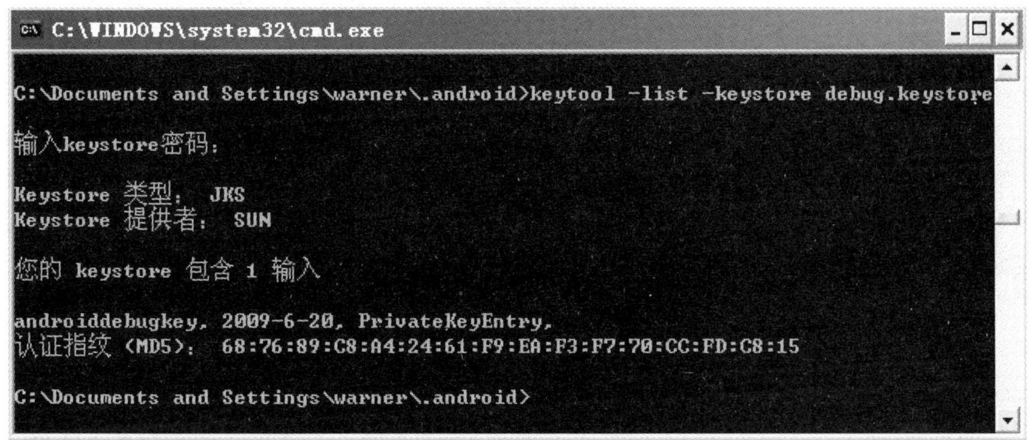

图 9-4　获取 Keystore 的 MD5 散列值图

Google Projects for Android: Google APIs

Home Docs Blog Group

Download
Terms

Getting Started
What is Google APIs Add-on?
Installing the Add-on

Maps
Overview
Obtaining a Maps API Key
Maps API Key Signup
API Reference

Resources
Android SDK »
Android Developer's Guide »

Maps API Key Signup

Use the form on this page to register with the Google Maps service and obtain a Maps API Key. Registration is free.

If you are developing an Android application that will display Google Maps data using the API provided in the Maps external library, you must register with the service and get an API Key.

A single Maps API key is valid for all applications signed by the corresponding developer certificate. For more information about signing, see Signing Your Applications on the Android Developers site.

Before you register, read Obtaining a Maps API Key to understand how your Maps API Key is used in your Android application, why it's needed, and how to generate an MD5 fingerprint based on your developer certificate.

To register for a Key, you also need a Google Account. Once you register, your Key will be associated with your Google Account.

```
Android Maps APIs Terms of Service

Last Updated: October 13, 2008

Thanks for your interest in the Android Maps APIs. The Android Maps
APIs are a collection of services (including, but not limited to, the
"com.google.android.maps.MapView" and "android.location.Geocoder"
classes) that allow you to include maps, geocoding, and other content
from Google and its content providers in your Android applications.
The Android Maps APIs explicitly do not include any driving directions
data or local search data that may be owned or licensed by Google.

1.  Your relationship with Google.
    1.1. Your use of any of the Android Maps APIs (referred to in this
```

☑ I have read and agree with the terms and conditions (printable version)

My certificate's MD5 fingerprint:

Generate API Key

图 9-5 获取 Maps API Key 页面图

输入 MD5 散列值后，单击 Generate API Key 按钮，将提示用户输入 Google 账户，正确输入 Google 账户后，将产生申请“地图密钥”的获取结果，如图 9-6 所示。

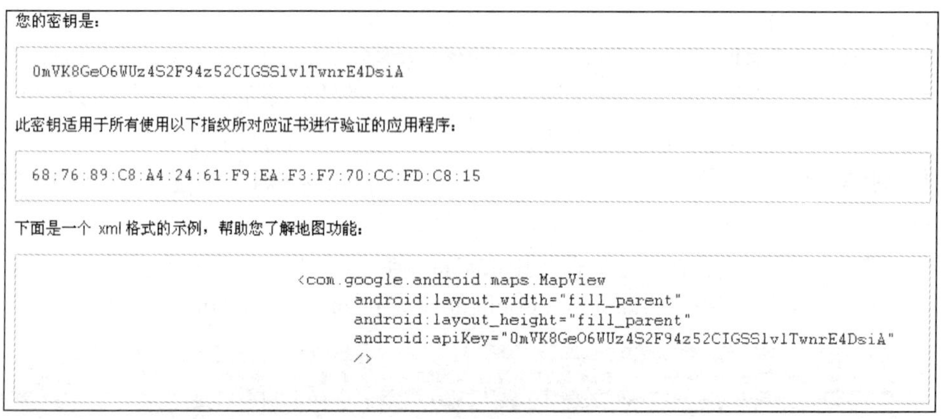

图 9-6 “地图密钥”获取结果图

笔者获取的“地图密钥”是 OmVK8Ge06WUz4S2F94z52CIGSSlvlTwnrE4DsiA，在以后使用到 MapView 的时候都要输入这个密钥。但需要注意的是，读者并不能使用这个密钥，而是需要根据自己的 Debug 证书的 MD5 散列值，重新到 Google 网站上申请一个用于调试程序的“地图密钥”。

9.2.2 使用 Google 地图

在申请到“地图密钥”后，下面考虑如何在 Android 系统中显示和控制 Google 地图。MapView 是地图显示控件，可以设置不同显示模式，例如卫星模式、街道模式或交通模式。而 MapController 则是 MapView 的控制器，可以控制 MapView 的显示中心和缩放级别等功能。

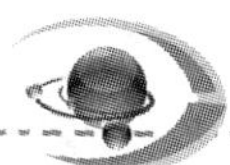

下面的内容以 GoogleMapDemo 为例，说明如何在 Android 系统中开发 Google 地图程序。这个示例将在程序内部设置一个坐标点，然后在程序启动时，使用 MapView 控件在地图上显示这个坐标点的位置。

因为普通的 Android 程序并不包含支持 Google 地图开发的扩展库，因此应在建立工程时将 com.google.android.maps 的扩展库添加到工程中，这样就可以使用 Google 地图的所有功能了，添加 com.google.android.maps 扩展库的扩展方式是在创建工程时，在 Buid Target 选项中选择 Google APIs，如图 9-7 所示。

Build Target

Target Name	Vendor	Platform	AP...
☐ Android 1.1	Android Open Source Project	1.1	2
☐ Android 1.5	Android Open Source Project	1.5	3
☑ Google APIs	Google Inc.	1.5	3

Android + Google APIs

图 9-7　引入 Google 地图扩展库图

创建工程后，修改\res\layout\main.xml 文件，在布局中加入一个 MapView 控件，并设置刚获取的“地图密钥”。Main.xml 文件的完整代码如下：

```
<?xml version="1.0" encoding="utf-8"?>
<LinearLayout
  xmlns:android="http://schemas.android.com/apk/res/android"
  android:orientation="vertical"
  android:layout_width="fill_parent"
  android:layout_height="fill_parent">
  <TextView android:layout_width="fill_parent"
      android:layout_height="wrap_content"
      android:text="@string/hello"/>
  <com.google.android.maps.MapView
      android:id="@+id/mapview"
      android:layout_width="fill_parent"
      android:layout_height="fill_parent"
      android:enabled="true"
      android:clickable="true"
      android:apiKey="0mVK8GeO6WUz4S94z52CIGSSlvlTwnrE4DsiA"/>
</LinearLayout>
```

仅在布局中添加 MapView 控件，还不能够直接在程序中调用这个控件，还需要将程序本身设置成 MapActivity(com.google.android.MapActivity)，MapActivity 类负责处理显示 Google 地图所需的生命周期和后台服务管理。下面先给出 GoogleMapDemo.java 文件的完整代码：

```
public class GoogleMapDemo extends MapActivity {
 private MapView mapView;
 private MapController mapController;
```

```
@Override
public void onCreate(Bundle savedInstanceState) {
    super.onCreate(savedInstanceState);
    setContentView(R.layout.main);
    mapView = (MapView)findViewById(R.id.mapview);
   //获取了 MapController
    mapController = mapView.getController();
    final int INITIAL_LATITUDE = 29494000; //设定经度
    final int INITIAL_LONGITUDE = 106542000; //设定纬度
//没有直接使用这个坐标的，而是将其转化为 GeoPoint 再使用
     GeoPoint point = new GeoPoint(INITIAL_LATITUDE,
         INITIAL_LONGITUDE);
         //设置 MapView 的“显示中点”
     mapController.setCenter(point);
     mapController.setZoom(11);// 设置放大层级
  //将 MapView 显示区域的中心移到设置的“显示中心”
     mapController.animateTo(point);
//设定 MapView 的地图显示模式是否为卫星模式，设置 true 则为卫星模式，设置 false 则为普通模式
     mapView.setSatellite(false)    }
 Override
//isRouteDisplayed()方法，是用来统计程序是否在 Google 地图中显示路径信心，默认为不显示
 protected boolean isRouteDisplayed() {
     // TODO Auto-generated method stub
     return false;  }
}
```

运行前还需在 AndroidManifest.xml 文件中添加允许访问互联网的许可，原因是获取 Google 地图是需要使用互联网的。AndroidManifest.xml 文件的添加代码如下：

```
<application android:icon="@drawable/icon"
android:label="@string/app_name">
  <activity android:name=".GoogleMapDemo"
      android:label="@string/app_name">
<intent-filter>
 <action android:name="android.intent.action.MAIN" />
<category android:name="android.intent.category.LAUNCHER" />
 </intent-filter>
   </activity>
 <uses-library android:name="com.google.android.maps"/>
  </application>
  <uses-sdk android:minSdkVersion="7" />
 <uses-permission android:name="android.permission.INTERNET"/>
 </manifest>
```

最后，程序运行时需要连接互联网，运行结果如图 9-8 和图 9-9 所示。

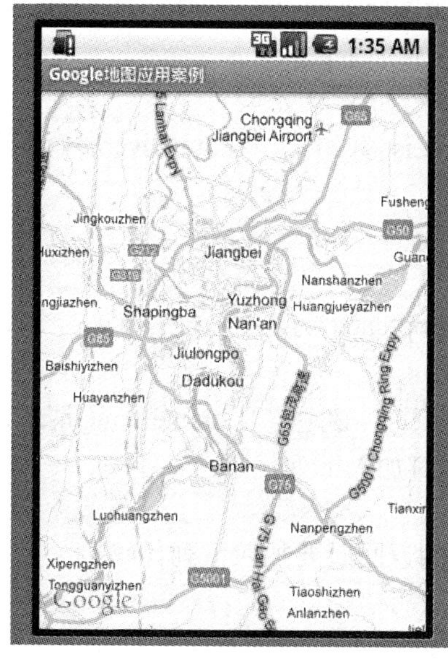

图 9-8 地图模式图

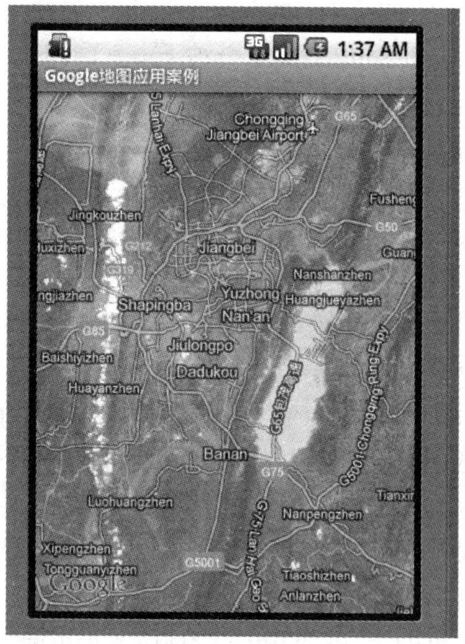

图 9-9 卫星模式图

9.2.3 Google 地图上贴上标记

本节讲一个在 Google 地图某一个固定的经纬度上贴上相应的标记。在前面的章节中，我们接触了如何与 Google 地图服务系统的互动，但是我们更希望在地图上贴上自己的标记和注释，这要求能在地图上覆盖一些对象。要做到这一点，首先要实现一个 ItemizedOverlay 类，它可以管理一套覆盖项目功能，下面我们通过一个简单的案例来学习这个知识点的应用。

建立一个 Android 项目，名称为 HelloMapView。需要的图标复制到相应的资源文件内，项目里面包含两个类文件，分别是 HelloItemizedOverlay.java 和 HelloMapView.java。下面对这个项目的完成进行详细的介绍。

实现一个 ItemizedOverlay 类程序 HelloItemizedOverlay.java 需要建立以下 4 个方法：

```
public HelloItemizedOverlay(Drawable defaultMarker)
public void addOverlay(OverlayItem overlay)
protected OverlayItem createItem(int i)
public int size()
```

请按下列步骤来实现 HelloItemizedOverlay.java 程序：

（1）建立一个新的 Android 应用程序 HelloMapView，并在 src 文件夹建立 Java 类 HelloItemizedOverlay 来实现 ItemizedOverlay。

（2）建立一个 OverlayItem ArrayList 数组，这个数组可以放置所有要贴在地图上的标记对象。

```
private ArrayList<OverlayItem> mOverlays = new
 ArrayList<OverlayItem>();
```

（3）构造方法会定义每一个被使用 OverlayItems 的默认标记，为了使 Drawable 标记可以贴上，必须界定它的范围。地图坐标上的标记希望贴在地图底部的中心点，使用 boundCenterBottom()方法来处理，defaultMarker 会被 super 调用。参考代码如下：

```
super(boundCenterBottom(defaultMarker));
```

（4）为了要增加一个新的 OverlayItem 到 ArrayList 数组上，在 addOverlay()方法中加入下列代码：

```
mOverlays.add(overlay);
  populate();
```

每一次增加一个新的 OverlayItem，一定要调用 populate()方法，它会读出每一个 OverlayItem 并使它可以被贴上。

（5）为了读取每一个 OverlayItem 会调用 populate()，必须定义 createItem()请求，确认是从 ArrayList 数组读出。在处理 createItem()方法的返回时加入下列代码：

```
return mOverlays.get(i);
```

还需要一个覆盖 size()方法，在处理 size()方法的返回时要加入下列代码：

```
return mOverlays.size();
```

一个完整的 HelloItemizedOverlay.java 程序的代码如下：

```
public class HelloItemizedOverlay extends ItemizedOverlay {
    //创建一个可以放置所有要贴在地图上的标记对象的数组
    private ArrayList<OverlayItem> mOverlays = new
ArrayList<OverlayItem>();
    //构造方法定义一个默认标记
    public HelloItemizedOverlay(Drawable defaultMarker) {
        super(boundCenterBottom(defaultMarker));
        // TODO Auto-generated constructor stub
    }
    public void addOverlay(OverlayItem overlay) {
        mOverlays.add(overlay);
        populate();
        }
    @Override
    protected OverlayItem createItem(int i) {
        return mOverlays.get(i);
        }
    @Override
    public int size() {
        return mOverlays.size();
    }
}
```

HelloMapView 主类的参考代码如下：

```
public class HelloMapView extends MapActivity {
//设置缩放 Zoom 的初始值
    static final int INITIAL_ZOOM_LEVEL = 14;
```

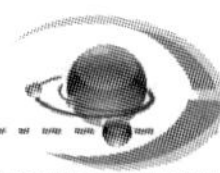

```
//设置纬度的初始值
    static final int INITIAL_LATITUDE = 29494000;
//设置经度的初始值
    static final int INITIAL_LONGITUDE = 106542000;
    MapView mapView;
    List<Overlay> mapOverlays;
    Drawable drawable;
    HelloItemizedOverlay itemizedOverlay;
//建立 HelloMapView 应用程序
    @Override
    public void onCreate(Bundle savedInstanceState) {
        super.onCreate(savedInstanceState);
        setContentView(R.layout.main);//设置显示窗体
        // 设定 MapView 可以缩放
        mapView = (MapView) findViewById(R.id.mapview);
        mapView.setBuiltInZoomControls(true);
        // 设定初始 Zoom 大小和地图的中心点经纬度
        MapController mc = mapView.getController();
        mc.setZoom(INITIAL_ZOOM_LEVEL);
        mc.setCenter(new GeoPoint(INITIAL_LATITUDE,
     INITIAL_LONGITUDE));
        /** 贴上标记**/
//实例化一个 List<Overlay>类对象 mapOverlays
        mapOverlays = mapView.getOverlays();
//从资源文件中读取图标文件
        drawable = this.getResources().getDrawable(R.drawable.marker);
        itemizedOverlay = new HelloItemizedOverlay(drawable);
//设置贴上图标的坐标及具体位置
        GeoPoint point = new GeoPoint(INITIAL_LATITUDE,
INITIAL_LONGITUDE);
        //实例化一个 OverlayItem 类对象 overlayitem 来保存标记的坐标
     OverlayItem overlayitem = new OverlayItem(point, "", "");
     //在 itemizedOverlay 对象上加入贴图的标记
        itemizedOverlay.addOverlay(overlayitem);
     //把 itemizedOverlay 对象加到 MapView 的 Overlay 上
        mapOverlays.add(itemizedOverlay);
    }
    //目前尚未实施且需要覆盖的方法
     @Override
     protected boolean isRouteDisplayed() {
         return false;
     }
}
```

运行该项目，运行效果如图 9-10 所示。

图 9-10　Google 地图贴标记图

9.3　利用 Google API 完成天气预报

天气预报就是未来时期内天气变化的预先估计和预告。它是根据大气科学的基本理论和技术对某一地区未来的天气作出分析和预测，这是大气科学为国民经济建设和人民生活服务的重要手段，准确及时的天气预报对于经济建设和国防建设的趋利避害、保障人民生命财产等有极大的社会和经济效益。按照天气预报的时限分：1～6 小时之内则为短临预报（近临预报），1～2 天短期天气预报，3～15 天为中期天气预报，月、季为长期天气预报。目前的天气学方法以天气图为主，配合气象卫星云图、雷达等资料：数值天气预报以计算机为工具，通过求解由流体力学、热力学、动力气象学组成的预报方程，来制作天气预报；统计预报以概率论数理统计为手段来制作天气预报。本节我们将学习如何在 Android 上完成一个天气预报的应用，以便随时随地获取最新的天气信息。项目名称为 ChinaWeather。

9.3.1　信息来源

要实现一个天气预报的应用，肯定需要找到一个提供天气信息的 Web 服务，当然现在提供天气预报的网站很多，方式也是多种多样。这里为了讲述 Android 平台有关 XML 解析的方法，选择了 Google 提供的天气预报，其获得发式有两种：

（1）通过经纬度来定位获得该地区的天气信息。

在浏览器中输入 http://www.google.com/ig/api?hl=zh_cn&weather=,,,30670000,104019996（30670000 和 104019996 分别表示经度和纬度的数据），即可获得一个 XML 形式的数据，其中

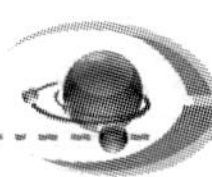

包括了天气信息，如有关天气信息的 XML 属性，如图 9-11 所示。通过网络获得这些信息之后，只需要将这些 XML 信息进行解析就可以得到我们需要的天气预报。

```
  <?xml version="1.0" ?>
- <xml_api_reply version="1">
  - <weather module_id="0" tab_id="0" mobile_row="0" mobile_zipped="1" row="0" section="0">
    - <forecast_information>
        <city data="Chengdu, Sichuan" />
        <postal_code data="chengdu" />
        <latitude_e6 data="" />
        <longitude_e6 data="" />
        <forecast_date data="2011-06-20" />
        <current_date_time data="2011-06-20 21:00:00 +0000" />
        <unit_system data="SI" />
      </forecast_information>
    - <current_conditions>
        <condition data="多云" />
        <temp_f data="77" />
        <temp_c data="25" />
        <humidity data="湿度： 78%" />
        <icon data="/ig/images/weather/cn_cloudy.gif" />
        <wind_condition data="风向： 北、风速：4 米/秒" />
      </current_conditions>
    - <forecast_conditions>
        <day_of_week data="周一" />
        <low data="20" />
        <high data="27" />
        <icon data="/ig/images/weather/chance_of_rain.gif" />
        <condition data="可能有雨" />
      </forecast_conditions>
    - <forecast_conditions>
        <day_of_week data="周二" />
```

图 9-11 有关天气信息的 XML 属性图

（2）通过城市的名字来获取天气信息。

在浏览器中输入 http://www.google.com/ig/api?hl=zh_cn&weather=chengdu（chengdu 代表城市的名字，这里输入的是成都），就可以获得与第一种方式一样的数据信息。可以看出，这两种方式获取得的数据都包含了 forecast-information，表示当前的一些信息，比如城市、时间等，current_conditons 表示当前的实时天气预报，还有 4 个 current_conditons 表示预报的最后四天的天气预报。

9.3.2 UI 设计

上面我们找到了天气预报的消息来源，也看到了是什么内容的信息。下面需要根据信息的内容来设计适合我们程序的界面。这些信息包括最高（低）温度、一个 ICNO 图标（表示天气信息的小图标）、一些附加的描述。也就是说，我们要用 ImageView 来显示这些图标，一个 TextView 来显示另外的温度和附加信息，因此先创建 SingleWetherInfoView 继承自 LinearLayout 来自定义一个线性布局以显示所获取的信息。其中包括一个 ImageView 和一个 TextView。只是需要注意这个 ImageView 是通过网络获取的图片，所以在 SingleWeatherInfoView 类中定义 setWeatherIcon 方法专门来设置图片，该类的代码如下：

```
public class SingleWeatherInfoView extends LinearLayout{
  private ImageView myWeatherImageView= null;
  private TextView  myTempTextView= null;
  public SingleWeatherInfoView(Context context)
  {super(context);}
  public SingleWeatherInfoView(Context context, AttributeSet attrs){
      super(context, attrs);
      this.myWeatherImageView = new ImageView(context);
      this.myWeatherImageView.setPadding(10, 5, 5, 5);
      this.myTempTextView = new TextView(context);
      this.myTempTextView.setTextColor(R.color.black);
      this.myTempTextView.setTextSize(16);
      this.addView(this.myWeatherImageView, new
LinearLayout.LayoutParams(LayoutParams.WRAP_CONTENT,
LayoutParams.WRAP_CONTENT));
  this.addView(this.myTempTextView, new
LinearLayout.LayoutParams(LayoutParams.WRAP_CONTENT,
LayoutParams.WRAP_CONTENT)); }
  public void setWeatherString(String aWeatherString){
      this.myTempTextView.setText(aWeatherString); }
  public void setWeatherIcon(URL aURL){
      try{
          URLConnection conn = aURL.openConnection();
          conn.connect();
          InputStream is = conn.getInputStream();
          BufferedInputStream bis = new BufferedInputStream(is);
          Bitmap bm = BitmapFactory.decodeStream(bis);
          bis.close();
          is.close();
          this.myWeatherImageView.setImageBitmap(bm);
      }catch (Exception e){} }
}
```

该例中，我们将分别使用两种方式来获得数据信息，因此需要一个 Spinner 来显示一些城市供用户选择，还需要一个 EditText 来供用户输入城市名字，具体布局如图 9-12 所示。

然后我们将获得天气预报信息，并通过自定义的 SingleWeatherInfoView 视图显示出来，如图 9-13 所示。

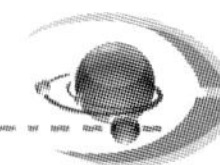

图 9-12　选择和输入城市图

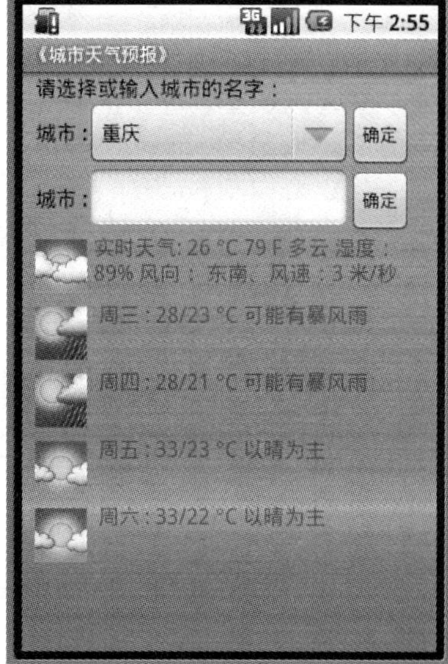

图 9-13　获取天气预报的效果图

根据项目获取到天气预报的效果图，图 9-14 显示项目结构图。其布局文件的代码如下：

```
<?xml version="1.0" encoding="utf-8"?>
<LinearLayout xmlns:android="http://schemas.android.com/apk/res/android"
  android:orientation="vertical" android:layout_width="fill_parent"
  android:layout_height="fill_parent"
   android:background="@drawable/bg">
<TextView android:id="@+id/TextView001"
android:layout_width="wrap_content"
     android:layout_height="wrap_content"
     android:text="@string/inputstr"
     android:textStyle="bold" android:textSize="16px"
     android:layout_marginLeft="10px"
 android:textColor="@color/black"></TextView>
<TableLayout android:id="@+id/TableLayout02"
    android:layout_height="wrap_content"
     android:layout_width="fill_parent">
    <TableRow android:id="@+id/TableRow01"
        android:layout_height="wrap_content"
        android:layout_width="fill_parent">
    <TextView android:id="@+id/TextView01"
            android:layout_width="wrap_content"
            android:layout_height="wrap_content"
            android:text="@string/msg"
            android:textStyle="bold" android:textSize="16px"
            android:layout_marginLeft="10px"
```

```
            android:textColor="@color/black"></TextView>
        <Spinner android:id="@+id/Spinner01"
            android:layout_height="wrap_content"
            android:layout_width="fill_parent"
            android:paddingLeft="10px"
            android:minWidth="200px"></Spinner>
        <Button android:id="@+id/Button01"
            android:layout_width="wrap_content"
            android:layout_height="wrap_content"
            android:text="@string/OK"
            android:paddingLeft="10px"></Button>
    </TableRow>
</TableLayout>
<TableLayout android:id="@+id/TableLayout002"
    android:layout_height="wrap_content"
    android:layout_width="fill_parent">
    <TableRow android:id="@+id/TableRow001"
        android:layout_height="wrap_content"
        android:layout_width="fill_parent">
        <TextView android:id="@+id/TextView002"
            android:layout_width="wrap_content"
            android:layout_height="wrap_content"
            android:text="@string/msg" android:textStyle="bold"
            android:textSize="16px"
            android:layout_marginLeft="10px"
            android:textColor="@color/black"></TextView>
        <EditText android:id="@+id/EditText001"
            android:layout_height="wrap_content"
            android:layout_width="fill_parent"
            android:paddingLeft="10px"
            android:minWidth="200px"></EditText>
        <Button android:id="@+id/Button001"
            android:layout_width="wrap_content"
            android:layout_height="wrap_content"
            android:text="@string/OK"
            android:paddingLeft="10px"></Button>
    </TableRow>
</TableLayout>
<TableLayout android:id="@+id/TableLayout01"
    android:layout_width="fill_parent"
    android:layout_height="wrap_content">
    <TableRow android:id="@+id/TableRow02"
```

```
        android:layout_width="wrap_content"
        android:layout_height="wrap_content">
        <com.yarin.android.CityWeather.SingleWeatherInfoView
            android:id="@+id/weather_0"
            android:layout_width="wrap_content"
            android:layout_height="wrap_content" />
    </TableRow>
    <TableRow android:id="@+id/TableRow03"
        android:layout_width="wrap_content"
        android:layout_height="wrap_content">
        <com.yarin.android.CityWeather.SingleWeatherInfoView
            android:id="@+id/weather_1"
            android:layout_width="wrap_content"
            android:layout_height="wrap_content" />
    </TableRow>
    <TableRow android:id="@+id/TableRow04"
        android:layout_width="wrap_content"
        android:layout_height="wrap_content">
        <com.yarin.android.CityWeather.SingleWeatherInfoView
            android:id="@+id/weather_2"
            android:layout_width="wrap_content"
            android:layout_height="wrap_content" />
    </TableRow>
    <TableRow android:id="@+id/TableRow05"
        android:layout_width="wrap_content"
        android:layout_height="wrap_content">
        <com.yarin.android.CityWeather.SingleWeatherInfoView
            android:id="@+id/weather_3"
            android:layout_width="wrap_content"
            android:layout_height="wrap_content" />
    </TableRow>
    <TableRow android:id="@+id/TableRow06"
        android:layout_width="wrap_content"
        android:layout_height="wrap_content">
        <com.yarin.android.CityWeather.SingleWeatherInfoView
            android:id="@+id/weather_4"
            android:layout_width="wrap_content"
            android:layout_height="wrap_content" />
    </TableRow>
  </TableLayout>
</LinearLayout>
```

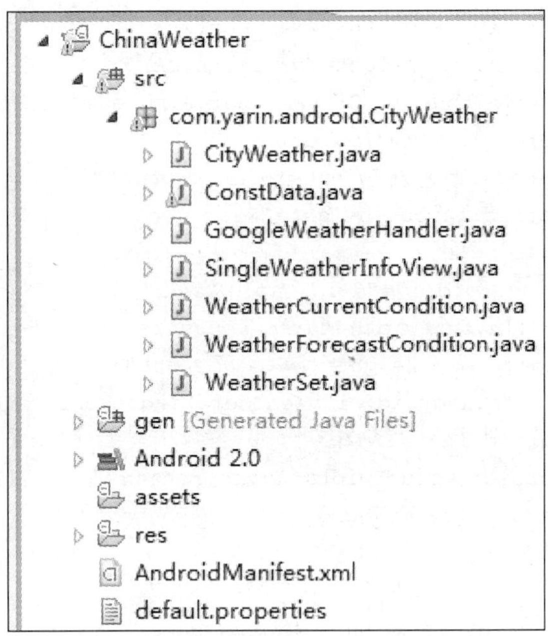

图 9-14　ChinaWeather 项目结构图

9.3.3　解析 XML

该实例的核心内容就在于如何解析 XML。Android 平台最大的一个优势在于它利用了 Java 编程语言。Android SDK 并未向标准 Java Runtime Environment（JRE）提供一切可用功能，但它支持其中很大一部分功能。Java 平台支持通过许多不同的方式来使用 XML，并且大多数与 XML 相关的 Java API 在 Android 上得到了完全的支持。举例来说，Java 的 Simple API for XML（SAX）和 Document Object Model（DOM）在 Android 上都是可用的。这些 API 多年以来一直都是 Java 技术的一部分，已经是相当稳定的技术。较新的 Streaming API for XML（StAX）在 Android 中并不可用。但是，Android 提供了一个相当功能的库。最后，Java XML Binding API 在 Android 中也不可用。这个 API 已确定可以在 Android 中实现，但是，它更倾向于是一个重量级的 API，需要使用许多不同类的实例来表示 XML 文档。因此，对于受限的环境，比如说 Android 针对的手持设备，不太理想。下面我们将先介绍 Android 中 3 种解析 XML 的方法，然后选择一种来完成天气预报程序。

1. 使用 DOM 方法来解析 XML。

DOM 是 Document Object Model 的缩写，即文档对象模型。XML 将数据组织为一棵树，所以 DOM 就是对这棵树的一个对象描述。通俗地说，就是通过解析 XML 文档，为 XML 文档在逻辑上建立一个树模型，树的一个节点是一个个对象。我们通过存取这些对象就能够存取 XML 文档的内容。DOM 解析器是通过将 XML 文档解析成树状模型并将其放入内存来完成解析工作的，而后对文档的操作都是在这个树状模型上完成。这个在内存中的文档树将是文档实

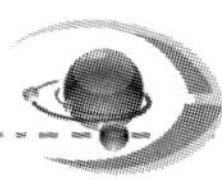

际大小的几倍。这样做的好处是结构清楚、操作方便，而带来的麻烦就是极其耗费系统资源。

要使用 DOM 方式来解析 XML，就需要引入如下两个包：

```
import javax.xml.parsers.*;
import org.w3c.dom.*;
```

其中 javax.xml.parsers 包含有 DOM 解析器和 SAX 解析器的具体实现。org.w3c.dom 包中定义了 W3C 所制定的 DOM 接口。

要将 XML 的内容解析到 Java 对象中去供程序使用，首先需要创建一个 DocumentBuilderFactory，代码如下：

```
DocumentBuilderFactory dbf=DocumentBuilderFactory.newInstance();
```

使用 DocumentBuilderFacotry 是为了创建与具体解析器无关的程序，当 DocumentBuilderFacotry 类的静态方法 newInstance()被调用时，它根据一个系统变量来决定具体使用哪一个解析器，又因为所有的解析器都服从于 JAXP 所定义的接口，所以无论具体使用哪一个解析器，代码都是一样的。所以当在不同的解析器之间进行切换时，只需要更改系统变量的值，而不用更改任何代码。这就是工厂对象所带来的好处。

当获得一个工厂对象后，使用它的静态方法 newDocumentBuilder()可以获得一个 DocumentBuilder 对象，这个对象代表了具体的 DOM 解析器，代码如下：

```
DocumentBuilder db=dbf. newDocumentBuilder();
```

现在我们就可以利用这个解析器来对 XML 文档进行解析了，DocumentBuilder 的 parse()方法接收一个 XML 文档名作为输入参数，返回一个 Document 对象，这个 Document 对象就代表了一个 XML 文档的树模型。以后所有对 XML 文档的操作，都与解析器无关，直接在这个 Document 对象上进行操作就可以了。而对 Document 操作的具体方法，就是由 DOM 所定义的了。Documet 对象的 normalize()可以去掉 XML 文档中作为格式化内容的空白而映射在 DOM 树中的不必要的 TextNode 对象，否则得到的 DOM 树可能并不如你所想象的那样。特别是在输出的时候，这个 normalize()更为有用。

```
Document doc=db.parse("xxxxx.xml");
doc.noramlize();
```

现在可以通过 Document 的 getElementsByTagName 方法来取得一个节点的 NodeList 对象，代码如下：

```
NodeList nodelist=doc.getElementsByTagName("节点");
```

然后用 item()方法提取想要的元素，并输入每个元素的数据，代码如下：

```
for(int i=0;i<nodelist.getLength();i++){
Element element=(Element)nodelist.item(i);
Srring str=element.getElementsByTagName("元素名称").item(0)
   .getFirstChind().getNodeValue();}
```

这样就可以得到我们需要的数据了，当然还可以修改一个 DOM 树中的值，然后重新写到一个 XML 文件中去。Document 对象中的 createElement 方法可以创建一个元素，如下代码即是创建一个节点，并对其进行赋值。

```
Element element=doc.createElement("节点");
Element elementtext=doc. createElement("text");
String textseg=doc.createTextNode(text);
```

```
elementtext.appendChild(textseg);
element.appendChild(elementtext);
```

在创建好节点之后，就苦于将修改的内容写入到XML文件中。Transformer类的transfrom方法接收两个参数、一个数据源Source和一个输出目标Result，这里数据源和输出目标使用的分别是DOMSource和StreamResult，这样就能够把DOM的内容输出到一个输出流中，当这个输出流是一个文件的时候，DOM的内容就被写入到文件中去了，代码如下：

```
TransformerFactory tFactory= TransformerFactory.newInstance();
Transformer transformer= tFactory.newTransformer();
DOMSource  source=new DOMSource(doc);
StreamResult result=new StreamResult(new java.io.File("text.xml"));
transformer.transform(source, result);
```

2. 使用XmlPullParser来解析XML。

Android并未提供对Java SAX API的支持。但是，Android确实附带了一个XmlPullParser解析器，其工作方式类似于SAX。它允许应用程序代码从解析器中获取事件，这与SAX解析器自动将事件推入处理程序相反。XmlPullParser是Android平台标准的XML解析器，它的正确名字应该是XML PULL PARSER，这项技术来自一个开源的XML解析API项目XMLPULL。下面是使用XmlPullParser来解析一个XML文件的方法，代码如下所示：

```
public void getXML(String url) throws
    XmlPullParserException,IOException,URISyntaxException{
   XmlPullParserFactory factory=XmlPullParserFactory.newInstance();
   factory.setNamespaceAware(true);
   XmlPullParser parser=factory.newPullParser();
   parser.setInput(new InputStreamReader(getUrlData(url)));
   XmlUtils.beginDocument(parser,"results");
   int eventType=parser.getEventType();
   do{ XmlUtils.nextElement(parser);
       parser.next();
       eventType=parser.getEventType();
       if(eventType==XmlPullParser.TEXT){
           String str="";
           str+=parser.getText(); }
   }while(eventType!=XmlPullParser.END_DOCUMENT);
}
```

XmlPullParser解析器的运行方式与SAX解析器相似。它提供了类似的事件（开始元素和结束元素），但需要使用parser.next()方法来提供它们。事件将作为数值代码被发送，因此可以根据不同的事件代码值来进行不同的处理。上面代码通过parser.getEventType()方法取得事件的代码值（如XmlPullParser.START_DOCUMENT、XmlPullParser.STRART_TAG、XmlPullParser.END_TAG），解析并未像SAX解析那样监听元素的结束，而是开始处完成了大部分处理。当某个元素开始时，可以调用parser.nextText()从XML文档中提取所有的字符数据。

同样，还可以使用XmlPullParser来创建XML文件，要创建XML文件需要使用一个StringBuilder

来创建 XML 字符串，下面代码就是用 XmlPullParser 解析器来创建一个 XML 文件并写入数据。

```
private String writexml(List<Message> messages){
    XmlSerializer serializer=Xml.newSerializer();
    StringWriter writer=new StringWriter();
    try{ serializer.setOutput(writer);
        serializer.startDocument("UTF-8", true);
        serializer.startTag("", "messages");
        serializer.attribute("", "number", String.valueOf(messages.size()));
        for(Message msg:messages){
            serializer.startTag("", "message");
            serializer.attribute("", "date", msg.getData());
            serializer.startTag("", "title");
            serializer.text(msg.getTitle());
            serializer.endTag("", "title");
            serializer.startTag("", "url");
            serializer.text(msg.getLink().toExternalForm());
            serializer.endTag("", "url");
            serializer.startTag("", "body");
            serializer.text(msg.getDescription());
            serializer.endTag("", "body");
            serializer.endTag("", "message");  }
        serializer.endTag("", "messages");
        serializer.endDocument();
        return writer.toString();
        }catch(Exception e){throw new RuntimeException(e);}
}
```

3. 使用 SAX 来解析 XML 文件。

在 Java 环境中，当需要一个速度快的解析器并且希望最大限度地减小应用程序的内存占用时，通常可以使用 SAX API，这非常适用于运行 Android 的移动设备。可以在 Java 环境中照原样使用 SAX API，不需要做任何修改地在 Android 上运行它。本例中我们将采用 SAX 来解析 XML 文件，获取天气预报信息。

SAX 是 Simple API for XML 的缩写，它并不是由 W3C 官方提出的标准。实际上，它是一种社区讨论的产物。虽然如此，在 XML 中对 SAX 的应用丝毫不比 DOM 少，几乎所有的 XML 解析器都支持它。与 DOM 比较而言，SAX 是一种轻量型的方法。我们知道，在处理 DOM 的时候，需要读入整个 XML 文档，然后在内存中创建 DOM 树，生成 DOM 树上的每个 Node 对象，当文档比较小的时候，这不会造成什么问题，但是一旦文档大起来，处理 DOM 就会变得相当费时、费力。特别是其对于内存的需求，也将是成倍地增长，以至于在某些应用中（比如在 applet 中）使用 DOM 是一件很不划算的事。这时候，一个较好的替代解决方法就是 SAX。

我们创建一个继承 DefaultHandler 的 GoogleWeatherHandler 类，要使用 SAX 来解析 XML，

首先需要引入以下几个包：

```
import javax.xml.parsers.*;
import org.xml.sax.*;
import org.xml.sax.helpers.DefaultHandler;
import org.xml.sax.helpers.*;
import java.util.*;
import java.io.*;
```

当遇到一个开始标签，在 startElement()方法中根据不同的标签来获取得不同的数据信息存放到一个我们自定义的用来保存获取的天气信息的 WeatherSet 类中，GoogleWeatherHandler 类具体代码如下所示：

```
public class GoogleWeatherHandler extends DefaultHandler{
   //天气信息
   private WeatherSet   myWeatherSet= null;
   //实时天气信息
   private boolean  is_Current_Conditions= false;
   //预报天气信息
   private boolean  is_Forecast_Conditions= false;
   private final String CURRENT_CONDITIONS= "current_conditions";
   private final String FORECAST_CONDITIONS= "forecast_conditions";
   public GoogleWeatherHandler(){}
    //返回天气信息对象
   public WeatherSet getMyWeatherSet(){
      return myWeatherSet;}
   @Override
   public void endDocument() throws SAXException{
      // TODO Auto-generated method stub
      super.endDocument();}
   @Override
   public void endElement(String uri, String localName, String name)
   throws SAXException{
      if (localName.equals(CURRENT_CONDITIONS)){
          this.is_Current_Conditions = false;}
      else if (localName.equals(FORECAST_CONDITIONS)){
          this.is_Forecast_Conditions = false;}  }
   @Override
   public void startDocument() throws SAXException{
      this.myWeatherSet = new WeatherSet();}
   @Override
   public void startElement(String uri, String localName, String name,
     Attributes attributes) throws SAXException{
      if (localName.equals(CURRENT_CONDITIONS)){//实时天气
         Log.i("localName+CURRENT", localName);
         this.myWeatherSet.setMyCurrentCondition(
```

```
                new WeatherCurrentCondition());
            Log.i("localName+CURRENT+1", localName);
            this.is_Current_Conditions = true; }
        else if (localName.equals(FORECAST_CONDITIONS)){//预报天气
            this.myWeatherSet.getMyForecastConditions().add(
            new WeatherForecastCondition());
            this.is_Forecast_Conditions = true; }
        else{
            //分别将得到的信息设置到指定的对象中
            if (localName.equals(CURRENT_CONDITIONS)){
                Log.i("localName+CURRENT", localName);}
            String dataAttribute = attributes.getValue("data");
            if (localName.equals("icon")){
                if (this.is_Current_Conditions){
                this.myWeatherSet.getMyCurrentCondition()
                    .setIcon(dataAttribute); }
                else if (this.is_Forecast_Conditions){
                    this.myWeatherSet.getLastForecastCondition()
                    .setIcon(dataAttribute); } }
            else if (localName.equals("condition")){
                if (this.is_Current_Conditions){
                    this.myWeatherSet.getMyCurrentCondition()
                    .setCondition(dataAttribute);}
                else if (this.is_Forecast_Conditions){
                    this.myWeatherSet.getLastForecastCondition()
                        .setCondition(dataAttribute); }}
            else if (localName.equals("temp_c")){
                this.myWeatherSet.getMyCurrentCondition()
                .setTemp_celcius(dataAttribute); }
            else if (localName.equals("temp_f")){
                this.myWeatherSet.getMyCurrentCondition()
                .setTemp_fahrenheit(dataAttribute); }
            else if (localName.equals("humidity")){
                this.myWeatherSet.getMyCurrentCondition()
                .setHumidity(dataAttribute); }
            else if (localName.equals("wind_condition")){
                this.myWeatherSet.getMyCurrentCondition()
                .setWind_condition(dataAttribute); }
            else if (localName.equals("day_of_week")){
                this.myWeatherSet.getLastForecastCondition()
                .setDay_of_week(dataAttribute); }
            else if (localName.equals("low")){
                this.myWeatherSet.getLastForecastCondition()
                .setLow(dataAttribute); }
```

```
            else if (localName.equals("high")){
                this.myWeatherSet.getLastForecastCondition()
                .setHigh(dataAttribute);  }}
            }
        @Override
        public void characters(char ch[], int start, int length){}
        }
```

WeatherSet 类的主要代码如下：

```
    public class WeatherSet{
        //实时天气信息
        private WeatherCurrentCondition myCurrentCondition = null;
        //预报后 4 天的天气信息
        private ArrayList<WeatherForecastCondition> myForecastConditions =
            new ArrayList<WeatherForecastCondition>();
        public WeatherSet(){}
        //得到实时天气信息的对象
        public WeatherCurrentCondition getMyCurrentCondition(){
            return myCurrentCondition;}
        //设置实时天气信息的对象
        public void setMyCurrentCondition(WeatherCurrentCondition
        myCurrentCondition){
            this.myCurrentCondition = myCurrentCondition;}
        //得到预报天气
        public ArrayList<WeatherForecastCondition>
          getMyForecastConditions(){
            return myForecastConditions;}
        //得到最后一个预报天气
        //这里我们每次添加一个数据都是在最后
        //所以得到最后一个
        public WeatherForecastCondition getLastForecastCondition(){
        return myForecastConditions.get(myForecastConditions.size() - 1);
          }
    }
```

在 startElement()方法中使用了同 DOM 一样的设计技巧，在创建 SAXParser 对象的时候，通过一个 SAXParserFactory 类来创建具体的 SAXParser 对象，这样，当需要使用不同解析器的时候，要改变的只是一个环境变量的值，而程序的代码可以保持不变。这就是 FactoryMethod 模式的思想。代码如下：

```
    SAXParseFactory spf= SAXParseFactory.newInstsance();
```

在获得了 SAXParserFactory 对象之后，要解析 XML 还需要一个 SAXParser 或者 XMLReader，SAXParser 是 JAXP 中对 XMLReader 的一个封装类，而 XMLReader 是定义在 SAX2.0 中的一个用来解析文档的接口。可以同样调用 SAXParser 或者 XMLReader 中的 parser()方法来解析文档，效果是完全一样的。不过 SAXParser 中的 parser()方法接收更多的参数，可以

对不同的 XML 文档数据源进行解析，因而使用起来要比 XMLReader 方便一些。

```
SAXParser sp=spf.newSAXParser();
XMLReader xr=sp.getXMLReader();
```

在创建了 XMLReader 之后，就可以使用上一步骤创建的 GoogleWeatherHandler 来解析 XML，具体操作代码如下：

```
GoogleWeatherHandler gwh = new GoogleWeatherHandler();
xr.setContentHandler(gwh);
InputStreamReader isr = new
    InputStreamReader(url.openStream(), "GBK");
    InputSource is = new InputSource(isr);
    xr.parse(is);
```

主逻辑界面 CityWeather 类的代码如下：

```
public class CityWeather extends Activity{
    /** Called when the activity is first created. */
    @Override
    public void onCreate(Bundle savedInstanceState){
        super.onCreate(savedInstanceState);
        setContentView(R.layout.main);
        init();}
    private void init(){
        Spinner city_spr = (Spinner) findViewById(R.id.Spinner01);
        ArrayAdapter<String> adapter = new ArrayAdapter<String>(this,
         android.R.layout.simple_spinner_item, ConstData.city);
        adapter.setDropDownViewResource(
         android.R.layout.simple_spinner_dropdown_item);
        city_spr.setAdapter(adapter);
        Button submit = (Button) findViewById(R.id.Button01);
        submit.setOnClickListener(new OnClickListener() {
             @Override
             public void onClick(View v){
                  // TODO Auto-generated method stub
                  Spinner spr = (Spinner) findViewById(R.id.Spinner01);
                  Long l = spr.getSelectedItemId();
                  int index = l.intValue();
                  String cityParamString = ConstData.cityCode[index];
                  try{ URL url = new URL(ConstData.queryString +
                     cityParamString);
                     getCityWeather(url); }
                  catch (Exception e){
                  Log.e("CityWeather", e.toString()); }}
        });
        Button submit_input = (Button) findViewById(R.id.Button001);
        submit_input.setOnClickListener(new OnClickListener(){
```

```
        public void onClick(View v) {
        EditText inputcity = (EditText) findViewById(R.id.EditText001);
        String tmp = inputcity.getText().toString();
        try{URL url = new URL(ConstData.queryString_intput + tmp);
            getCityWeather(url);}
        catch (Exception e){ Log.e("CityWeather", e.toString());} }
        }); }
    // 更新显示实时天气信息
    private void updateWeatherInfoView(int aResourceID,
    WeatherCurrentCondition aWCC) throws MalformedURLException{
        URL imgURL = new URL("http://www.google.com/" + aWCC.getIcon());
        ((SingleWeatherInfoView) findViewById(aResourceID))
          .setWeatherIcon(imgURL);
        ((SingleWeatherInfoView)findViewById(aResourceID))
        .setWeatherString(aWCC.toString());}
    // 更新显示天气预报
    private void updateWeatherInfoView(int aResourceID,
     WeatherForecastCondition aWFC) throws MalformedURLException{
    URL imgURL = new URL("http://www.google.com/" + aWFC.getIcon());
      ((SingleWeatherInfoView) findViewById(aResourceID))
         .setWeatherIcon(imgURL);
    ((SingleWeatherInfoView) findViewById(aResourceID))
       .setWeatherString(aWFC.toString()); }
    //获取天气信息
    //通过网络获取数据
    //传递给XMLReader解析
    public void getCityWeather(URL url){
    try{  SAXParserFactory spf = SAXParserFactory.newInstance();
            SAXParser sp = spf.newSAXParser();
            XMLReader xr = sp.getXMLReader();
            GoogleWeatherHandler gwh = new GoogleWeatherHandler();
            xr.setContentHandler(gwh);
            InputStreamReader isr = new
             InputStreamReader(url.openStream(), "GBK");
            InputSource is = new InputSource(isr);
            xr.parse(is);
            WeatherSet ws = gwh.getMyWeatherSet();
            updateWeatherInfoView(R.id.weather_0,
            ws.getMyCurrentCondition());
            updateWeatherInfoView(R.id.weather_1,
            ws.getMyForecastConditions().get(0));
            updateWeatherInfoView(R.id.weather_2,
            ws.getMyForecastConditions().get(1));
            updateWeatherInfoView(R.id.weather_3,
```

```
            ws.getMyForecastConditions().get(2));
            updateWeatherInfoView(R.id.weather_4,
            ws.getMyForecastConditions().get(3)); }
        catch (Exception e){ Log.e("CityWeather", e.toString());
            }
        }
    }
```

项目中剩下几个类 ConstData.java、WeatherCurrentCondition.java 和 Weather ForecastCondition.java 的参考代码公布如下。

ConstData.java 类的代码如下：

```
public class ConstData{
  public static final String
  queryString="http://www.google.com/ig/api?hl=zh-cn&weather=,,,";
  public static final String queryString_intput=
  "http://www.google.com/ig/api?hl=zh_cn&weather=";
  public static final String [] cityCode={
      "39930000,116279998",//北京   "31399999,121470001",//上海
      "39099998,117169998",//天津   "29520000,106480003",//重庆
      "39669998,118150001",//唐山   "38029998,114419998",//石家庄
      "38900001,121629997",//大连   "45750000,126769996",//哈尔滨
      "20030000,110349998",//海口   "43900001,125220001",//长春
      "28229999,112870002",//长沙   "30670000,104019996",//成都
      "26079999,119279998",//福州   "23129999,113319999",//广州
      "26579999,106720001",//贵阳   "30229999,120169998",//杭州
      "31870000,117230003",//合肥   "40819999,111680000",//呼和浩特
      "36680000,116980003",//济南   "25020000,102680000",//昆明
      "29657589,91132050",//拉萨    "36040000,103879997",//兰州
      "28600000,115919998",//南昌   "32000000,118800003",//南京
      "22819999,108349998",//南宁   "36069999,120330001",//青岛
      "22549999,114099998",//深圳   "41770000,123430000",//沈阳
      "37779998,112550003",//太原   "43779998,87620002",//乌鲁木齐
      "30620000,114129997",//武汉   "34299999,108930000",//西安
      "36619998,101769996",//西宁   "24479999,118080001",//厦门
      "34279998,117150001",//徐州   "38479999,106220001",//银川
      "34720001,113650001"//郑州  };
  public static final String [] city ={
      "北京","上海","天津","重庆","唐山",
      "石家庄","大连","哈尔滨","海口","长春","长沙","成都","福州","广州",
      "贵阳","杭州","合肥","呼和浩特","济南","昆明","拉萨","兰州","南昌",
      "南京","南宁","青岛","深圳","沈阳","太原","乌鲁木齐","武汉","西安",
      "西宁","厦门","徐州","银川","郑州"
      };  }
```

WeatherCurrentCondition.java 类的代码如下：

```
public class WeatherCurrentCondition{
    private String  condition;          // 多云
    private String  temp_celcius;       // 摄氏温度
    private String  temp_fahrenheit;    // 华氏温度
    private String  humidity;           // 湿度:58%
    private String  wind_condition;     // 风向...
    private String  icon;               // 图标
    public WeatherCurrentCondition(){}
    //得到Condition（多云）
    public String getCondition(){   return condition;}
    //设置Condition（多云）
    public void setCondition(String condition){ this.condition =
condition;}
    //得到设置温度
    public String getTemp_c(){
        return temp_celcius;}
    //得到华氏温度
    public String getTemp_f(){
        return temp_fahrenheit;}
    //设置摄氏温度
    public void setTemp_celcius(String temp_celcius){
        this.temp_celcius = temp_celcius;}
    //设置华氏温度
    public void setTemp_fahrenheit(String temp_fahrenheit){
        this.temp_fahrenheit = temp_fahrenheit;}
    //得到（湿度:58%）
    public String getHumidity(){
        return humidity;}
    //设置（湿度:58%）
    public void setHumidity(String humidity){
        this.humidity = humidity;}
    //得到风向指示
    public String getWind_condition(){
        return wind_condition;}
    //设置风向指示
    public void setWind_condition(String wind_condition){
        this.wind_condition = wind_condition;}
    //得到图标地址
    public String getIcon(){
        return icon;}
    //设置图标地址
    public void setIcon(String icon){
        this.icon = icon;}
    //得到一个封装打包的字符串，包括除icno外的所有东西
```

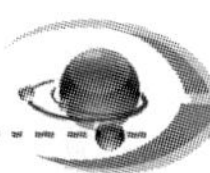

```
    public String toString(){
        StringBuilder sb = new StringBuilder();
        sb.append("实时天气: ").append(temp_celcius).append(" ¡ãC");
        sb.append(" ").append(temp_fahrenheit).append(" F");
        sb.append(" ").append(condition);
        sb.append(" ").append(humidity);
        sb.append(" ").append(wind_condition);
        return sb.toString();}
}
```

WeatherForecastCondition.java 类的代码如下：

```
public class WeatherForecastCondition {
    private String day_of_week;        //星期
    private String low;                //最低温度
    private String high;          //最高温度
    private String icon;          //图标
    private String condition;          //提示
    public WeatherForecastCondition(){}
    public String getCondition(){
        return condition;}
    public void setCondition(String condition)
    {this.condition = condition;}
    public String getDay_of_week(){return day_of_week;}
    public void setDay_of_week(String day_of_week){this.day_of_week =
day_of_week;}
    public String getLow(){  return low;}
    public void setLow(String low){this.low = low;}
    public String getHigh(){return high;}
    public void setHigh(String high){this.high = high;}
    public String getIcon(){return icon;}
    public void setIcon(String icon){this.icon = icon;}
    public String toString(){
        StringBuilder sb = new StringBuilder();
        sb.append(" ").append(day_of_week);
        sb.append(" : ").append(high);
        sb.append("/").append(low).append(" ¡ãC");
        sb.append(" ").append(condition);
        return sb.toString();}
    }
```

9.3.4 AndroidManifest.xml（限设置）

由于我们的信息是通过网络获取的，所以需要在 AndroidManifest.XML 中注册访问网络的权限：

```
<uses-permission android:name="android.permission.INTERNET">/<use—
permission>。
```

最后将解析 XML 得到的数据更新到我们自定义的视图上，展示给用户即可完成从网络获取天气信息的程序。

本节的重点内容在于如何解析 XML 文件，该例只是使用了 Android 中的其中一种方式，大家可以根据我们所讲述的其他两种方法来完成这个天气程序。这三种方式都有各自的优势，究竟使用哪一种方式，还需要根据应用程序的实际需要来确定。大多数情况下，使用 SAX 是比较安全的，并且 Android 提供一种传统的 SAX 使用方法以及一个便捷的 SAX 包装器。如果文档比较小，那么 DOM 可能是一种比较简单的方法，如果文档比较大，但只需要文档的一部分，则 XML PULL 解析器可能是更为有效的方法。

第10章

综合案例设计与开发

本章将以“手机相册服务软件”作为示例，综合运用前面章节所学到的知识和技巧，从需求分析、界面设计、模块设计和程序设计等几个方面，详细介绍 Android 应用程序的设计思路与开发方法。本章提供的“手机相册服务软件”是 2012 年“全国软件杯”软件设计大赛 Android 开发项目的本地相册内容，是一个综合的案例。

10.1 需求分析

通过前面章节的学习，读者应该已经掌握了较多 Android 应用程序开发的知识和方法，但如何能够综合地运用这些知识和方法，解决实际开发中所遇到的问题，还需要继续学习和探讨。本章的设计就是希望读者能够根据实际项目的需求，准确地分析出 Android 应用程序可能涉及到的知识点，并学会如何通过分析软件的需求，快速地设计出应用程序的用户界面和功能模块，并最终完成应用程序的开发和调试。

在这个综合案例中，打开软件，就可以浏览相册里面的相片，并且也可以浏览到每张相片的具体信息；用户可以对每张相片进行编辑和删除等操作；还可以选择照相模块实现照相等功能。其项目的具体架构图如图 10-1 所示。

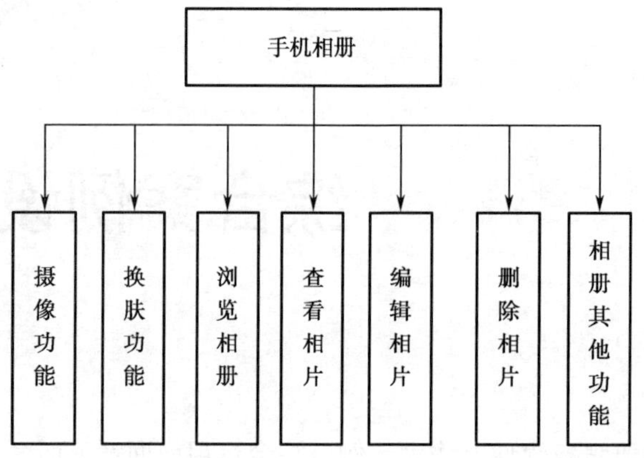

图 10-1　手机相册总体架构图

从总体结构图可以基本了解到该软件的功能需求，用户登录相册就直接进入浏览相册界面，即浏览相册界面为主界面。浏览相册界面设计如图 10-2 所示。主界面的右上角有两个图标，分别是摄像和换肤功能的进入图标；界面主要用 GridView 控件显示 SD 卡里面的相片。当用户点击某一张图片，则会看到该图片的放大图；长时间按某一张图片，在弹出提示对话框，可以对图片进行删除、编辑和查看详情等操作。当用户按 MENU 键，将弹出“关于”和“帮助”的菜单。

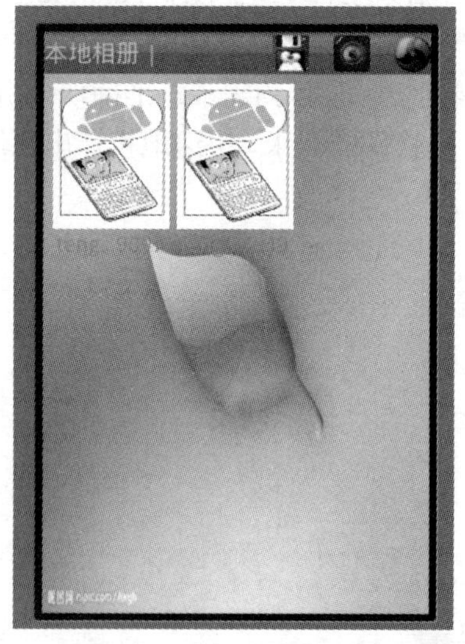

图 10-2　浏览相册界面图

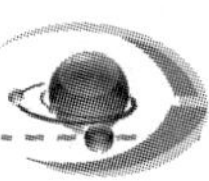

10.2　策划与准备

10.2.1　图片资源的准备

在本项目中，需要涉及到一些图片，尤其在换肤功能的实现中，本项目所涉及到的图片如图 10-3 所示。

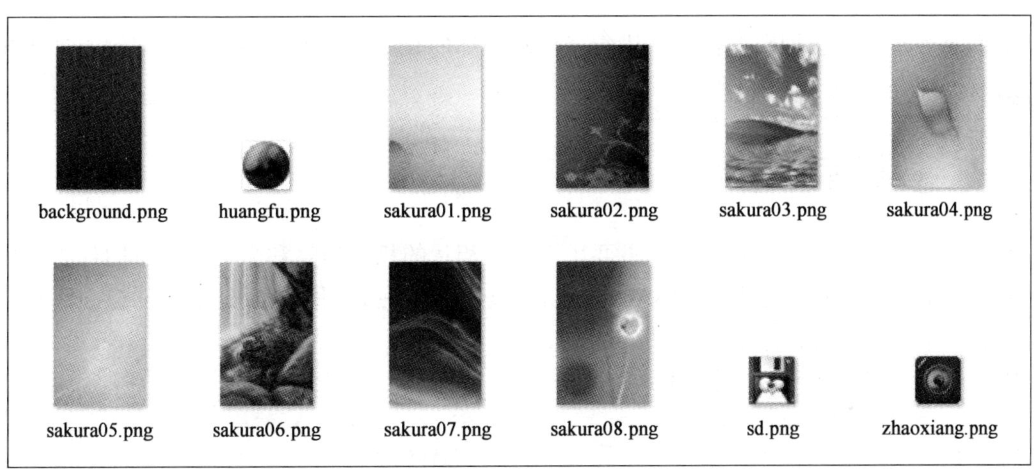

图 10-3　项目图片资源图

10.2.2　数据库设计

本案例中主要涉及到两种数据需要存储，一个是相片的信息，另一个是换肤图片信息的存储。由于数据量都很小，从 Android 支持的存储方式上分析，可以保存在 SharePreference、文件和 SQLite 数据库中。我们这个项目中都是用 SQLite 数据库对相关信息进行保存。保存图片信息的表名为 photo，表结构如表 10-1 所示；对换肤功能所涉及到的换肤图片信息保存在表 IdXu 中，表结构如表 10-2 所示。

表 10-1　photo 表

字 段 名	键　值	类　型	长　度	默 认 值	备　注
photoName	PK	Varchar	50	实际图片的值	唯一的键值
photoDescribe		Varchar	50	NULL	图片描述
photoPath		Varchar	50	NULL	图片的存放路径

表 10-2　IdXu 表

字段名	类　型	长　度	备　注
id	Varchar	50	图片编号
tuhao	Varchar	50	换肤图片的名称

注意：本案例在 SD 卡根目录下，创建了一个名为 XiangCe 的目录文件夹，用于储存用户拍照的图片。

10.3　程序设计

前面对该案例的策划和架构设计进行了比较详细的介绍，接下来我们对本案例的具体开发过程进行详细介绍。

10.3.1　数据库适配器

数据库适配器是最底层的模块，主要用于保存相片的相关信息和主界面背景图片的信息。数据库适配器的核心代码主要在 DBAdapter.java 文件中，在介绍数据库适配器的核心代码前，首先了解一下保存相片信息的类文件 Photo.java 文件。

Photo.java 文件的全部代码如下：

```
package thi.xc.sqlite.com;
public class Photo {
    private String photoName = "";
    private String photoDescribe = "";
    private String photoPath = "";
    public String getPhotoName() {
        return photoName;  }
    public void setPhotoName(String photoName) {
        this.photoName = photoName;  }
    public String getPhotoDescribe() {
        return photoDescribe;  }
    public void setPhotoDescribe(String photoDescribe) {
        this.photoDescribe = photoDescribe;  }
    public String getPhotoPath() {
        return photoPath;  }
    public void setPhotoPath(String photoPath) {
        this.photoPath = photoPath;  }
    }
```

DBAdapter 类与以往介绍过的数据库适配器类相似，都具有继承 SQLiteOpenHelper 的帮助类 DBOpenHelper。DBOpenHelper 在建立数据库时，同时建立两张数据库表，并对表里面的信息进行了初始化。

DBOpenHelper.java 文件的全部代码如下：

```
package thi.xc.sqlite.com;
```

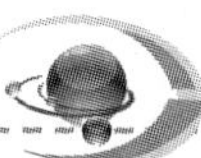

```
import android.content.Context;
import android.database.sqlite.SQLiteDatabase;
import android.database.sqlite.SQLiteOpenHelper;
public class DBOpenHelper extends SQLiteOpenHelper {
    private static final int VERSION = 1;
    private static final String DBNAME = "picture.db";
    private static final String DBTABLE = "photo";
    public DBOpenHelper(Context context) {
        super(context, DBNAME, null, VERSION);
    }
    public void onCreate(SQLiteDatabase db) {
        db.execSQL("create table " + DBTABLE + " (photoName varchar(50),
photoDescribe varchar(50),photoPath varchar(50))");
        db.execSQL("create table IdXu (id varchar(50) ,tuhao varchar(50))");
    }
    public void onUpgrade(SQLiteDatabase db, int oldVersion, int newVersion) { }
 }
```

核心类 DBAdapter.java 文件的全部代码如下：

```
package thi.xc.sqlite.com;
import java.util.ArrayList;
import java.util.List;
import android.content.ContentValues;
import android.content.Context;
import android.database.Cursor;
import android.database.sqlite.SQLiteDatabase;
public class DBAdapter {
    private SQLiteDatabase db = null;
    private DBOpenHelper dboper;
    private static final String PHOTO_NAME = "photoName";
    private static final String PHOTO_DESCRIBE = "photoDescribe";
    private static final String PHOTO_PATH = "photoPath";
    private static final String DBTABLE = "photo";
    public DBAdapter(Context context) {
        dboper = new DBOpenHelper(context);
        try {
            db = dboper.getWritableDatabase();
        } catch (Exception e) {
            db = dboper.getReadableDatabase(); }
    }

    /** 添加相片信息*/
    public long add(Photo photo) {
        ContentValues newValues = new ContentValues();
        newValues.put(PHOTO_NAME, photo.getPhotoName());
```

```
        newValues.put(PHOTO_DESCRIBE, photo.getPhotoDescribe());
        newValues.put(PHOTO_PATH, photo.getPhotoPath());
        return db.insert(DBTABLE, null, newValues);
    }
    /* * 更新相片信息 */
    public long update(String photoName, Photo photo) {
        ContentValues newValues = new ContentValues();
        newValues.put(PHOTO_NAME, photo.getPhotoName());
        newValues.put(PHOTO_DESCRIBE, photo.getPhotoDescribe());
        return db.update(DBTABLE, newValues, PHOTO_NAME + "='" + photoName
                 + "'", null);
    }
    /* * 查找相片信息 */
    public Photo find(String photoName) {
        Photo photo = new Photo();
        Cursor cursor = db.query(DBTABLE, new String[] { PHOTO_NAME,
                  PHOTO_DESCRIBE, PHOTO_PATH }, PHOTO_NAME + "='" + photoName
                 + "'", null, null, null, null);
        cursor.moveToFirst();
        if (cursor != null) {
            photo.setPhotoName(cursor.getString(cursor
                    .getColumnIndex(PHOTO_NAME)));
            photo.setPhotoDescribe(cursor.getString(cursor
                    .getColumnIndex(PHOTO_DESCRIBE)));
            photo.setPhotoPath(cursor.getString(cursor
                    .getColumnIndex(PHOTO_PATH)));  }
        return photo;
    }
    /* * 查找全部相片信息 */
    public List<Photo> find() {
        List<Photo> photolist = new ArrayList<Photo>();
        Cursor cursor = db.query(DBTABLE, new String[] { PHOTO_NAME,
                  PHOTO_DESCRIBE, PHOTO_PATH }, null, null, null, null, null);
        cursor.moveToFirst();
        for (int i = 0; i < cursor.getCount(); i++) {
            if (cursor != null) {
                Photo photo = new Photo();
                photo.setPhotoName(cursor.getString(cursor
                        .getColumnIndex(PHOTO_NAME)));
                photo.setPhotoDescribe(cursor.getString(cursor
                        .getColumnIndex(PHOTO_DESCRIBE)));
                photo.setPhotoPath(cursor.getString(cursor
                        .getColumnIndex(PHOTO_PATH)));
                photolist.add(photo);
```

```
                photo = null;
                cursor.moveToNext();
            }  }
        cursor.close();
        if (photolist.size() == 0) {
            return null;  }
        return photolist; }
    /**  删除相片信息 */
    public long detele(String photoName) {
        return db.delete(DBTABLE, PHOTO_NAME + "='" + photoName + "'", null);
    }
    public void close() {
        if (db != null) {
            db.close();
            db = null;   }
    }
    /* * 查找换肤图号  */
    public String CheckTuhao() {
        Cursor cursor = db.query("IdXu", new String[] { "id", "tuhao" }, null,
                null, null, null, null);
        if (cursor == null) {
            return null;  }
        cursor.moveToFirst();
        String tuhao = null;
        for (int i = 0; i < cursor.getCount(); i++) {
            if (cursor != null) {
                tuhao = cursor.getString(cursor.getColumnIndex("tuhao"));
                cursor.moveToNext();  }
        }
        cursor.close();
        return tuhao;
    }
    /*  更新换肤图号  */
    public int updataTuhao(String tuhao) {
        ContentValues newValues = new ContentValues();
        newValues.put("tuhao", tuhao);
        return db.update("IdXu", newValues, "id='one'", null);
    }

    /*  初始化换肤数据 */
    public long insertTuhao() {
        if (CheckTuhao() != null) {
            return 0; }
        ContentValues newValues = new ContentValues();
```

```
            newValues.put("id", "one");
            newValues.put("tuhao", "-1");
            return db.insert("IdXu", null, newValues);
        }
    }
```

10.3.2　主界面类 PhotographActivity.java 的实现

进入相册软件，首先能够浏览到该相册里面的所有相片，能够查看手机里面的所有图片资源，并且可以照相和换肤等功能。PhotographActivity.java 类作为整个项目的入口类，也是主界面类，运行效果如图 10-4 所示。从图中我们可以看出相册的主界面用了自定义的布局方式，在界面上有查看手机所有图片资源的图标按钮，照相功能的图标按钮，换肤功能的图标按钮以及刷新相册相片的按钮。当用户单击“本地相册”就可以达到对相册刷新的功能；当用户单击查看手机所有图片资源的图标按钮，就会看到手机上所有的图片资源，如图 10-5 所示；当用户单击照相功能的图标按钮，就会出现照相的界面，如图 10-6 所示；当用户单击换肤功能的图标按钮，就可以看到我们可以更换哪几种皮肤，如图 10-7 所示，当用户单击其中任意一种换肤图片，都可以实现换肤的功能，并把换肤图片名称保存到相应的数据库中，以后打开都一直有效，直到下次换肤。

在相册软件里面还设置了“帮助”和“关于”内容。当用户单击模拟器的 MENU 按钮，就会弹出“帮助”和“关于”的子菜单选项，如图 10-8 所示。当用户选择“帮助”子菜单，界面会弹出软件的帮助内容，如图 10-9 所示；当用户选择“关于”子菜单，界面会弹出软件的关于内容，如图 10-10 所示。

在相册软件中还设置了对每张相片的操作功能实现，当用户长按某张相片时，界面会弹出相片操作的功能选项图，图如 10-11 所示。当用户选择“删除”功能，则会删除该张相片，返回主界面；当用户选择“相片详情”功能，则界面弹出如图 10-12 所示，把该张相片的详细信息显示给用户；当用户选择“编辑”功能，则弹出如图 10-13 所示，用户可以根据自己的需要对相片的名称和描述进行修改。这些信息修改以后将保存到 SQLite 数据库中。

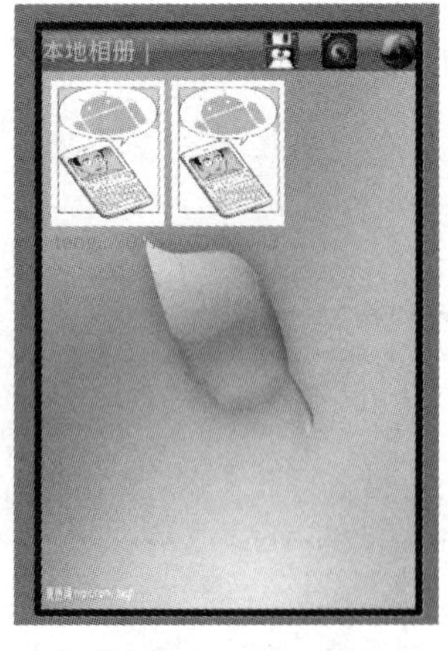

图 10-4　主界面图

图 10-5　查看所有相片图

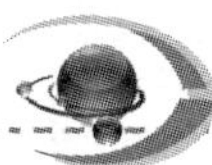

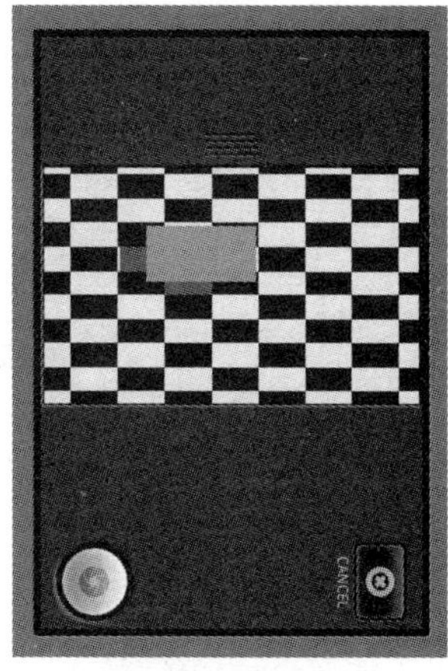

图 10-6　照相功能图

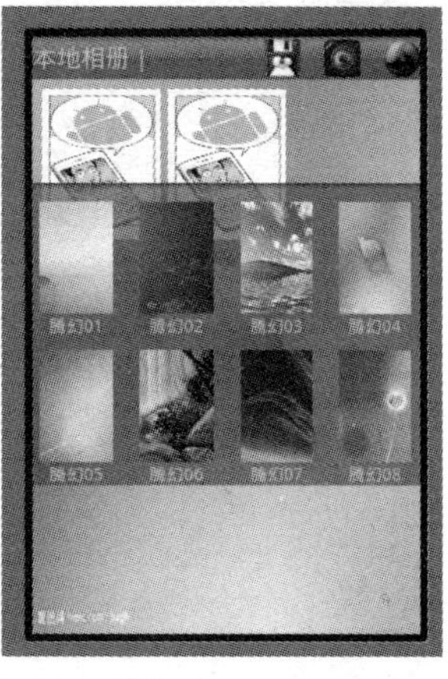

图 10-7　换肤功能图

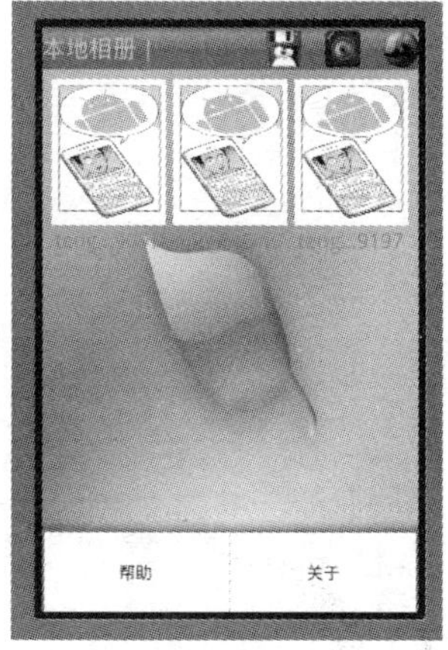

图 10-8　子菜单图

图 10-9　帮助子菜单图

图 10-10　关于子菜单图图

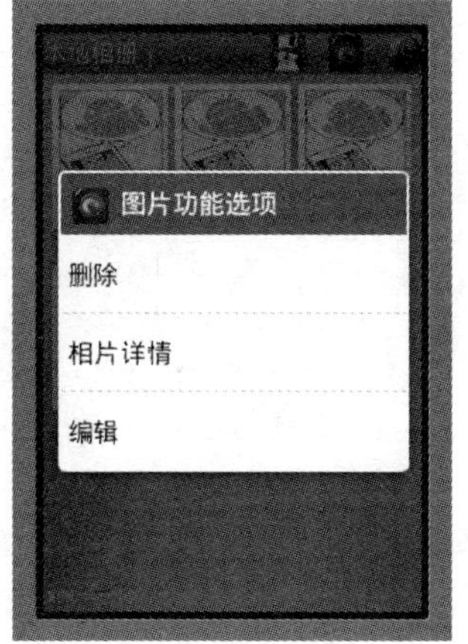

图 10-11　相片功能选项图

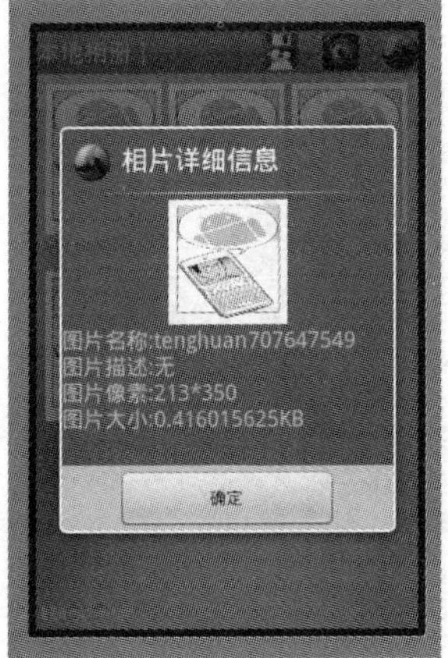

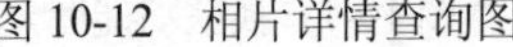
图 10-12　相片详情查询图

图 10-13　相片编辑功能图

前面对主界面的功能进行了详细的描述，在理解主界面的各个功能模块的时候，首先需要理解 PhotographActivity.java 类显示的布局文件 index.xml。

index.xml 文件的完整代码如下：

```
<?xml version="1.0" encoding="utf-8"?>
<LinearLayout xmlns:android="http://schemas.android.com/apk/res/android"
    android:layout_width="fill_parent"
    android:layout_height="fill_parent"
    android:orientation="vertical" >
    <RelativeLayout
        android:layout_width="fill_parent"
        android:layout_height="wrap_content"
        android:background="@drawable/back" >
        <LinearLayout
            android:layout_width="fill_parent"
            android:layout_height="wrap_content"
            android:layout_centerVertical="true"
            android:orientation="horizontal" >
            <TextView
                android:id="@+id/bendi"
                android:layout_width="wrap_content"
                android:layout_height="wrap_content"
                android:text="本地相册"
                android:textSize="20dip" />
        </LinearLayout>
        <AbsoluteLayout
            android:layout_width="wrap_content"
            android:layout_height="wrap_content"
            android:layout_alignParentRight="true"
            android:gravity="center_vertical" >
```

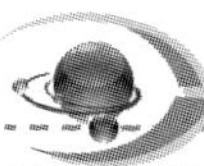

```
            <ImageView
                android:id="@+id/sd"
                android:layout_width="wrap_content"
                android:layout_height="wrap_content"
                android:layout_x="0dip"
                android:layout_y="3dip"
                android:src="@drawable/sd" />
            <ImageView
                android:id="@+id/zhaoxiang"
                android:layout_width="wrap_content"
                android:layout_height="wrap_content"
                android:layout_x="50dip"
                android:layout_y="3dip"
                android:src="@drawable/zhaoxiang" />
            <ImageView
                android:id="@+id/huangfu"
                android:layout_width="wrap_content"
                android:layout_height="wrap_content"
                android:layout_x="100dip"
                android:layout_y="3dip"
                android:src="@drawable/huangfu" />
        </AbsoluteLayout>
    </RelativeLayout>
    <LinearLayout
        android:id="@+id/linear01"
        android:layout_width="fill_parent"
        android:layout_height="fill_parent"
        android:background="@drawable/background"
        android:visibility="visible" >
        <GridView
            android:id="@+id/sd01"
            android:layout_width="fill_parent"
            android:layout_height="fill_parent"
            android:gravity="center"
            android:numColumns="3" />
    </LinearLayout>
    <LinearLayout
        android:id="@+id/linear02"
        android:layout_width="fill_parent"
        android:layout_height="fill_parent"
        android:background="@drawable/background"
        android:gravity="center"
        android:orientation="vertical"
        android:visibility="gone" >
```

```
<WebView
   android:id="@+id/webviewid"
   android:layout_width="fill_parent"
   android:layout_height="fill_parent"
   android:visibility="visible" />
<GridView
   android:id="@+id/gridviewss"
   android:layout_width="fill_parent"
   android:layout_height="fill_parent"
   android:columnWidth="85dip"
   android:gravity="center"
   android:horizontalSpacing="10dp"
   android:numColumns="1"
   android:stretchMode="columnWidth"
   android:verticalSpacing="10dp"
   android:visibility="gone" />
</LinearLayout>
</LinearLayout>
```

PhotographActivity.java 类的全部代码如下：

```
package th.xc.main.com;
import java.io.File;
import th.xc.adapter.com.BenDiBaseAdapter;
import th.xc.file.com.FileSD;
import th.xc.yibu.com.BenDiTuAdd;
import thi.xc.sqlite.com.DBAdapter;
import thi.xc.sqlite.com.Photo;
import android.app.Activity;
import android.app.AlertDialog;
import android.app.AlertDialog.Builder;
import android.content.DialogInterface;
import android.content.Intent;
import android.graphics.Bitmap;
import android.graphics.BitmapFactory;
import android.net.Uri;
import android.os.Bundle;
import android.provider.MediaStore;
import android.view.*;
import android.view.ContextMenu.ContextMenuInfo;
import android.view.View.OnClickListener;
import android.widget.*;
import android.widget.AdapterView.OnItemClickListener;
import android.widget.AdapterView.OnItemLongClickListener;
public class PhotographActivity extends Activity implements
      View.OnClickListener, OnItemLongClickListener, OnItemClickListener {
```

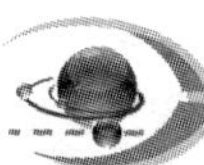

```
    // 相关控件
    private TextView bendi = null;
    private ImageView sd = null;
    private ImageView zhaoxiang = null;
    private ImageView huangfu = null;
    private LinearLayout linear01 = null;
    private FileSD sdfile = null;
    private GridView gridview = null;
    private GridView gv = null;
    private String txt = null;
    private HuanFuDialog my01 = null;
    private DBAdapter db = null;

    public void onCreate(Bundle savedInstanceState) {
        super.onCreate(savedInstanceState);
        requestWindowFeature(Window.FEATURE_NO_TITLE);
        getWindow().setFlags(WindowManager.LayoutParams.FLAG_FULLSCREEN,
                WindowManager.LayoutParams.FLAG_FULLSCREEN);
        setContentView(R.layout.index);
        db = new DBAdapter(this);
        bendi = (TextView) findViewById(R.id.bendi);
        bendi.setOnClickListener(this);
        linear01 = (LinearLayout) findViewById(R.id.linear01);
        sd = (ImageView) findViewById(R.id.sd);
        sd.setOnClickListener(this);
        zhaoxiang = (ImageView) findViewById(R.id.zhaoxiang);
        zhaoxiang.setOnClickListener(this);
        huangfu = (ImageView) findViewById(R.id.huangfu);
        huangfu.setOnClickListener(this);
        gridview = (GridView) findViewById(R.id.sd01);
        this.registerForContextMenu(gridview);
        gridview.setOnItemLongClickListener(this);
        gridview.setOnItemClickListener(this);
        gv = (GridView) findViewById(R.id.gridviewss);
        gv.setOnItemClickListener(this);
        my01 = new HuanFuDialog(this);
        my01.addLinearLayout(linear01);
        my01.jiazaiGround();
        sdfile = new FileSD();
        sdfile.createFile();
        BenDiAdd();
    }
//选择相册软件各个功能的按钮事件方法
    public void onClick(View v) {
```

```
        if (v.getId() == R.id.bendi) {
            linear01.setVisibility(LinearLayout.VISIBLE);
            BenDiAdd();
        }
        else if (v.getId() == R.id.sd) {
            Intent intentUpload = new Intent();
            intentUpload.setAction(Intent.ACTION_GET_CONTENT);
            intentUpload.setType("image/*");
            ((Activity) this).startActivityForResult(intentUpload, 2);
        }
        else if (v.getId() == R.id.zhaoxiang) {
            zhaoxiang();
        } else if (v.getId() == R.id.huangfu) {
            my01.show();
        } else if (v.getId() == R.id.sd01) {}
    }

    /* 完成照相功能的方法*/
    public void zhaoxiang() {
        Intent intent = new Intent(MediaStore.ACTION_IMAGE_CAPTURE);
        File ifile = new File(sdfile.morenPATH, "tenghuan"
                + sdfile.checkTuNumber() + ".jpg");
        intent.putExtra(MediaStore.EXTRA_OUTPUT, Uri.fromFile(ifile));
        Photo photo = new Photo();
        // 本地数据库中的图片 id
        String name = ifile.getName().substring(0,
                ifile.getName().lastIndexOf("."));
        // 图片名(可以编辑,显示时用的是这个名字,同时是唯一标识)
        photo.setPhotoName(name);
        // 本地数据库中的图片描述(开始图片描述没有,要在编辑里面进行编写)
        photo.setPhotoDescribe("");
        // 图片路径(储存图片的路径)
        photo.setPhotoPath(ifile.getAbsolutePath());
        db.add(photo);
        photo = null;
        startActivityForResult(intent, 1);
    }

    /* 获取请求返回结果的方法*/
    protected void onActivityResult(int requestCode, int resultCode, Intent
data) {
        super.onActivityResult(requestCode, resultCode, data);
        if (requestCode == 1) {
            File picture = new File(sdfile.morenPATH);
```

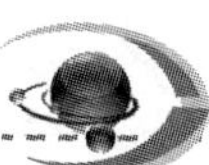

```
            startPhotoZoom(Uri.fromFile(picture));
        }
        if (data == null) {
            return; }
    }

    /** 储存照相相片的方法*/
    public void startPhotoZoom(Uri  uri) {
        Intent intent = new Intent("com.android.camera.action.CROP");
        intent.setDataAndType(uri, "image/*");
        intent.putExtra("crop", "true");
        // aspectX aspectY 是宽高的比例
        intent.putExtra("aspectX", 1);
        intent.putExtra("aspectY", 1);
        // outputX outputY 是裁剪图片宽高
        intent.putExtra("outputX", 64);
        intent.putExtra("outputY", 64);
        intent.putExtra("return-data", true);
        BenDiAdd();
        startActivityForResult(intent, 3);
    }

    /* 刷新或获取本地图片的方法  */
    public void BenDiAdd() {
        gridview = (GridView) findViewById(R.id.sd01);
        this.registerForContextMenu(gridview);
        gridview.setOnItemLongClickListener(this);
        new BenDiTuAdd(this, gridview).execute();
    }
    //创建关于相片的上下文菜单
    public void onCreateContextMenu(ContextMenu menu, View v,
            ContextMenuInfo menuInfo) {
        menu.setHeaderIcon(R.drawable.zhaoxiang);
        menu.setHeaderTitle("图片功能选项");
        menu.add(0, 1, 0, "删除");
        menu.add(0, 2, 0, "相片详情");
        menu.add(0, 4, 0, "编辑");
    }
    //处理上下文菜单的选项
    public boolean onContextItemSelected(MenuItem item) {
        switch (item.getItemId()) {
        case 1:// 删除相片
            deletTuPian(txt);    break;
        case 2: // 查看相片详细信息
```

```
            detailsTuPian(txt);   break;
        case 4:// 编辑
            bianji(txt);          break; }
        return super.onContextItemSelected(item);
    }

    /*  创建menu菜单  */
    public boolean onCreateOptionsMenu(Menu menu) {
        menu.add(0,11, 0, "帮助");
        menu.add(0, 12, 0, "关于");
        return super.onCreateOptionsMenu(menu); }
  //处理帮助和关于选项菜单的方法
    @Override
    public boolean onOptionsItemSelected(MenuItem item) {
        AlertDialog.Builder builder = new AlertDialog.Builder(this);
        switch (item.getItemId()) {
        case 11:
            builder.setTitle("帮助");
            TextView textView=new TextView(this);
            textView.append("1、具有拍摄、编辑、查看、共享及分发的功能； \n3、能够
            绑定 SNS 社区账户，第一次访问需要授权及认证，以后便可直接访问； \n4、具有
            换肤功能，提供换肤模板库供用户选择；\n5、网络异常时能够提示或禁止相关网络
            相册功能； \n6、自动切换 3G、WIFI 等网络通道，优先使用 WIFI。 ");
            builder.setView(textView);
            break;
        case 12:
            builder.setTitle("关于");
            TextView textView1=new TextView(this);
            textView1.append("指导员：向守超\n组    长：贺华川\n程序员：郑传庆、李
            虹锋\n测试员：牟兰、杨怡菡\n");
            builder.setView(textView1);
            break; }
        builder.setPositiveButton("OK",new
        android.content.DialogInterface.OnClickListener() {
            public void onClick(android.content.DialogInterface dialog,int
            whichButton) {
                setResult(RESULT_OK);
            }
        });
        builder.create();
        builder.show();
        return super.onOptionsItemSelected(item);
    }
    //  长按某张相片的事件
```

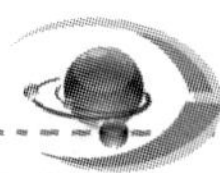

```
    public boolean onItemLongClick(AdapterView<?> arg0, View arg1, int arg2,
            long arg3) {
        txt = null;
        @SuppressWarnings("unused")
        BenDiBaseAdapter a = (BenDiBaseAdapter) arg0.getAdapter();
        LinearLayout liner = (LinearLayout)
         (arg0.getTouchables().get(arg2));
        TextView ts = (TextView) liner.getChildAt(1);
        txt = ts.getText().toString();
        a = null;
        liner = null;
        ts = null;
        return false;
    }
//删除相片的方法
    public void deletTuPian(String tupianming) {
        Photo photo = db.find(tupianming);
        if (photo != null) {
            File file = new File(photo.getPhotoPath());
            if (file.isFile()) {
                if (file.delete()) {
                    new AlertDialog.Builder(PhotographActivity.this).setTitle(
                "相片删除成功!").setPositiveButton("确定",
                new DialogInterface.OnClickListener() {
                public void onClick(DialogInterface dialog,
                int which) {
                BenDiAdd(); }
                }).setIcon(R.drawable.huangfu).show();
                db.detele(photo.getPhotoName());}
            } else {new
AlertDialog.Builder(PhotographActivity.this).setTitle(
"没有此图片，请刷新界面").setNegativeButton("确定", null).show(); }
            file = null;
            photo = null; }
    }
//查看相片详细信息的方法
    @SuppressWarnings("static-access")
    public void detailsTuPian(String tupianming) {
        Photo photo = db.find(tupianming);
        if (photo != null) {
            File file = new File(photo.getPhotoPath());
            if (file.isFile()) {
                Bitmap a =
                BitmapFactory.decodeFile(file.getAbsolutePath());
```

```
                Bitmap b = a.createScaledBitmap(a, 100, 100, true);
                LinearLayout liner = new LinearLayout(this);
                liner.setOrientation(LinearLayout.VERTICAL);
                liner.setGravity(Gravity.CENTER_HORIZONTAL);
                ImageView image = new ImageView(this);
                image.setImageBitmap(b);
                TextView tst = new TextView(this);
                tst.setTextSize(18);
                tst.append("图片名称:" + photo.getPhotoName() + "\n");
                if (photo.getPhotoDescribe().equals("")) {
        tst.append("图片描述:无\n");
            } else {
        tst.append("图片描述:" + photo.getPhotoDescribe() + "\n");}
        tst.append("图片像素:" + a.getWidth() + "*" + a.getHeight() + "\n");
        tst.append("图片大小:" + a.getRowBytes() / 1024d + "KB\n");
        liner.addView(image);
        liner.addView(tst);
        Builder c = new AlertDialog.Builder(PhotographActivity.this);
            c.setIcon(R.drawable.huangfu);
            c.setTitle("相片详细信息");
            c.setNegativeButton("确定", null);
            c.setView(liner);
            c.show();
            a = null;
            b = null;
            liner = null;
            image = null;
            tst = null;
        } else {
        new AlertDialog.Builder(PhotographActivity.this).setTitle(
    "没有此图片，请刷新界面").setNegativeButton("确定", null).show();
            }
            file = null;
        }
    }
//编辑相片信息的方法
    public void bianji(String tupianming) {
        if (tupianming != null) {
            final String name = tupianming;
            Photo photo = db.find(tupianming);
            LinearLayout a0 = new LinearLayout(this);
            a0.setOrientation(LinearLayout.VERTICAL);
            a0.setGravity(Gravity.CENTER_HORIZONTAL);
            a0.setPadding(10, 10, 10, 10);
```

```
        LinearLayout a1 = new LinearLayout(this);
        a1.setOrientation(LinearLayout.HORIZONTAL);
        TextView t0 = new TextView(this);
        t0.setText("图片名称:");
        final EditText eo = new EditText(this);
        eo.setWidth(200);
        eo.setMaxLines(1);//设置最大的行数是 1
        eo.setHorizontallyScrolling(true);//设置为可以水平移动
        eo.setText(photo.getPhotoName());
        a1.addView(t0);
        a1.addView(eo);
        LinearLayout a2 = new LinearLayout(this);
        a2.setOrientation(LinearLayout.HORIZONTAL);
        TextView t1 = new TextView(this);
        t1.setText("图片描述:");
        final EditText e1 = new EditText(this);
        e1.setWidth(200);
        e1.setMaxLines(1);//设置最大的行数是 1
        e1.setHorizontallyScrolling(true);//设置为可以水平移动
        e1.setText(photo.getPhotoDescribe());
        a2.addView(t1);
        a2.addView(e1);
        LinearLayout a3 = new LinearLayout(this);
        a3.setOrientation(LinearLayout.HORIZONTAL);
        a3.setGravity(Gravity.CENTER_HORIZONTAL);
        Button button01 = new Button(this);
        button01.setText("确定");
        button01.setWidth(100);
        Button button02 = new Button(this);
        button02.setText("取消");
        button02.setWidth(100);
        a3.addView(button01);
        a3.addView(button02);
        a0.addView(a1);
        a0.addView(a2);
        a0.addView(a3);
        Builder builder = new AlertDialog.Builder(this);
        builder.setIcon(R.drawable.huangfu);
        builder.setTitle("编辑");
        builder.setView(a0);
        final AlertDialog alertDialog = builder.show();
        button01.setOnClickListener(new OnClickListener() {
           public void onClick(View v) {
              Photo photo = new Photo();
```

```
                photo.setPhotoName(eo.getText().toString());
                photo.setPhotoDescribe(e1.getText().toString());
                db.update(name, photo);
                alertDialog.dismiss();
                BenDiAdd(); }
        });
        button02.setOnClickListener(new OnClickListener() {
            public void onClick(View v) {
                alertDialog.dismiss(); }
        }); } }
    protected void onDestroy() {
        super.onDestroy();
        this.finish();
        System.exit(0);
    }
//处理某张相片的单击事件
    public void onItemClick(AdapterView<?> arg0, View arg1, int arg2, long
    arg3) {
    @SuppressWarnings("unused")
    BenDiBaseAdapter a = (BenDiBaseAdapter) arg0.getAdapter();
    LinearLayout liner = (LinearLayout) (arg0.getTouchables().get(arg2));
        TextView ts = (TextView) liner.getChildAt(1);
        String txt01 = ts.getText().toString();
            a = null;
            liner = null;
            ts = null;
        Photo photo=db.find(txt01);
        Intent intent = new Intent(PhotographActivity.this,
        FangDaActivitys.class);
            intent.putExtra("path",photo.getPhotoPath());
            startActivity(intent);
    }
}
```

10.3.3 辅助类的设计

1. FileSD 的实现。

该类主要自动创建 SD 卡的默认路径，并有检查本机 SD 卡中相片的张数。其开发代码如下：

```
package th.xc.file.com;
import java.io.File;
import android.os.Environment;
```

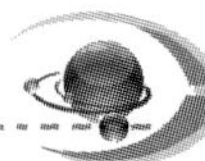

```
/* 文件处理 */
public class FileSD {
 public String PATH = null;
 public String morenPATH = null;// 默认路径
public FileSD() {
// 得到当前设备的目录
PATH = Environment.getExternalStorageDirectory() + "/";
morenPATH = PATH + "XiangCe";
 }
 /*创建文件夹 本软件图片默认路径 */
 public void createFile() {
     File file = new File(morenPATH);
     if (!file.exists()) {
         @SuppressWarnings("unused")
         boolean a = file.mkdir();
     }
}

/*  检查图片有多少张  */
public int checkTuNumber() {
    int nmuber = 0;
    nmuber =(int)System.currentTimeMillis();
    return nmuber;
    }
}
```

2. BenDiBaseAdapter.java 类的开发。

该类主要是本地相册图片操作的适配器类。该类的开发代码如下：

```
package th.xc.adapter.com;
import java.util.ArrayList;
import th.xc.main.com.PhotographActivity;
import android.graphics.Bitmap;
import android.graphics.Color;
import android.text.TextUtils.TruncateAt;
import android.view.Gravity;
import android.view.View;
import android.view.ViewGroup;
import android.widget.BaseAdapter;
import android.widget.Gallery;
import android.widget.ImageView;
import android.widget.LinearLayout;
import android.widget.TextView;
/* 本地相册图片适配器 */
```

```
public class BenDiBaseAdapter extends BaseAdapter {
 @SuppressWarnings("unchecked")
 private ArrayList name = null;
 @SuppressWarnings("unchecked")
 private ArrayList image01 = null;
 private PhotographActivity activity = null;
 @SuppressWarnings("unchecked")
 public BenDiBaseAdapter( PhotographActivity activity,
 ArrayList name, ArrayList image01) {
     this.name=name;
     this.image01=image01;
     this.activity=activity;
 }
 public int getCount() {
    return image01.size();   }
 public Object getItem(int position) {
    return null; }
 public long getItemId(int position) {
    return 0;    }
public View getView(int position, View convertView, ViewGroup parent) {
    // 布局1
    LinearLayout lv2 = new LinearLayout(activity);
    lv2.setOrientation(LinearLayout.VERTICAL);
    lv2.setGravity(Gravity.CENTER);
    lv2.setPadding(3, 3, 3, 3);
    // 创建图片对象
    ImageView image = new ImageView(activity);
    image.setScaleType(ImageView.ScaleType.FIT_XY);
    image.setLayoutParams(new Gallery.LayoutParams(100, 120));
    image.setImageBitmap((Bitmap)image01.get(position));
    // 获取数据
    TextView text = new TextView(activity);
    text.setTextSize(18);
    text.setTextColor(Color.GRAY);
    text.setGravity(Gravity.CENTER);
    text.setEllipsize(TruncateAt.MIDDLE);
    text.setText(""+name.get(position));
    // 添加
    lv2.addView(image);
    lv2.addView(text);
    return lv2;
 }
}
```

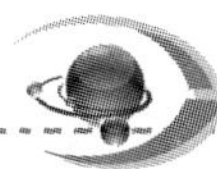

3. BenDiTuAdd.java 类的实现。

该类负责对本地图片添加到界面的功能。该类实现的只要代码如下：

```
    package th.xc.yibu.com;
    import java.io.File;
    import java.util.ArrayList;
    import java.util.List;
    import th.xc.adapter.com.BenDiBaseAdapter;
    import th.xc.main.com.PhotographActivity;
    import thi.xc.sqlite.com.DBAdapter;
    import thi.xc.sqlite.com.Photo;
    import android.app.ProgressDialog;
    import android.graphics.Bitmap;
    import android.graphics.BitmapFactory;
    import android.os.AsyncTask;
    import android.view.Window;
    import android.view.WindowManager;
    import android.widget.GridView;
    /* 本地图片加载类*/
public class BenDiTuAdd extends AsyncTask<String, BenDiBaseAdapter, String> {
    private GridView gridview = null;
    private List<Photo> photo = null;
    private PhotographActivity activity = null;
    private ProgressDialog myDialog = null;
    private DBAdapter db=null;
    public BenDiTuAdd(PhotographActivity activity, GridView gridview) {
        this.gridview = gridview;
        this.activity = activity;
        db=new DBAdapter(activity);
        photo=db.find();
    }
    protected void onPreExecute() {
        super.onPreExecute();
        myDialog = new ProgressDialog(activity);
        myDialog.setTitle("正在加载本地图片...");
        Window window=myDialog.getWindow();
        WindowManager.LayoutParams lp = window.getAttributes();
        lp.alpha = 0.4f;// 透明度
        myDialog.show();
    }
    @SuppressWarnings("static-access")
    protected String doInBackground(String... params) {
        if(photo==null){
```

```
            return null;
        }
        ArrayList<Bitmap> Image01 = new ArrayList<Bitmap>();
        ArrayList<String> name01 = new ArrayList<String>();
        if (photo.size() != 0) {
            for (int i = 0; i < photo.size(); i++) {
                Photo photo01=photo.get(i);
                File file=new File(photo01.getPhotoPath());
                String name =file.getName();
                name = name.substring(name.indexOf(".") + 1);
                if (file.isFile()
                        & (name.equals("png") || name.equals("jpg"))) {
                    Bitmap bitmap = null;
        try {
                bitmap = BitmapFactory.decodeFile(file.getAbsolutePath());
                bitmap = bitmap.createScaledBitmap(bitmap, 100, 120,true);
                        } catch (Exception e) {
                            e.getMessage();
                        }
                        Image01.add(bitmap);
                        name01.add(photo01.getPhotoName());
                        bitmap = null;
                    }
                }
                // 实例化一个适配器
    BenDiBaseAdapter a = new BenDiBaseAdapter(activity, name01, Image01);
        publishProgress(a);
        }
        return null;
    }
    protected void onProgressUpdate(BenDiBaseAdapter... values) {
        super.onProgressUpdate(values);
        gridview.setAdapter(values[0]);
    }
    protected void onPostExecute(String result) {
        super.onPostExecute(result);
        myDialog.dismiss();
     }
}
```

4. FangDaActivitys.java 类的实现。

该类实现照片放大的功能。在主界面图 10-4 中，当用户点击其中任意一张相片，就会看到该照片的放大图。放大效果如图 10-14 所示。

图 10-14　相片放大图

FangDaActivitys.java 类的整体代码如下：

```
package th.xc.main.com;
import java.io.File;
import android.app.Activity;
import android.content.Intent;
import android.graphics.Bitmap;
import android.graphics.BitmapFactory;
import android.os.Bundle;
import android.view.Window;
import android.view.WindowManager;
import android.widget.ImageView;
public class FangDaActivitys extends Activity {
  ImageView image=null;
public void onCreate(Bundle savedInstanceState) {
    super.onCreate(savedInstanceState);
    requestWindowFeature(Window.FEATURE_NO_TITLE);
 getWindow().setFlags(WindowManager.LayoutParams.FLAG_FULLSCREEN,
          WindowManager.LayoutParams.FLAG_FULLSCREEN);
    setContentView(R.layout.fangdaphot);
    image=(ImageView)findViewById(R.id.switchters_fangda);
    Intent intent=getIntent();
    String path=intent.getStringExtra("path");
    File file=new File(path);
    String name =file.getName();
```

```
name = name.substring(name.indexOf(".") + 1);
if (file.isFile()&(name.equals("png") || name.equals("jpg"))) {
Bitmap bitmap = null;
try {
bitmap = BitmapFactory.decodeFile(file.getAbsolutePath());
} catch (Exception e) {
 e.getMessage();   }
   if(image!=null){
      image.setImageBitmap(bitmap);
   } } } }
```

5. HuanFuDialog.java 类的实现。

该类实现相册的换肤功能。该类实现的效果如图 10-7 所示。该类的全部代码如下：

```
 package th.xc.main.com;
import java.util.ArrayList;
import java.util.HashMap;
import thi.xc.sqlite.com.DBAdapter;
import android.app.Dialog;
import android.view.View;
import android.widget.AdapterView;
import android.widget.AdapterView.OnItemClickListener;
import android.widget.GridView;
import android.widget.LinearLayout;
import android.widget.SimpleAdapter;
/* * 自定义 Dialog 用于换肤 */
public class HuanFuDialog extends Dialog implements
OnItemClickListener {
 private PhotographActivity dialog = null;
 private GridView menuGrid = null;
 private String myMenuStr[] = { "腾幻 01", "腾幻 02", "腾幻 03",
 "腾幻 04", "腾幻 05", "腾幻 06", "腾幻 07", "腾幻 08" };
 private int myMenuBit[] = { R.drawable.sakura01,
R.drawable.sakura02, R.drawable.sakura03, R.drawable.sakura04,
R.drawable.sakura05, R.drawable.sakura06, R.drawable.sakura07,
R.drawable.sakura08, };
 private ArrayList<LinearLayout> linear = null;
 private DBAdapter adapter = null;
 public HuanFuDialog(PhotographActivity dialog) {
     super(dialog, R.style.dialog_fullscreen);
     setContentView(R.layout.mymenu);
     linear = new ArrayList<LinearLayout>();
     this.dialog = dialog;
     menuGrid = (GridView)this.findViewById(R.id.GridView_toolbar);
```

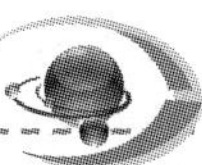

```
    menuGrid.setAdapter(getMenuAdapter(myMenuStr, myMenuBit));
    menuGrid.setOnItemClickListener(this);
    View view = findViewById(R.id.mainlayout);
    view.getBackground().setAlpha(180);// 120 为透明的比率
    this.setCanceledOnTouchOutside(true);
    adapter = new DBAdapter(dialog);
}

/*  设置窗体属性  */
public SimpleAdapter getMenuAdapter(String[] menuNameArray,
        int[] imageResourceArray) {
ArrayList<HashMap<String, Object>> data =
    new ArrayList<HashMap<String, Object>>();
    for (int i = 0; i < menuNameArray.length; i++) {
        HashMap<String, Object> map = new HashMap<String, Object>();
        map.put("itemImage", imageResourceArray[i]);
        map.put("itemText", menuNameArray[i]);
        data.add(map);
    }
    SimpleAdapter simperAdapter = new SimpleAdapter(dialog, data,
    R.layout.item_menu, new String[] { "itemImage", "itemText" },
    new int[] { R.id.item_image, R.id.item_text });
    return simperAdapter;
}
public void onItemClick(AdapterView<?> parent, View view, int position,
long id) {
    if (this.linear != null) {
        for (int i = 0; i < linear.size(); i++) {
            if (linear.get(i) != null) {

linear.get(i).setBackgroundResource(myMenuBit[position]);
                adapter.updataTuhao(position + "");
            } } }
      this.dismiss();  }
   public void addLinearLayout(LinearLayout linear) {
      this.linear.add(linear);
   }

   /**  * 加载保存的图片号  */
   public void jiazaiGround() {
      adapter.insertTuhao();
      String a = adapter.CheckTuhao();
      if (a == null) {
          return;
```

```
            }
            int tuhao = Integer.parseInt(a);
            for (int i = 0; i < linear.size(); i++) {
                if (linear.get(i) != null) {
                    if(tuhao!=-1){
                linear.get(i).setBackgroundResource(myMenuBit[tuhao]);
                    }
            }
        }
      }
    }
```

该项目开发过程中涉及到的布局文件和清单文件等，就不在这里赘述了，相信读者经过这段时间的学习，根据相关图片的显示界面，能够独立自主地完成其他相关文件的代码编写。